In the Box
Music Production

In the Box Music Production

Advanced Tools and Techniques for Pro Tools

Mike Collins

Focal Press
Taylor & Francis Group

NEW YORK AND LONDON

First published 2015
by Focal Press
70 Blanchard Road, Suite 402, Burlington, MA 01803

and by Focal Press
2 Park Square, Milton Park, Abingdon, Oxon OX14 4RN

Focal Press is an imprint of the Taylor & Francis Group, an informa business

Notices
Knowledge and best practice in this field are constantly changing. As new research and experience broaden our understanding, changes in research methods, professional practices, or medical treatment may become necessary.

Practitioners and researchers must always rely on their own experience and knowledge in evaluating and using any information, methods, compounds, or experiments described herein. In using such information or methods they should be mindful of their own safety and the safety of others, including parties for whom they have a professional responsibility.

Product or corporate names may be trademarks or registered trademarks, and are used only for identification and explanation without intent to infringe.

Library of Congress Cataloging-in-Publication Data

Collins, Mike, 1949– author.
  In the box music production : advanced tools and techniques for Pro Tools / Mike Collins.
      pages cm
  1. Pro Tools.   2. Digital audio editors.   I. Title.
    ML74.4.P76C64   2014
    781.3'4536—dc23
    2014013446

ISBN: 978-0-415-81460-7 (pbk)
ISBN: 978-0-203-06636-2 (ebk)

Typeset in Myriad Pro
By Apex CoVantage, LLC

MIX
Paper from
responsible sources
FSC® C014174
www.fsc.org

Printed and bound in the United States of America by Sheridan Books, Inc. (a Sheridan Group Company).

Bound to Create

You are a creator.

Whatever your form of expression — photography, filmmaking, animation, games, audio, media communication, web design, or theatre — you simply want to create without limitation. Bound by nothing except your own creativity and determination.

Focal Press can help.

For over 75 years Focal has published books that support your creative goals. Our founder, Andor Kraszna-Krausz, established Focal in 1938 so you could have access to leading-edge expert knowledge, techniques, and tools that allow you to create without constraint. We strive to create exceptional, engaging, and practical content that helps you master your passion.

Focal Press and you.

Bound to create.

We'd love to hear how we've helped you create. Share your experience:
www.focalpress.com/boundtocreate

Contents

Acknowledgments

First of all, I would like to thank Anaïs Wheeler at Focal Press for commissioning this book. I am also grateful for the efforts of Carlin Bowers and all the staff at Focal Press who are involved with publishing and marketing my books.

Louise Wells at Red Lorry Yellow Lorry PR—www.rlyl.com—was enormously helpful throughout the time I was writing this book, liaising constantly with Avid to provide review products and information. Avid UK Solutions Specialist Simon Sherbourne, as usual, was extremely helpful, supplying much-needed clarifications of various technical points.

Colin McDowell provided lots of useful information together with his entire suite of McDSP AAX plug-ins—www.mcdsp.com. Matt Ward, president at Universal Audio, and Erik Hanson, director of marketing at Universal Audio—www.uaudio.com—kindly supplied a UAD-2 system with its extremely comprehensive suite of plug-ins for Pro Tools.

Special thanks to Thomas Lund of TC Electronic (www.tcelectronic.com); Marco Alpert and Andy Hildebrand at Antares Audio Technologies (www. antarestech.com); Arjen van der Schoot and Aram Verwoest at Audio Ease (www.audioease.com); Guillaume Jeulin at Blue Cat Audio (www.bluecataudio. com); Keith Malin at KMR Audio (www.kmraudio.com), who kindly supplied the Crane Song Phoenix plug-in; Simon Stock of Music Track (www.musictrack. co.uk), who supplied a copy of Digital Performer 8; Alex Theakston of Source Distribution (www.sourcedistribution.co.uk), who supplied the Eventide AAX plug-ins; Frederik Slijkerman of FabFilter Software Instruments, who supplied his TotalBundle of plug-ins; Felix Niklasson of Flux:: sound and picture development (www.fluxhome.com); Kim Pfluger of iZotope Audio Signal Processing (www.izotope.com); Denis Goekdag of Zynaptiq (www.zynaptiq. com); Alex Siegel of Slate Digital (www.slatedigital.com); Mattias Danielsson of Softube (www.softube.com); Nathan Eames of Sonnox Oxford Plug-ins (www. sonnox.com); Mitch Thomas of SoundToys Professional Effects Plug-ins (www. soundtoys.com); and Paul J. de Benedictis, who supplied the Spectrasonics Virtual Instruments (www.spectrasonics.net).

To all my regular musical collaborators and partners, including Jim Mullen, Lyn Dobson, John McKenzie, Winston Blissett, José Joyette, Marc Parnell, Vivienne

McKone, Rouhangeze Baichoo, and David Philips, my wholehearted thanks for the musical inspirations and motivations with which they have kept me sane and fulfilled while I have developed my ideas for this book.

To Ernest Ranglin, Jamaica's foremost guitarist, who has not only been a major source of musical inspiration but has also acted as a personal and musical mentor—providing much-welcomed encouragement and positive feedback since we met back in 2009—I offer my greatest respects and appreciation—see www.ernestranglin.co.uk for more information.

I would also like to thank all the members of my family, most importantly my father, Luke Collins (who passed away peacefully on September 28, 2013 aged 94), my mother, Patricia Collins, my brothers, Anthony and Gerard Collins, and all my close friends, in particular Sia Duma, Dario Marianelli, Barry Stoller, Anthony Washington, Clive Mellor, and Keith O'Connell, for their continuing support throughout.

Mike Collins
November 2013

Advanced *In the Box* Music Production with Pro Tools

Introduction

I have divided this book into three logical parts: Part 1: 'Working with Loops and Beats'; Part 2: 'Working with Virtual Instruments'; and Part 3: 'Studio Techniques'.

Part 1 aims to equip the reader with a more in-depth understanding of how to bring files into Pro Tools, how to work with clips and loops, how to edit these using Beat Detective, and how to time stretch audio using Elastic Audio. These are all essential techniques for popular music production, and can also be very useful when working in other genres such as jazz, folk, blues, or 'world music'.

Part 2 is all about working with plug-in 'virtual' instruments including drum machines, synthesizer, and samplers. This part of the book will not only be of interest to all those involved in popular music production, but will also be relevant for those working on other genres who can use these to provide temporary sounds that will later be replaced by acoustic instruments. It will also be useful for composers working to picture who often need to produce electronic versions of their scores—even if they do eventually hire an orchestra!

Part 3 covers recording and processing vocals, signal processing, and effects, and finishes off with a chapter all about mastering audio before final delivery. So—how to capture vocals (and instruments) 'into the box', how to manipulate this audio using the panoply of technical and creative tools available 'inside the box', and how to transfer audio 'outside of the box' to deliver this to its target audience.

Part 1

Very often you will start a project in Pro Tools by bringing in already existing audio files or loops. This is common enough in music production, especially in dance music where many projects are almost totally loop based. It is also the primary way of working in postproduction and multimedia production, where lots of individual audio files containing music, sound effects, or dialog may be needed.

A popular way of working on music is to start out by assembling loops of audio sourced from commercially available collections in a variety of formats. REX files, for example, contain audio sliced up into individual beats so that when these are played back using a tick-based sequencer the tempo can be increased or decreased from the original tempo and the beats in the REX file will follow. ACID files behave similarly, with the tempo increased or decreased using time stretching. Files analyzed with Pro Tools Elastic Audio work in a similar way. You can conveniently audition these various file types using the Pro Tools browsers then bring these directly into tick-based (or sample-based) tracks during your Pro Tools session.

Working in Pro Tools, you can duplicate or repeat these short 'loops' as you develop your arrangement using the basic commands in the Edit menu that let you repeatedly play back regions to form 'loops'. You can also loop audio clips or clip groups using the Clip menu's 'Loop . . .' command. If you want to edit these, a special loop trimmer tool streamlines this process. A TCE trimmer tool lets you time compress or time expand any audio region in the Edit window simply by dragging the beginning or end of the region to make it shorter or longer. This is particularly useful for adjusting short audio 'loops' to fit your arrangement exactly.

One of the keys to creating successful loops in Pro Tools is being completely familiar with how to make accurate selections in the Edit window and how to trim regions using the Edit menu commands. You also need to make sure that you are confident about making Timeline selections so that you can loop playback while you are auditioning regions to make sure that they will 'loop' correctly. The Dynamic Transport feature can be very useful here, letting you audition a series of prospective 'loops' without stopping and restarting playback. It also lets you loop playback around one section of your song while you do some edits on the fly, then drag the timeline selection to another part of the song and loop playback around this section so you can do more edits there.

Pro Tools itself allows you to manipulate loops and beats directly in the Edit window in many useful ways. Short 1-, 2-, 4-, or 8-bar recordings of drums, percussion, or of any other suitable material can be played back repeatedly so that the short pattern 'loops'. In other words, the pattern plays through, then plays through again, and continues to play repeatedly for as long as you have decided to 'loop' it. This terminology originated with the use of MIDI samplers that could be used to 'loop' playback of samples internally, but even with these, a MIDI sequencer was typically used to retrigger, that is, replay, the sample every bar, every two bars, or whatever. So 'looping' really is being used to mean 'playing back this short audio recording repeatedly'.

The ability to change sample-based audio tracks into 'tick-based' tracks is extremely useful when you want to change the tempo of a Pro Tools session and have your audio follow this tempo. You can achieve this in various ways, including the use of REX, ACID, and Elastic Audio-analyzed files. You can also cut any audio recording into 'slices' using Beat Detective to create audio clips that act similarly to REX files. Sometimes you will want to make the tempo of your Pro Tools session follow the tempo of audio that you have recorded. Beat Detective and Elastic Audio can be used for these purposes as well. The Identify Beat capability allows you to define any location along the timeline as corresponding to a particular beat location. Beat Detective takes this concept much further, providing more sophisticated tools to analyze transients within the audio that may (or may not) correspond to individual beats. Elastic Audio provides even more sophisticated transient analysis algorithms along with tools that allow you to 'warp' (stretch or compress) even the smallest region of audio—'massaging' this until it plays back exactly the way that you want it to do.

Mastery of all these tools and techniques in Pro Tools will put you completely 'in charge' of every aspect of time, tempo, beat creation and manipulation, looping, slicing, stretching and compressing, and much, much more. If you also master the most important third-party plug-ins introduced here, you will have control of the most popular and advanced beat manipulation tools 'on the planet' available at your fingertips!

Part 2

When you are starting out on a new songwriting project or working ideas out for a new arrangement, it can be very useful to have a drum machine with a standard beat, such as pop, Latin, country, rock, or blues, to play along to. When you get further along with your project, you might write your own specific drum patterns using MIDI, or use sampled beats, or record real drummers and percussionists. AIR Music Technology offers Strike, a virtual instrument that has lots of preset patterns with popular beats to suit everything from pop ballads to rap, funk, and even reggae. You can edit the patterns to make them fit your own music and you can also play the individual drum sounds using MIDI—so you can write your own drum parts into the MIDI or Instrument tracks in Pro Tools using Strike's drum kit sounds.

If you plan on creating lots of drum parts using MIDI, AIR Music Technology's Structure sampler plug-in library includes a very useful selection of acoustic and electronic drum kits including vintage kits and jazz brushes, hip-hop loops, R & B loops, and lots of percussion instruments. Structure can also play back REX files and has lots of other useful features. For example, its sample looping feature lets you set up a single loop within any sample, with a crossfade at the loop point. Just click the Loop button in the wave editor, drag the edit points to where you want the loop to start and end, zooming in for more accuracy, then click Edit Xfade and create a crossfade at the loop point. It's very quick and easy to work with—and beautifully integrated with Pro Tools.

If you are really serious about working with loops and beats in Pro Tools, the AIR Music Technology Transfuser plug-in lets you slice, dice, shuffle, trigger, rephrase, or resequence your audio loops and phrases. It works with WAV or AIFF, ACID or Apple Loops, and REX 1 or 2 formats and comes with a library of REX files and loops to get you started. Drag and drop an audio file into Transfuser and it 'asks' you how you want it to handle this. There are three main options: First, Transfuser can chop the audio into slices that you can work with—much like using Reason's Dr. REX or Ableton Live's slicing capabilities. The slices are linked to a slice sequencer with a preset pattern that plays back the slices in sequence to sound as they would sound if you played back the original audio clip. You can then change the tempo so that the slices play back more slowly or quickly and you can manipulate the slices in various ways, applying effects in Transfuser. The second option time stretches and beat matches the audio and lets you play this using the phrase sequencer—pitching, stretching, and vocoding your phrases and loops. The third method slices the audio and assigns the slices to a set of 12 virtual drum pads, again linking this with a sequencer module that lets you play the samples using either a 12-note step sequencer or a more conventional piano roll-type sequencer. You can also insert Transfuser on any audio track and use it to process the audio using its Audio Input synthesizer. This passes audio from disk or the audio input of the track on which Transfuser is inserted so that you can directly process Pro Tools audio using Transfuser without having to actually import the audio into Transfuser.

Other virtual instruments from AIR Music Technology include Hybrid subtractive/wavetable synthesizer, Velvet electric pianos, Vacuum Pro vintage-style polyphonic synthesizer, and Loom additive synthesizer.

Spectrasonics offers its powerful trio of virtual instrument plug-ins: Stylus RMX for drums, Trilian for basses, and Omnisphere with its wide choice of pads, strings, voices, and other synthesizer textures.

Arturia offers its V-Collection, which focuses on recreating classic analog and early digital synthesizers and drum machines. V-Collection includes 10 software instruments: Mini V, Modular V, CS-80V, ARP2600 V, Prophet V and Prophet VS, Jupiter 8-V, Oberheim SEM V, Wurlitzer V, and Spark Vintage—a comprehensive collection of 30 classic drum machines (including the TR-808, TR-909, LinnDrum, and other favorites).

Other popular third-party software applications that work well with Pro Tools include Propellerhead Reason and Ableton Live. Reason is a stand-alone software application that integrates with Pro Tools using a technology called ReWire that feeds the audio outputs from Reason into the Pro Tools mixer, allows MIDI communication between Pro Tools and Reason, and provides transport synchronization. Reason contains a rack of devices including a REX file player, sample playback, synthesizer, sound effects, and sequencing features. Ableton Live, another stand-alone application that integrates with Pro Tools via ReWire, is the leading software application for beat slicing and for manipulating loops, and also includes a selection of virtual instruments.

Part 3

Vocals are unquestionably the most important element when it comes to popular music, so it is appropriate to look at how to record and process these to the highest standards. Although some projects may only involve importing existing audio files into Pro Tools, most will involve recording at least 'live' vocals—especially in pop music. Of course, classical and orchestral music and most jazz and world music genres solely involve recording musicians, bands, or orchestras 'live'. For this you need microphones and preamplifiers—and possibly various hardware signal processors—to 'capture' these 'live' performances. Once you have these recordings 'inside the box', you can get to work on these using software to hone these to perfection.

'In the box' is, of course, where most of the 'action' will be throughout your music production once you have recorded all the 'live' elements. A large part of many projects involves signal processing of one type or another, so this section covers the different types in some depth and offers detailed overviews of popular products that you can add to your system.

How to get it 'out of the box' when you have finished your project is obviously of crucial importance as well! Once all the mixing is done the question of 'mastering' arises. This is the process of preparing the final mixes as the last stage of the journey toward the distribution medium that will be used. Will the audio files be transferred to vinyl discs, cassette tapes, CDs, or USB 'sticks', broadcast on radio, downloaded, or streamed via iTunes or Spotify, or whatever? The mastering engineer can take all these possibilities into account and can also add a creative element if this is desired. Attaching detailed metadata to files is becoming increasingly important, and it may fall to the mastering engineer or to producers involved in this stage of a project to ensure that this is done correctly. Then, of course, you (or someone) will need to get on with promoting and selling the recordings—and in today's 'wired' world, that involves making sure that information about your recordings is available in all the various databases used throughout the music industry, such as those associated with iTunes, GooglePlay, HD Radio, Amazon Cloud Player, and others.

You will also need to consider how to back up and archive your projects when you are all done—after all, always remember that computers and hard drives are not 'forever'; they are frail creatures that can easily die or malfunction such that your data is lost and gone (unless you had this backed up)! Also, there are many reasons why you should carefully archive your master recordings—after all you may want to rerelease them 50 or even 70 years after they were first made (which is the case with the Beatles' 1963 recordings, for example, which are still popular and selling well today)!

And if you get involved in reissues of older recordings, the question of audio restoration often arises. A working knowledge of music rights and copyrights, contracts, and legal issues is also essential for everyone working professionally in music and recording.

PART 1

Working with Loops and Beats in Pro Tools

Chapter 1—Bringing in Audio and Loops—covers how to bring in audio files and loops using the Workspace browser in Pro Tools, how to preview and audition files, how to search for files, how to relink missing files, how to import audio from CD, and how to work with REX and ACID files.

Chapter 2—Working with Clips—is about what happens after you have brought audio and loops into Pro Tools and need to move audio clips around in the Edit window so that these can be edited and corrected. So, for example, it covers how to select clips, how to make a timeline and edit selections, how to loop clips, and how to use Dynamic Transport mode.

Chapter 3—Beat Detective—is all about identifying the beats within an audio recording using either the basic 'Identify Beat' command or the more advanced Beat Detective feature in Pro Tools.

Chapter 4—Elastic Audio—explains how to apply Time Compression and Expansion (TCE) in real time to individual audio tracks. It also features beat and tempo analysis and TCE processing algorithms that allow you to quantize audio, to follow tempo variations in the audio, or to conform the tempo of the audio to that of Pro Tools.

in this chapter

CHAPTER 1

Bringing in Audio and Loops

How to Find Your Files

Using the Browsers

The first thing you need to do before bringing any audio or other media file (video, for example) into Pro Tools is to find the file—yes, I know this sounds obvious, but it is worth looking at the various ways you can do this. On the Mac you could use the appropriately named Finder for this and on Windows machines you could use Windows Explorer. You would look through the files and folders available on your system until you find what you are looking for and bring this into your session in some way.

But there is a better way! Pro Tools actually has its own file browser that you can use to work with your disk volumes, folders, and files as an alternative to using the Mac's Finder or Windows Explorer. This so-called Workspace browser has specialized features to speed up this process and makes things much easier for you to access the various media files that Pro Tools uses. You can set up multiple browsers and customize and arrange these to suit the ways that you like to work.

> **NOTE**
> Previous versions of Pro Tools used multiple types of browsers for media management: Catalog browsers, the Project browser, and Workspace browsers. The new Workspace uses a single-pane grid view that provides easier navigation and combines the Catalog, Project, and Workspace browsers in one window.

To get started, just choose New Workspace from the Window menu or press Option-I (Mac) or Alt-I (Windows)—see Figure 1.1.

> **TIP**
> If multiple Workspace browsers are open, these key commands bring successive Workspace browsers to the foreground.

A Workspace browser, as its name suggests, lets you browse through all the available folders and files that form your 'workspace'. Workspace browsers will display every type of computer file, even including unknown file types, as well as aliases and desktop folders. However, to protect vital system software components, Workspace browsers do not display the System folder on the Mac or the WU Temp or System Volume Information folders in Windows. This prevents these from being indexed, searched, sorted, or affected in any way when you are using the Workspace.

Files that you can search for, sort, audition, and import to your session include audio, MIDI, video, (.txf) plug-in settings, other session files, and Guitar Rig settings for the Eleven Rack if you have one of these. You can drag audio, video, or session files directly from browsers into your current Pro Tools session.

The Workspace browser window is divided into two sections—the Locations 'pane' at the left and the Workspace 'pane' to the right of this. In the Workspace browser, you can use the Locations pane to look through the 'volumes', that is, the local and networked drives, on your system or to look through the indexed catalogs of media or the system user directory. If you are looking for folders or files on your desktop, for instance, you can find these in the user directory's desktop folder.

Figure 1.1

Choosing a new Workspace from the Windows menu

The Locations Pane

The default Workspace browser—see Figure 1.2—has a Locations pane (in the window) at the left that will contain a folder called Volumes, an icon representing the session file that you are currently working with, an item named Catalogs, and the User folder. If you click on the revealing arrows to the left of each of these items, their 'contents' will be revealed.

> **NOTE**
>
> The Locations pane is used to navigate the volumes (local and networked drives) on your system, the currently open session, indexed catalogs of media, and the system user directory.

The Volumes folder contains all the disk drives available in your system; the Session file opens to reveal the Session Audio Files folder, which can be opened

Figure 1.2
A default Workspace browser

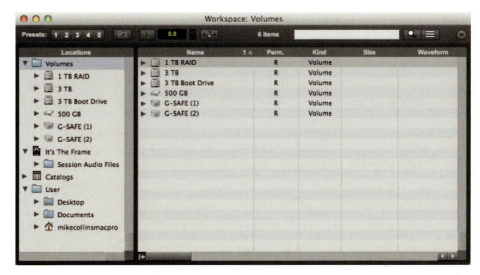

in turn to reveal the audio files used during this session. No catalogs are shown here yet, but these can be set up later on. The User folder shows all the folders available in your computer's user account.

You can show or hide the Locations pane at the left of the Workspace browser window by clicking on the Show/Hide icon, see Figure 1.3, in the lower right corner of the left-hand pane.

Figure 1.3
Show/Hide the Locations pane

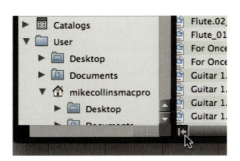

NOTE: PERFORMANCE AND TRANSFER VOLUMES

Workspace browsers let you view, manage, and import sessions and media from both performance and transfer volumes.

Performance volumes are storage volumes (hard drives) that are suitable for playback and have been designated in the Workspace browser as Record and Playback (R) or Playback Only (P) of media files in a Pro Tools session.

(Continued)

Transfer volumes are volumes that are not supported for media playback in Pro Tools (such as shared network volumes or CD-ROMs), or storage volumes (hard drives) that have been designated in the Workspace browser as Transfer (T) volumes. Transfer volumes cannot be used to record or play back media during a Pro Tools session. Designated Transfer (T) volumes can be useful for transferring session and media files between different Pro Tools systems.

Workspace Browser Pane

The main part of the Workspace browser window, the Browser pane, displays the contents of items selected in the Locations pane or displays search results.

The browser pane is subdivided into columns in which the volumes, folders, or file names are displayed along with metadata for file size and kind, duration, creation date, sample rate, and so forth.

Customizing the Browser Pane

You can show or hide individual columns in the Browser pane to fit your workflow. The default column display includes the items you would use in most sessions.

- ■ To show or hide individual columns: Control-click (Mac), Start-click (Windows), or Right-click a column label and select or deselect the column name from the pop-up menu.

- ■ To show or hide all columns: Control-click (Mac), Start-click (Windows), or Right-click a column header and choose ALL or NONE from the pop-up menu. (Only the Name column remains when you choose NONE.)

- ■ To show the default set of columns: Control-click (Mac), Start-click (Windows), or Right-click a column header and choose DEFAULT from the pop-up menu.

- ■ To move a column: drag the column to a new location in the Workspace.

- ■ To resize a column: drag the right column boundary to a new width.

- ■ To scroll the active pane up or down: press the Page Up or Page Down key.

- ■ To scroll to the top or bottom of the active pane: press Home (for the top) or End (for the bottom).

- To move items up or down in the current Workspace browser: select items in a Workspace browser and make sure that window is in the foreground, then press the Up Arrow or Down Arrow key.

- Columns can be sorted in ascending or descending order, and multiple sorts can be applied (up to four levels).

- To sort by columns: click the column title header.

- To toggle the current sort order between ascending and descending: click the column title header a second time.

- To sort by multiple columns: Option-click (Mac) or Alt-click (Windows) successive column headers.

Workspace Browser Toolbar

The toolbar at the top of the Workspace browser window lets you conveniently access various functions.

View Presets

Five small buttons at the left of the toolbar let you store and recall preset browser configurations: arrange the Workspace browser any way you like, then Command-click (Mac) or Control-click (Windows) on any of these view presets to store this arrangement for future recall. To recall a saved preset, just click on the appropriate View Preset button.

Folder-Hierarchy Icon

To the right of the Presets buttons, there is a folder-hierarchy icon that becomes active whenever you select any items that are revealed below the Volumes, Session, Catalogs, or User folders in the Locations pane. For example, with the Session Audio Files folder revealed and selected below the Session folder, as in Figure 1.4, this toolbar folder icon becomes activated.

Figure 1.4
Activated folder-hierarchy icon

If you click on the folder-hierarchy icon, the selection in the Locations pane will move up to the next higher level in the hierarchy of folders—in this case up to the top level of the Session folder—and the folder-hierarchy icon will become deactivated, as in Figure 1.5.

Preview Button

To the right of the folder-hierarchy icon, there is a Preview button to start and stop preview with associated volume controls and meters, and an 'Audio Files Conform to Session Tempo' button—see Figure 1.6.

Figure 1.5
Deactivated folder-hierarchy icon

Figure 1.6
Preview button, volume control, meters, and 'Audio Files Conform to Session Tempo' button

Search Functions

To the right of the Preview controls on the toolbar, every Workspace browser provides a set of search tools for finding files on your system quickly and easily.

A simple text search using the Browser Search field—see Figure 1.7—lets you quickly search all volumes for items by file name as well as any other text field available in the Workspace browser (such as File Comment, Database Comment, Scene, Take, Plug-In Name, and so on).

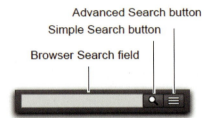

Figure 1.7
Search tools

Using the Search Functions

In the Workspace browser window, press Command-F (Mac) or Control-F (Windows) to highlight the Browser Search field so this is ready for you to type the names of files, dates, durations, or whatever. As soon as you start typing text into this field, the search will commence. Search results will be displayed in the Browser pane. Very often the results will appear before you have even finished typing the full name, as the database searches immediately for the first characters that you type.

> **TIP**
> To search within a particular volume, catalog, session, or folder, select this first in the Locations pane. If you do not make a selection, the entire system is searched (including shown offline volumes).

You can also click the Search button to initiate a simple search before typing any text into the Browser Search field—see Figure 1.8.

The Simple Search button changes from a magnifying glass icon to an 'X' and immediately initiates a search. Click the Search button again to end the search (or press the Escape (Esc) key).

Figure 1.8
Initiating a simple
search of the
Session folder with
no search criteria

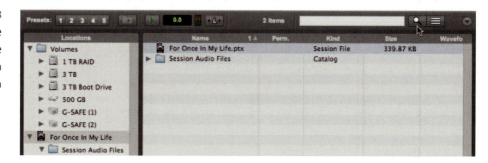

Figure 1.8
Initiating a simple
search of the
Session folder with
no search criteria

Initiating a search in this way, with no search criteria, will find all the items available on the system that you have selected. In the example shown in Figure 1.9, the search has found all the audio files in the Session Audio Files folder associated with the session folder 'For Once in My Life'.

Figure 1.9
Results returned by
a simple search of
the Session folder
with no search
criteria

Pro Tools 11 also provides advanced tools for searching files and volumes by specific categories. To initiate an advanced search, click the Advanced Search button or, if you prefer to use a keyboard command, press Command-F (Mac) or Control-F (Windows).

In the Locations pane, select the volume, catalog, session, or folder that you wish to search, select the column type and the search criteria, then click in the Search Text field and type the text for your search. When you press Enter, the search results will be displayed in the Browser pane.

In the example shown in Figure 1.10, the session is selected in the Locations pane and the first two letters of the word 'flute' have just been typed into the Advanced Search field. The three flute recordings, located inside the Session Audio Files folder, have been found and are displayed in the Browser pane.

Figure 1.10
Advanced search for
flute recordings

You can further refine your search using up to eight rows of search criteria. Click the Add Row button (the '+' sign to the right of the search field) to add an additional search field with associated search criteria pop-ups—see Figure 1.11.

Figure 1.11
Additional search
criteria

Pop-up Browser Menu

All Workspace browsers provide a Browser menu in the upper right corner of their toolbar. Workspace Browser menus provide commands specific to the type of items selected in the Locations pane or the Browser pane. You can also Right-click on items in the Locations pane or in the Browser pane to access relevant Browser menu commands.

For example, if you select a file in the Browser pane and click on the Browser menu, or simply Right-click the item, you can access commands such as Reveal in Finder (Mac) or Reveal in Explorer (Windows)—see Figure 1.12.

Similarly, if you select any volume in the Locations pane and click on the Browser menu, or simply Right-click the item, you can access commands such as Update Index for Selected Items or Delete Index for All Offline Volumes—see Figure 1.13.

Figure 1.12
Browser Menu 1

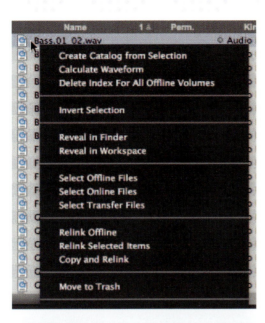

Figure 1.13
Browser Menu 2

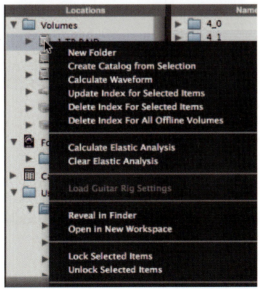

Previewing Audio in the Browsers

To audition an audio file, navigate to a folder containing audio files, click on any of the files within the folder, then click on the Preview button in the toolbar at the top of the window.

13

To the right of the loudspeaker icon there is a numerical display that defaults to 0.0 indicating that the playback level is at 0 dB—that is neither boosted nor attenuated from the level in the audio file. If you click on this, a small fader appears that you can use to increase or decrease the playback level of the audio in the file—see Figure 1.14.

TIP

If you press the down arrow on your computer keyboard while one file within a folder is being previewed, the next file down in the folder will start playing instead. This lets you go through a whole folder of audio files very quickly until you find the one you want.

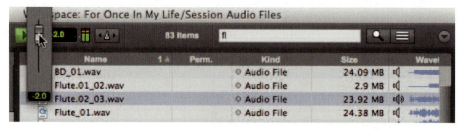

Figure 1.14
Auditioning
audio files in the
Workspace browser

As usual, Pro Tools offers different ways to do the same thing, so you can also preview any audio file in the browsers by clicking the Waveform Preview button (the speaker icon to the left of the waveform display at the right of the browser window)—see Figure 1.15. Just click again to stop auditioning.

If you prefer to use keyboard commands, when a browser is in front of the other Pro Tools windows you can use Command-P (Mac) or Control-P (Windows) to start and stop previewing—and you can use the Escape (Esc) key to stop previewing.

Figure 1.15
Previewing a file
in the Workspace
browser

> **TIP**
> If you want to play from somewhere within the file, just click in the waveform display at the point you wish to play from.

Preview Modes

Pro Tools browsers offer three preview modes: Normal Preview, Loop Preview, and Auto Preview. These are particularly useful when you are looking through your libraries of loops or sound effects. You can go through your files in turn using Normal mode, looping them in Loop mode if you need to get an idea of how they will sound looped, or letting Auto mode play through them for you.

Normal Preview simply plays the selected audio file then stops at the end of the file. If you select more than one file, it plays through each file in order.

Using the Browser menu, you can select Loop Preview or Auto Preview and choose whether to enable or disable the spacebar toggle and conform to session tempo options—see Figure 1.16.

Figure 1.16
Previewing items in
the Browser menu

Figure 1.17
Loop Preview
Mode icon

Loop Preview plays the selected file by looping playback of the file. If more than one file is selected, only the first will be played. When Loop Preview mode is enabled, the Preview button shows a Loop Preview Mode icon to remind you that you are in this mode—see Figure 1.17.

Figure 1.18
Auto Preview
Mode icon

Auto Preview plays the audio file as soon as it is selected, and the Preview button shows an Auto Preview icon to remind you that this mode is active—see Figure 1.18.

Figure 1.19
Preview button
Right-click menu

You can also Right-click the Preview button to change the preview mode and set whether or not the spacebar toggles file preview—see Figure 1.19.

> **TIP**
>
> If 'Spacebar Toggles File Preview' is selected, you can use the spacebar to start and stop auditioning. When this option is disabled, you can use the spacebar to start and stop session playback. Using this configuration, and with the 'Audio Files Conform to Session Tempo' option enabled, you can preview the selected audio file in time with the session.

> **NOTE**
>
> You can always use the Preview button to start and stop preview. Also, when a Workspace browser is front most, Control-P (Windows) or Command-P (Mac) starts and stops preview and the Escape (Esc) key stops preview.

Conforming Audio Files to the Session Tempo

When you preview 'sample-based' audio files in a browser, they will play at their original tempo—and if you import them they will stay at that tempo. When you preview or import 'tick-based' REX and ACID files, these will adjust automatically to the tempo of your session.

The Pro Tools Elastic Audio feature can be used to analyze any other standard audio files so that these will also be automatically adjusted to the session's tempo while being previewed or after being imported.

Make sure that the 'Audio Files Conform to Session Tempo' option is enabled in the Browser menu or click on the 'Audio Files Conform to Session Tempo' button in the browser's toolbar. With this feature enabled, Pro Tools will analyze any audio files that have not yet been analyzed by Elastic Audio when you preview or import these.

> **NOTE**
>
> When the 'Audio Files Conform to Session Tempo' option is disabled, any tick-based audio files with Elastic Audio analysis, and REX and ACID files, preview and import at their native tempo.

Elastic Audio Plug-In Selector

The Workspace browser's toolbar has a convenient Elastic Audio Plug-In selector that lets you select any real-time Elastic Audio plug-in as the default plug-in for previewing and importing Elastic Audio files.

You can access the Elastic Audio Plug-In selector by Right-clicking on the Audio Files Conform to Session Tempo button in the toolbar—see Figure 1.20.

Figure 1.20
Accessing the
Elastic Audio
Plug-In selector in
the toolbar

NOTE
Changing the Elastic Audio plug-in in any Workspace browser also affects the 'Elastic Audio Default Plug-In' option in the processing preferences.

Elastic Audio in the Workspace

You can use Elastic Audio to analyze WAV or AIFF audio files in Workspace browser—MP3 files or files in other formats must be converted to WAV or AIFF first.

When a tempo has been detected in an analyzed file, this will be updated as tick based and will also display its duration in Bars|Beats and its native tempo in the Tempo column.

NOTE
MIDI, REX, and ACID files are always tick based and contain their own tempo and Bar|Beat metadata.

Analyzed files in which no tempo was detected remain sample based. These files typically contain only a single transient (such as a snare hit) or they are longer files without a readily identifiable regular tempo (such as entire songs).

> **NOTE**
> On the Mac, if you do not see the Elastic Audio Analysis icon or the duration does not change to tick based during the preview process, check the permissions for the folder in the Mac Finder. You must have write access to the directory for this feature to work.

Identifying Elastic Audio–Analyzed Files in the Browsers

Workspace browsers provide two different icons for sample-based and tick-based files:

The Sample-Based File icon, which looks like a small clock face to the left of the legend 'Audio File' in the browser's 'Kind' column, indicates that the file is sample based. The file's duration is displayed in minutes and seconds. Sample-based WAV and AIFF files without Elastic Audio analysis data display a small grey clock icon. When sample-based files have been analyzed with Elastic Audio, they display a small blue clock icon.

The Tick-Based File icon, which looks like a small metronome to the left of the legend 'Audio File' in the browser's 'Kind' column, indicates that the file is tick based. The file's duration is displayed in Bars|Beats and the file's native tempo is displayed in the tempo column. Tick-based WAV and AIFF files with Elastic Audio analysis display a green metronome icon.

Elastic Audio 'Batch' Analysis

When you select a file to preview that has not already been analyzed with Elastic Audio, you will have to wait until the analysis is complete before you can preview this at the session tempo. To avoid this, select the files you are interested in—a whole folder if you like—and choose 'Calculate Elastic Analysis' from the Browser menu to analyze the batch of files—see Figure 1.21.

In the example shown in Figure 1.22, the analyzed files have been converted to tick based with green metronome icons, their durations are shown in Bars|Beats, and the tempos are displayed in BPM.

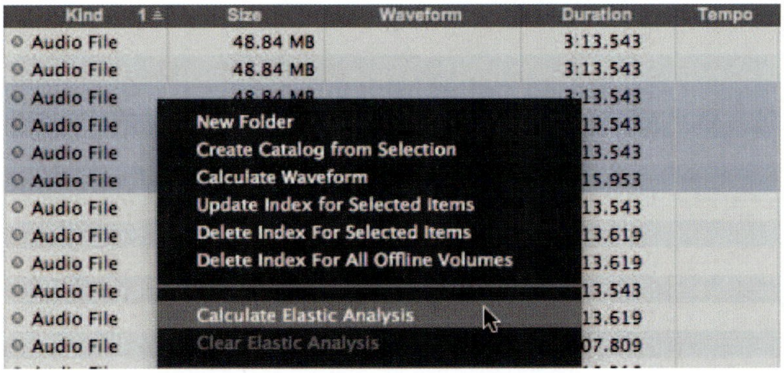

Figure 1.21
Elastic Audio 'batch'
analysis

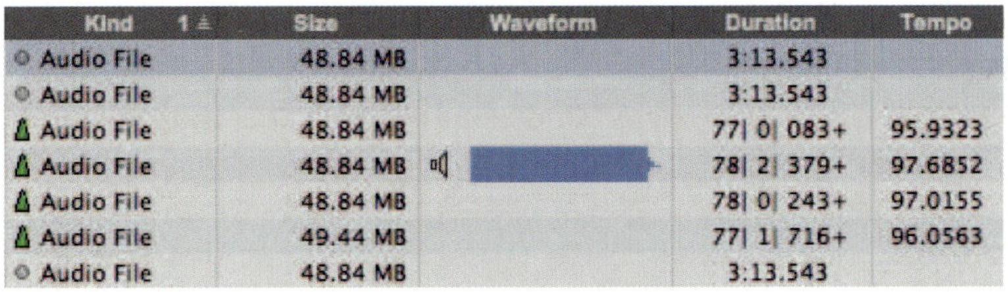

Kind	1 ≜	Size	Waveform	Duration	Tempo
○ Audio File		48.84 MB		3:13.543	
○ Audio File		48.84 MB		3:13.543	
○ Audio File		48.84 MB		3:13.543	
○ Audio File		New Folder		13.543	
○ Audio File		Create Catalog from Selection		13.543	
○ Audio File		Calculate Waveform		15.953	
○ Audio File		Update Index for Selected Items		13.543	
○ Audio File		Delete Index For Selected Items		13.619	
○ Audio File		Delete Index For All Offline Volumes		13.619	
○ Audio File				13.543	
○ Audio File		Calculate Elastic Analysis		13.619	
○ Audio File		Clear Elastic Analysis		07.809	

Figure 1.22
Elastic Audio–
analyzed file
identifier icons

Kind	1 ≜	Size	Waveform	Duration	Tempo
○ Audio File		48.84 MB		3:13.543	
○ Audio File		48.84 MB		3:13.543	
⌀ Audio File		48.84 MB		77| 0| 083+	95.9323
⌀ Audio File		48.84 MB		78| 2| 379+	97.6852
⌀ Audio File		48.84 MB		78| 0| 243+	97.0155
⌀ Audio File		49.44 MB		77| 1| 716+	96.0563
○ Audio File		48.84 MB		3:13.543	

> **NOTE**
> The 'Clear Elastic Analysis' command can be used to clear Elastic Audio analysis data from selected audio files. Files cleared of Elastic Audio analysis data revert to being sample based, display their duration in minutes and seconds, and do not report a tempo.

The Audition Path

The audio that you are auditioning in the browsers will play back via the master Audition path that you can select in the Output pane of the Pro Tools I/O Setup dialog. This defaults to the first pair of stereo outputs on your interface, but you can choose any outputs that are available on your system. See Figure 1.23.

There are some rules that apply to previewing the different file types. For example, multi-mono files must be previewed one file at a time; interleaved

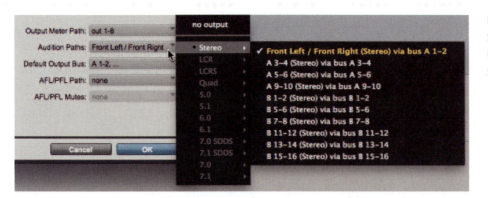

Figure 1.23
Setting the Audition
Path in the I/O
Setup window

files with more than two channels will be summed to mono for auditioning; and if you select one of a pair of Split Stereo files, such as 'Drums.L' and 'Drums.R', both files will play together.

> **TIP**
> If you just want to preview the selected channel of a split stereo pair, press and hold the Shift key while you initiate the preview.

Preview in Context

Sometimes, you may actually be working in your Pro Tools session when you decide that you would like to find a suitable loop or sound effect or other audio material to place at a particular bar and beat location. You want to hear the audio you are auditioning play back at the same time as the session—and you want these both to play from the appropriate bar and beat location.

Pro Tools will let you do this: first make sure that the 'Audio Files Conform to Session Tempo' option is enabled. In the Edit window, place the insertion point at the location where you want to preview the file, then click the Preview button in the browser. Any tick-based audio files with detected tempos will play back in tempo, and at the corresponding bar and beat location, while the session plays back.

You can check whether the audio file is tick based by checking its file icon in the browser and you can see that it has been analyzed if this is a green metronome icon.

TIP
If you want to use the spacebar to start and stop session playback instead of starting and stopping auditioning in the browser, you will need to deselect the 'Spacebar Toggles File Preview' option in the Browser menu. To start and stop preview in the front-most browser during session playback you can always use Command-p (Mac) or Control-p (Windows) from your computer keyboard.

Another thing to watch out for is that the 'Preview in Context' feature needs to use a number of disk voices for playback. The actual number depends on the number of channels used for (in other words, the channel width of) the selected Audition Path. So a stereo audition path uses two voices for preview in context and a 5.1 audition path uses six!

To make sure voices are available, Pro Tools provides an option in the Operation Preferences window to 'Reserve Voices for Preview in Context'. When this option is enabled, the number of reserved voices will always correspond exactly to the channel width of the selected Audition Paths in the I/O Setup.

NOTE
If no voices are available for Preview in Context, perhaps because they are all being used to play back tracks from disk, the Preview button in Workspace browsers will not work while you are playing back the session.

Task Manager Window

The Task Manager window is a utility for viewing and managing all of the background tasks that you initiate with Pro Tools. You can use the Task Manager window to monitor, pause, or cancel background tasks such as file copies, searches, fade regeneration, and indexing.

Relink Window

The Relink window provides tools and features for relinking sessions and catalogs to media files. You can use these Relink tools to search for and reacquire missing files for use in the current session. Files may be missing because they cannot be found where the session expects them, either because they have been moved, or because they are stored on volumes that are not currently mounted (offline volumes).

When you open a session, if it finds files on a volume unsuitable for playback, a dialog appears prompting you to copy them to a suitable volume.

> **TIP**
>
> The Workspace browser's 'Copy and Relink' command provides a convenient way to copy files and relink the session or catalog to the copies rather than to the originals.

If files are missing when you open a session, Pro Tools opens the session with all the files that it can find (if any), then opens the Missing Files dialog, which shows you how many files are missing and asks how you want to proceed—see Figure 1.24.

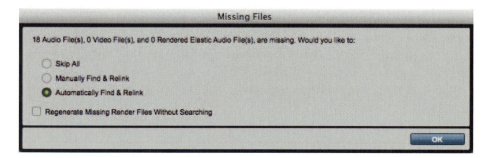

Figure 1.24
The Missing Files dialog

'Automatically Find and Relink' is the simplest method to relink sessions to required media, but it provides no way to compare files or verify links.

The 'Automatically Find and Relink' option searches all performance volumes for all missing items with matching name, ID, format, and length, then links missing items to the first matches found and commits these links.

If some (or all) of the files remain unlinked, the Task Manager window opens and any failed tasks, such as files not found, will appear in the window—as in Figure 1.25.

Figure 1.25
The Task Manager
window

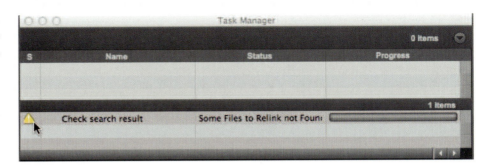

In the Task Manager window you can double-click the triangular yellow Task icon to open the Relink window (see Figure 1.26), which allows you to search for, compare, verify, and relink missing files.

Figure 1.26
The Relink window

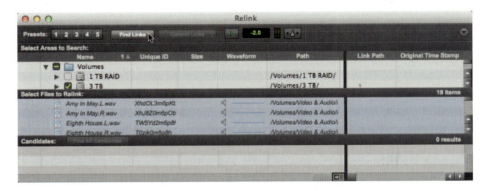

Figure 1.27
The Linking Options
dialog

To relink missing files in batches, select one or more items in the 'Select Files to Relink' list then click the 'Find Links' button that becomes available in the toolbar at the top of the window. The Linking Options dialog appears, allowing you to refine the search according to various criteria—see Figure 1.27.

Pro Tools searches, matches, and links each missing file in the 'Select Files to Relink' list and a link icon appears next to each file that Pro Tools has found—see Figure 1.28.

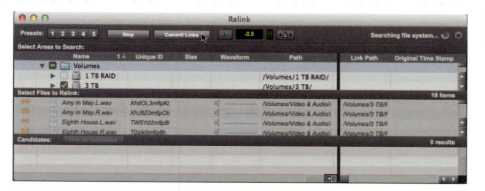

Figure 1.28
Commit Links

To view a candidate for a missing file, select to highlight the missing file. That file's candidate appears in the Candidates pane, where you can view its information and waveform.

When you are satisfied that Pro Tools has found and linked the correct files, click on the Commit Links button that becomes active in the Relink window's toolbar.

If you are not satisfied with the candidate for a particular missing file, you can click 'Find All Candidates' for a selected file to search for this individual file—see Figure 1.29.

Figure 1.29
Find all candidates
for a selected file

One or more potential 'candidates' will be listed in the Candidates pane after the search has been completed. Click the Link icon next to the appropriate candidate to relink this to the item currently selected in the Files to Relink list—see Figure 1.30. To complete the process, click the Commit Links button in the toolbar.

Figure 1.30
Relinking an
individual file

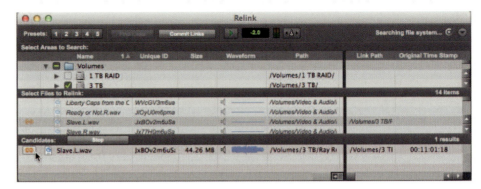

Importing Files into Pro Tools Sessions

You can import AIFF or WAV or BWF (.WAV) (including WAV Extensible and RF64) audio files into Pro Tools sessions without conversion.

Other types of audio files can be imported into Pro Tools sessions, but will be converted to the current session file format on import. The list includes AAC audio (including audio with AAC, mp4, and m4a file extensions); ACID files; MP3; MXF audio; QuickTime (Mac only); ReCycle (REX 1 and 2) files; SD II; SD I; and Sound Resource (AIFL–Mac only).

Importing Audio from Audio CDs

Pro Tools lets you import tracks from audio CDs using the same methods that you can use to import audio files: so you can use the 'Import Audio' command, or drag CD audio from the CD folder, or drag files from a Workspace browser. Because the transfer is made in the digital domain, there is no possibility of any loss in quality—unless a sample rate conversion is required, in which case it is possible for some small loss of quality.

The sample rate for audio CDs is 44.1 kHz, so, if your session's sample rate is set to 48 kHz or higher, Pro Tools will need to convert the sample rate for the imported audio. Before importing audio tracks from CD, you should set the Sample Rate Conversion Quality in the Import preferences—see Figure 1.31. I always choose 'TweakHead' to get the highest quality. If you are using a slow computer and you are in a hurry, you might want to choose a lower-quality conversion.

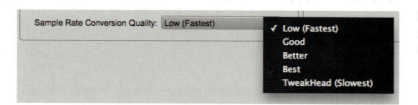

Figure 1.31
Sample Rate
Conversion Quality

Importing Elastic Audio, REX, and ACID Files

Pro Tools can import REX files created using ReCycle software, ACID files created using ACID software, and any files that have been analyzed using Elastic Audio.

> **NOTE**
> When the Elastic Audio feature analyzes audio files, it marks these with a tempo in a similar way to ACID files so that Pro Tools can control the playback tempo of these files.

REX files are created using Propellerhead's ReCycle software. They contain audio 'loops' edited into small slices and stored in a type of AIFF file along with information about the original loop tempo. You can play these back using software such as Reason or Cubase and if you change the tempo, these audio 'loops' immediately follow the tempo changes without any need for time stretching or other processing. Pro Tools audio tracks can do this if you change the time base from sample based to tick based. Simply drag and drop a REX file from a Workspace browser or from the desktop to a track, the Tracks list, the Timeline, or the Clips list.

> **NOTE**
> ReCycle software lets you slice up an audio file containing, for example, four bars of a drum kit playing a pattern that could be repeated or 'looped' to form the basis of a rhythm track. The results can be saved into a REX format file that contains all these slices, each of which may only contain a single drum 'hit'. When these slices are played back in turn from within a sequencer at the original tempo, you hear the original drum pattern at the original tempo. If you increase or decrease the tempo of the sequencer, the slices will simply be played back more quickly or more slowly, so the tempo of the drum pattern will follow the tempo of the sequencer.

Depending on the settings that you choose in the Import section of the Processing Preferences window (see Figure 1.32), a REX file may be converted to a tick-based clip group on import, with all its slices converted into individual audio files and clips. REX files can be imported either as sample-based clips or as tick-based Elastic Audio clips. During the Elastic Audio analysis process, Pro Tools uses the slice data in the REX files to create event markers with 100 percent confidence, and the REX files are converted to the session's audio file type.

Figure 1.32
Import section
of the Processing
Preferences window

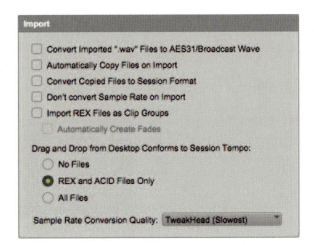

ACID files are a 'stretchable' and 'tunable' format created using Sony's ACID software for the PC. They contain metadata about tempo, number of bars, pitch, and slice information along with the audio in a type of WAVE file. You can simply drag and drop ACID files, like REX files, from a Workspace browser or from the desktop to the Timeline, a track, the Tracks list, or the Clips list.

> **NOTE**
> While they are being created using Sony's ACID software, ACID files are analyzed to discover the tempo of the audio they contain. The files are then marked with this tempo in a way that can be read by various software applications, including ACID itself—and Pro Tools. After importing into ACID or Pro Tools, when the tempo of the sequence is increased or decreased, time stretching can automatically be applied to these files, starting from the known tempos marked in the ACID files and moving these to the new tempo set in the sequencer.

Pro Tools users upgrading from older versions will recall that ACID files used to have to be imported into Pro Tools as Region (now Clip) Groups. ACID files can now be imported either as sample-based clips or as tick-based Elastic Audio clips. During the process, ACID files are analyzed for Elastic Audio events just like WAV files, so events are detected with varying degrees of confidence, and ACID files are converted to the session's audio file type.

Typically, REX and ACID files will be exactly 1-, 2-, 4- or 8-bars long and are used as basic 'building blocks' for musical arrangements. To build up 16- or 32-bar verses, choruses, or other sections of the musical arrangement, they are repeated or 'looped'. Often they contain drumbeats, but can also contain bass lines, keyboard pads, guitar licks, or any type of sounds.

How to Import a REX or ACID File

You can drag and drop a REX or ACID file into any existing audio track, but unless the Time Base Selector is set to Ticks (see Figure 1.33), the audio will stay at the original tempo when you change the tempo in Pro Tools.

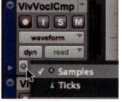

Figure 1.33
Click and hold the Time Base Selector to open the pop-up selector and choose 'Ticks'.

If you drag and drop a REX or ACID file into an audio track that is in Ticks mode, you can make the tempo faster or slower and the audio in the REX or ACID file will automatically follow the tempo changes.

> **NOTE**
> If you drag and drop a REX or ACID file into the empty space in a new session or below any existing tracks in the Edit window, Pro Tools automatically creates a new track for this and sets it to tick based.

Setting the Preferences for Importing REX Files

REX files can be imported as Clip groups, which makes sense as this keeps all the files together in a grouped Clip that is easier to manipulate in the Edit window in Pro Tools. For this to happen, the 'Import REX Files as Clip Groups' option must be enabled in the Processing Preferences page. All the individual slices are then imported as individual Clips contained within the Clip group.

> **NOTE**
> When 'Import REX Files as Clip Groups' is not selected, importing REX files into a session converts them to the session's audio file format. The individual slices are then consolidated and Elastic Audio analysis is applied to the slices so that they become tick-based files.

If you are working with REX files in Pro Tools, you are also very likely to want to apply fades and crossfades between the slices as you import these into Pro Tools. This can be done automatically if you enable the 'Automatically Create Fades' option in the Processing Preferences page—see Figure 1.34.

Figure 1.34
Part of the
Processing
Preferences page
showing the
'Import REX Files
as Clip Groups'
and 'Automatically
Create Fades'
options

In this case, in the Fades section of the Editing Preferences page you should also select 'REX' as the default envelope shape for the fades and crossfades that will be applied automatically between the slices in the imported REX files.

To change the default fade settings for REX files, click the REX button in the Default Fade Settings section of the Editing Preferences page. This opens the REX Auto-Crossfades dialog that lets you 'tweak' the default envelope shape for the fades and crossfades that will be automatically applied. See Figure 1.35.

Setting the Processing Preferences to Conform Audio to the Session Tempo

The 'Drag and Drop from Desktop Conforms to Session Tempo' option lets you choose whether or not REX, ACID, and audio files will be imported as tick-based Elastic Audio and conformed to (so that they will play back at) the session tempo when you use drag and drop. There are three options:

When 'No Files' is enabled, no audio files are conformed to the session tempo when imported by drag and drop from the Mac Finder or Windows Explorer. All audio files are imported as sample-based files and converted to the session's audio file format.

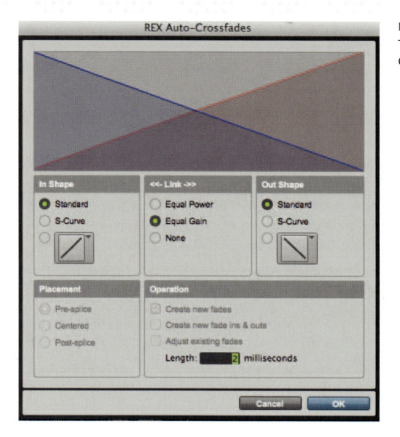

Figure 1.35
The REX Auto-
Crossfades dialog

When 'REX and ACID Files Only' is enabled, only REX and ACID files are conformed to the session tempo when imported by drag and drop from the Mac Finder or Windows Explorer. REX files are imported either as tick-based Elastic Audio or (if the 'Import REX Files as Clip Groups' option is enabled) as tick-based Clip groups.

When 'All Files' is enabled, all audio files are imported as tick-based Elastic Audio that will run at the session tempo—see Figure 1.36.

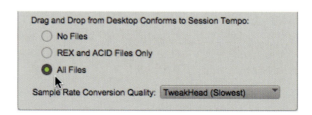

Figure 1.36
The 'Drag and Drop from Desktop Conforms to Session Tempo' options in the Import section of the Processing Preferences page

Importing Elastic Audio Files

When importing any audio file by drag and drop to an Elastic Audio–enabled track, the audio file is imported and analyzed (if not already analyzed).

If the imported audio file is tick based, it is conformed to tempo using the track's Elastic Audio plug-in.

If it is a sample-based audio file, such as a single drum hit or short sound effect, it maintains its original duration. Longer audio files with no detectable tempo also remain sample based.

You can also import audio by drag and drop to create new tick-based Elastic Audio–enabled tracks.

If you are importing an audio file by drag and drop from Windows Explorer or Mac Finder to the Track list or to empty space in the Edit window, files are imported as sample-based or as tick-based Elastic Audio depending on the 'Drag and Drop from Desktop Conforms to Session Tempo' setting in the Import section of the Processing Preferences page. If you want audio files to be conformed, you need to select the 'All Files' option here—see Figure 1.36.

When importing tick-based audio by drag and drop from Workspace browsers to the Tracks list or empty space in the Edit window, Pro Tools creates a new tick-based, Elastic Audio–enabled track using the default Elastic Audio plug-in selected in the Processing Preferences page—see Figure 1.37.

Elastic Audio Processing Preferences

Three processing preferences need to be set correctly to control the way you import Elastic Audio files. The first of these lets you choose the default plug-in that will be used for previewing and importing Elastic Audio. The second preference lets you adjust the input gain for the signal entering the plug-in. The third preference lets you choose whether new tracks are created with Elastic Audio enabled. If you tick this option, these new tracks will use the selected default plug-in.

Figure 1.37
Part of the Processing Preferences page showing the Elastic Audio preferences

Default Plug-in

By default, the Polyphonic plug-in is selected for previewing and importing from Workspace browsers or the Clips list. This is a good choice because the Polyphonic plug-in uses a general, all-purpose algorithm that is effective with a wide range of material including complex loops and multi-instrument mixes.

The other choices include Rhythmic, Monophonic, and Varispeed. The Rhythmic plug-in is best suited to material with clear attack transients, such as drums. The Monophonic plug-in is best suited to monophonic material where you want to keep the formant relationships intact, such as with vocals. The Monophonic plug-in is also well suited to monophonic instrumental lines, such as a bass track. With the Varispeed plug-in, if you make the audio play back faster, the pitch rises, and if you play the audio back more slowly, the pitch falls—just like when you vary the speed of a tape recorder.

NOTE

The Elastic Audio Plug-in selector in Workspace browsers inherits the selected default plug-in. Likewise, changing the selected plug-in in Workspace browsers updates the selected default plug-in in the Processing Preferences page.

Default Input Gain

The 'Default Input Gain' setting in the Elastic Audio Preferences window lets you lower the gain of the signal going into the Elastic Audio plug-ins by up to 6 dB when previewing and after importing. Why might you want to do this? If you find that the audio is clipping while you are previewing or after importing the file, you can simply lower the Input Gain until the clipping disappears, then preview or import the Elastic Audio processed file again.

NOTE

This preference also applies to any audio imported to an Elastic Audio–enabled track.

Normally, you would set the default preference for Input Gain based on your initial experiences with the audio for the project that you are working on. You would then make further adjustments to this Input Gain setting using the Elastic Properties window. This 'inherits' the Default Input Gain preference that you have chosen in the Processing Preferences. To apply further clip-based Input Gain for Elastic Audio processing, just select the clip and adjust the Input Gain setting directly in the Elastic Properties window.

Enable Elastic Audio on New Tracks

When 'Enable Elastic Audio on New Tracks' is selected in the Elastic Audio Preferences window, new tracks are created with Elastic Audio enabled and the selected default plug-in is used.

> **TIP**
> If you have chosen to use the 'Enable Elastic Audio on New Tracks' option, it makes sense to also choose the 'New Tracks Default to Tick Timebase' option in the Editing Preferences page—see Figure 1.38.

Figure 1.38
Editing Preferences:
Tracks

Importing Tick-based Audio Files from the Workspace to the Clips List

The Elastic Audio feature is very useful when bringing files from the Workspace into a session. Here's how this works:

When you have the 'Audio Files Conform to Session Tempo' option enabled, dragging and dropping tick-based audio from Workspace browsers to the Clips list (or to the Tracks list) creates two separate clips in the Clips list.

Take a look at OttoBass.08 in Figure 1.39 where you will see that this is shown as a single entry in the Workspace that becomes two items when dragged into the Clips list.

The first is a sample-based whole file clip and the second is a tick-based copy of the same clip. This can be seen more clearly after these two clips have been dragged into the Edit window, as shown at the left of Figure 1.39.

Figure 1.39
Dragging files from
the Workspace to
the Clips list, then
from the Clips list to
the Edit window

Importing Audio Files and Loops

You can always drag and drop audio files from the desktop into a Pro Tools
session and a track will be automatically created—or you can drop files onto
an existing track. If it contains audio that you want to run at the same tempo
as your session, simply change the track to tick based and use Elastic Audio to
conform it to the session tempo. You can then apply grooves to your imported
loops.

Importing Tempo from Tick-based Audio Files

If there are no tracks in the session, and you import a tick-based audio file to the
Clips list, Tracks list, or empty space in the Edit window, you will be prompted
either to import the tempo from the file or to use the default session tempo.
To keep the default session tempo, and have the loop conform to the session
tempo, click 'Don't Import'.

If the session already contains at least one track (regardless of track type), you
will not be prompted to import the tempo from the file and the file will be
conformed to the session tempo.

Summary

How to find and organize your files and how to bring in existing audio files to a
Pro Tools session are some of the most basic tasks that you need to be familiar
with when working on any project.

It is easy enough to drag and drop audio from the desktop, but using the
Workspace browsers lets you see much more information about these files
and, most important, lets you audition them.

If you are working with loops, then you should make sure that you fully understand how to work with Elastic Audio–analyzed, REX, and ACID files, and how to set all the preferences for these.

It won't take you long to master all this stuff, and you will definitely reap the benefit when things are busy and you need to be as organized as possible!

in this chapter

CHAPTER 2

Working with Clips

After you have brought audio and loops into Pro Tools it is vital to gain familiarity with how to move audio clips around in the Edit window so that these can be edited and corrected as you develop your arrangements. So, for example, you need to know how to make timeline and edit selections, how to loop clips, and how to use Dynamic Transport mode.

Making Accurate Selections in the Edit Window

Making an Edit Selection

To make an edit selection, you can simply drag the mouse over any clips or tracks in the Edit window when the selector tool is enabled. If you are using Grid mode, your selections will be constrained to the grid settings.

If you are using Slip mode, you may prefer to specify the edit selection numerically to make an accurate selection using the Edit Selection indicators in the Event Edit area at the top of the Edit window. The Edit Selection indicators use whichever time format you have selected for the Main Time Scale, so if you want to work with Bars|Beats, you should make sure that you have selected this format first—see Figure 2.1.

Figure 2.1
Setting the time
format for the Main
Counter

To make your edit selection, click with the selector tool in the track that you want to select, then click in the Edit Selection Start field in the Event Edit area at the top of the Edit window. Type in the start point for the selection and press the Forward Slash key (/) to enter the value and automatically move to the Edit Selection End field where you can type in the end point for the selection. Press Enter to accept the value and you will see your edit selection made—see Figure 2.2.

If you prefer to use the mouse, with the selector tool active, you can simply drag the mouse in the Edit window to highlight the selection—using Grid mode if you want to start and end on grid lines.

Figure 2.2 Edit window showing the edit selection made using the edit selection indicators

Working with Multiple Tracks

To make edits across multiple tracks or in all tracks, you must select the tracks first. The easiest way to do this is to drag the insertion cursor across all the tracks you want to select, dragging upward or downward to include adjacent tracks and dragging horizontally to define the time range. Choose the selector tool and point your mouse at any track of interest in the Edit window. Click and hold the mouse button, then drag horizontally to select a clip. Without letting go of the mouse button, drag vertically upward or downward to select additional tracks of whatever type. Once you have made your selection, you can let go of the mouse button. See Figure 2.3.

Figure 2.3
Make a selection
across multiple
tracks by dragging
across these using
the selector tool.

Figure 2.3
Make a selection across multiple tracks by dragging across these using the selector tool.

After you have made a selection and let go of the mouse button, if you change your mind and want to extend the selection to include more tracks or to shorten or lengthen the selection, press and hold the Shift key while you drag using the selector tool. See Figure 2.4.

Figure 2.4
Extend the selection by holding the Shift key and dragging to the right using the selector tool

If you click in the Edit window without holding the Shift key (and you will, sooner or later), you will lose your original selection. If this happens, you can use the 'Restore Last Selection' command from the Edit menu to get your selection back. See Figure 2.5.

Figure 2.5
'Restore Last Selection' command

TIP

To select and edit all the tracks simultaneously, you can drag with the selector tool in any of the Timebase rulers with the Timeline and Edit Selections linked, or with the 'All' edit group in the Groups List enabled—see Figure 2.6.

Figure 2.6
Enabling the ALL edit group in the groups list

If you expect to regularly make the same edits to particular groups of tracks, you can group these tracks together so that edits applied to one will apply to all. Typically you might do this with your drum tracks or with a brass section or a set of guitar or banjo tracks—see Figure 2.7.

To create a group of tracks, click on the track names that you want to group and use the 'Group . . .' command from the Track menu.

Figure 2.7
Four guitar tracks grouped together so they can be edited together

NOTE

If a group of tracks in the Edit window contains tracks that are hidden (i.e., deselected in the Tracks show/hide list at the left of the Edit window), these will not be affected by any edits made to the members of the group that are visible in the Edit window.

To paste to multiple tracks, engage the selector tool, then place the insertion point in each of the destination tracks by Shift-clicking in them. If you want to place the insertion point into all tracks, click in any of the Timebase rulers.

Timeline and Edit Selections

To make a timeline selection (as opposed to an edit selection among the tracks), you can either type your selection into the Transport window or drag with the selector or time grabber tool in any Timebase ruler.

The timeline selection is indicated in the Main Timebase ruler by blue Playback Markers, which turn red if a track is record enabled. The start, end, and length for the timeline selection are displayed in the corresponding fields in the Transport window.

You can also set the timeline selection by dragging the timeline selection markers. Using either the selector or time grabber tool, drag the Timeline Selection In Point (the downward-pointing arrow) to set the start or in point and drag the Timeline Selection Out Point (the up arrow) to set the end or out point. See Figure 2.8. You can also create selection in and out points in the timeline by pressing the Down and Up arrow keys, respectively, during playback.

Figure 2.8
Setting the
Timeline Selection
In and Out Points
by dragging the
timeline selection
markers

If you press the Option key (Mac) or the Alt key (Windows), you can drag either the Timeline Selection Start or End marker and the whole timeline selection will move forward or backward along the Main Timebase ruler as you drag. See Figure 2.9.

Figure 2.9
Dragging the
Timeline Selection
In Point while
holding the Option
(Alt) key to move
the whole timeline
selection

Edit Selection Markers

If the timeline and edit selections are unlinked, separate gold-colored edit selection start and end markers appear in the main Timebase ruler to let you define the edit selection separately from the timeline selection. These work similarly to the timeline selection markers. So, for example, you can drag either the Edit Selection Start (In Point) or End (Out Point) marker while holding the Option (Alt) key and the whole edit selection will move backward or forward along the timeline as you drag. See Figure 2.10.

Figure 2.10
Dragging the
gold-colored Edit
Selection In Point
marker while
holding the Option
(Alt) key to move
the whole edit
selection

Linked Timeline and Edit Selections

By default, Pro Tools sessions are set up so that the timeline and edit selections are linked.

With the timeline and edit selections linked, when you make an edit selection in a track the same time range automatically becomes selected in the Timebase rulers and a small blue arrow is placed at each side of the timeline selection in the Main Timebase ruler. See Figure 2.11.

Figure 2.11
Edit window showing
a highlighted edit
selection in a track
with the corresponding
timeline selection
made in the Bars|Beats
Timebase ruler

To toggle 'Link Timeline and Edit Selection' on and off you can either use the button in the toolbar (see Figure 2.12) or press Shift and Forward Slash (/) on your computer keyboard.

Figure 2.12
Enabling the Link
Timeline and Edit
Selection button in
the toolbar

Unlinked Timeline and Edit Selections

When you unlink the timeline and edit selections, you can make different selections in the Edit display and in the rulers for the timeline.

To make a timeline selection, choose the selector tool and drag the cursor along the Main Timebase ruler to set the playback and recording range. You can edit this selection by dragging the blue arrows (the timeline selection markers) that appear in the Timebase ruler to mark the beginning and end of your timeline selection. If a track is record enabled, the timeline selection markers will be red instead of blue.

To make an edit selection, choose the selector tool and drag the cursor across the Edit window, or you can enter the start and end locations into the edit selection indicators in the Event Edit area in the toolbar at the top of the Edit window. You can edit this selection by dragging the gold-colored brackets (the edit markers) that appear in the Timebase ruler to mark the beginning and end of your edit selection.

So why would you want to unlink the timeline and edit selections? Take a look at Figure 2.13 in which the timeline selection (the darkened area in the Bars|Beats ruler) is defining the range to be looped during playback while an edit selection containing a MIDI clip within the loop is being moved around using the separation grabber. During playback, the edit selection can be moved around like this, or quantized or transposed or whatever, while the loop plays back without interruption.

Figure 2.13
Timeline selection
defines the
playback loop while
the edit selection
in the MIDI track
can be moved or
otherwise edited
during playback`

Here's another example. Suppose you are working on chorus vocals at the end of a song, and you are trying to decide whether to copy the last two choruses to replace the first two choruses. You have made your edit selection to play the first two choruses, then you *unlink* the timeline and edit selections so that you can make a timeline selection around the two end choruses without losing your edit selection—see Figure 2.14.

Now you can audition the end choruses, then relink the timeline and edit selections so that the timeline selection corresponds to the edit selection around the first two choruses once more, listen again, and decide which to use.

Figure 2.14
Different timeline
and edit selections
in the Edit window

Even more convenient, two commands in the Edit menu let you play the edit selection instead of playing the timeline selection (which is what happens when you use the normal 'Play' command or hit the spacebar) or vice versa—see Figure 2.15. The current timeline selection stays in place when you play the edit selection using this command, so you can search through the Edit window, selecting and auditioning clips of potential interest. When you find material that you like, copy that selection, then go back to the timeline selection and paste the material that you have found.

Figure 2.15
'Play Edit Selection'
command in the
Edit menu

TIP
You can also use keyboard commands to play the timeline or edit selections: hold the Option key (Mac) or Alt key (Windows) and press the left bracket key to play the edit selection or press the right bracket key to play the timeline selection.

Selection Commands

When you have made separate edit and timeline selections, two other selection commands in the Edit menu become very useful.

If you select 'Change Timeline to Match Edit' from the Edit menu or press Option-Shift-5 (Mac) or Alt-Shift-5 (Windows), this will change the timeline selection to match the current edit selection.

If you select 'Change Edit to Match Timeline' or press Option-Shift-6 (Mac) or Alt-Shift-6 (Windows), this will change the edit selection to match the timeline selection. See Figure 2.16.

Figure 2.16
Selection
commands

Also, you can play either the timeline or the edit selection by choosing the relevant command from the Edit menu or by holding down the Option key (Mac) or the Alt key (Windows) and pressing the left or right bracket, respectively.

The Playback and Edit Markers

Timeline selections are indicated in the Main Timebase ruler by playback markers. These appear as blue arrows normally and as red arrows when any track is record enabled. If you have entered and enabled Pre- or Post-Roll amounts in the Transport window, these will be indicated in the Main Timebase ruler by a pair of yellow flags.

Take a look at the accompanying screenshot, Figure 2.17, to see how the Edit window might look with the timeline and edit selections linked and with Pre- and Post-Roll amounts set and enabled. Notice that the edit selection is represented in the Main Timebase ruler by the blue playback markers that also indicate the timeline selection. In this example the Main Timebase ruler is set to Bars|Beats.

Take a look at the accompanying screenshot, Figure 2.18, to see how the Edit window might look with the timeline and edit selections unlinked and different selections made in the Timebase ruler and waveform display.

Figure 2.17
Blue playback markers and yellow Pre- and Post-Roll flags indicate the timeline selection and the Pre- and Post-Roll selection, respectively.

Figure 2.18
Red playback markers indicate the timeline selection with a track enabled for record, while gold-colored brackets indicate where the edit selection is. White flags indicate disabled Pre- and Post-Roll amounts

When the timeline and edit selections are unlinked, edit selections are displayed in the ruler with edit markers, which appear as gold-colored brackets. In this example, a track's Record Enable button has been engaged so the playback markers are shown in red.

You can see the edit selection at the right of the waveform display with two gold-colored brackets in the Main Timebase ruler above this. To the left, you can see the timeline selection indicated by the red arrows in the Main Timebase ruler. The Pre- and Post-Roll amounts are set but disabled, so although you still see the flags in the Main Timebase ruler, these flags are colored white to indicate that the Pre- and Post-Roll amounts are not active.

Linked Track and Edit Selections

If you want your edit selection to follow when you select a new track, you can link the track and edit selections by choosing 'Link Track and Edit Selection' from the Options menu or by clicking the 'Link Track and Edit Selection' button in the toolbar at the top of the Edit window.

With the track and edit selections linked, if you make an edit selection within a track or across multiple tracks, each associated track will automatically become selected, with the track names highlighted. To select a new track, simply click the track name, and the edit selection will move from the previous track (or tracks) to the newly selected track (or tracks). To select multiple tracks, Shift-click on additional tracks.

In the accompanying screenshot, Figure 2.19, the 'Link Track and Edit Selection' button is enabled and an edit selection has been made in the Edit window's waveform display. This has caused the track name at the left of the Edit window to become highlighted, confirming that this track is selected.

Figure 2.19 Edit window with Link Track and Edit Selection button enabled, with an edit selection in the waveform and the track name at the left highlighted to indicate that this is also selected

Now, if you click on any other track name, this will become selected and the edit selection will be moved to the newly selected track—see Figure 2.20.

Figure 2.20 A click on any other track name selects the track and automatically moves the edit selection to the newly selected track

Finding the Ends of the Selection

If you have made a selection in the Edit window and then you zoom in to fine-tune your edit points, the display will zoom in around the start point of the selection. If you want to jump to the end point, you can use the right arrow on the computer keyboard to move the display so that the right-hand edge of the selection, that is, the end point, is in the center of the screen. To get back to the start point of the selection, just press the left arrow.

To see how this works, make a selection in the Edit window then zoom the display in horizontally so that both the beginning and the end of the selection are no longer visible—as in Figure 2.21.

Figure 2.21
The beginning and the end of the selection are both hidden from view

Press the left arrow to bring the start point of the selection into view in the center of the window—as in Figure 2.22.

Figure 2.22
Edit window showing the start point of the edit selection centered in the Edit window

Press the right arrow to bring the end point of the selection into view in the center of the window—as in Figure 2.23.

Figure 2.23
Edit window showing the end point of the edit selection centered in the Edit window

Time Compression and Expansion (TCE) Techniques

TCE Edit to Timeline Selection

The 'TCE Edit to Timeline' command can be used on multichannel selections and on selections across multiple tracks. All clips are compressed or expanded equally by the same percentage value, based on edit selection range. This ensures that the rhythmic relationship between the different channels or tracks is maintained.

On audio tracks, the 'TCE Edit to Timeline Selection' command uses whichever TCE AudioSuite plug-in you have selected in the Processing Preferences—see Figure 2.24.

Figure 2.24
Setting the TC/E
plug-in preferences

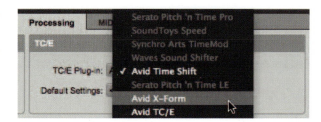

On Elastic Audio–enabled tracks, TCE Edit to Timeline Selection uses the track's selected Elastic Audio plug-in.

When the edit and timeline selections are unlinked, you can compress or expand an audio selection to fit the timeline selection using the Edit menu's 'TCE Edit to Timeline Selection' command—see Figure 2.25. If you prefer a keyboard command, you can use Option-Shift-U (Mac) or Alt-Shift-U (Windows).

Figure 2.25
Edit menu 'TCE
Edit to Timeline
Selection' command

TCE Edit to Timeline Selection ⌥⇧U

> **TIP**
> The standard Avid TC/E plug-in works, but doesn't sound too good, and Avid's Time Shift plug-in is only marginally better than this. The Elastic Audio real-time plug-ins are not very good either. These plug-ins are only

(Continued)

usable to give you a rough idea of how things will sound as you are working ideas out. You will need to buy a much higher-quality plug-in, such as Serato's Pitch 'n' Time Pro, if you want to achieve truly professional results. Avid also offers X-Form, which is absolutely excellent. This takes quite a lot longer to process the audio, but you can really hear the difference!

To fit an edit selection to the timeline: disable 'Link Timeline and Edit Selection'. With the selector tool, select the audio material to compress or expand—see Figure 2.26. In any Timebase ruler, select the time range where you want to fit the audio material, then choose 'TCE Edit to Timeline Selection' from the Edit menu.

Figure 2.26
An edit selection ready to be fitted to a timeline selection

The edit selection is compressed or expanded to the length of the timeline selection—see Figure 2.27.

Figure 2.27
The edit selection has been compressed and moved to fit into the timeline selection using the 'TCE Edit to Timeline Selection' command

Fitting an Audio Clip to an Edit Selection

Clips can be dragged from the Clip List and made to automatically fit within an edit selection using time compression or expansion. On audio tracks, fitting an audio clip to the edit selection uses whichever TCE AudioSuite plug-in you have selected in the Processing Preferences. On Elastic Audio–enabled tracks, this uses whichever Elastic Audio plug-in you have selected for the track.

It couldn't be much easier to use! First you make an edit selection in an audio track using the selector tool—see Figure 2.28.

Figure 2.28
An empty edit selection is made at the left of the Edit window. The clips list is open at the right

Then you drag the clip from the Clip List to the track on which you have made the edit selection while holding down the Command and Option keys (Mac) or the Control and Alt keys (Windows)—see Figure 2.29.

Figure 2.29
Dragging a clip from the list using the modifier keys immediately fits this to the edit selection.

The start of the clip you are dragging automatically snaps to the selection start, and the clip is time compressed or expanded to exactly match the length of the selection—see Figure 2.30.

Figure 2.30
The clip is expanded or compressed to fit the edit selection

NOTE
The 'Fit to Selection' command will let you drag multiple clips from the Clip List to multiple tracks or to multichannel tracks. However, all dragged clips are compressed or expanded equally by the same percentage value, based on length of the clip last clicked before dragging.

Looping Clips in Pro Tools

The most basic way to loop audio in Pro Tools is to select a clip and play this repeatedly, 'looping' around the selection in Loop Playback mode. This is ideal for checking out short sections while you are working on them, but it is not the kind of looping that you need when you are constructing your arrangement. This is where the other kind of 'looping' becomes more appropriate: taking an audio or a MIDI clip and playing this back repeatedly throughout sections of your music—or even throughout the whole piece. You can always create 'loops' (repeated rhythms or phrases) by duplicating or repeating selected clips, but there is a more flexible way, using clip looping.

Loop Playback Mode

There are lots of occasions when you will need to loop playback so that playback starts at the beginning of a range selection and jumps back from the end of the selection to start playing from the beginning again. The smallest selection that you can loop is half a second.

If you are working with short selections of one or two bars containing rhythmic material that you want to repeat, Loop Playback mode lets you check to see if you have selected exactly the right length and edit points for it to repeat cleanly without any glitches.

To loop playback of a selection, make sure that 'Link Timeline and Edit Selection' is enabled, use the selector tool to select the track range you want to loop, and select Loop Playback from the Options menu. Alternatively, you can Right-click the Play button in the Transport window and select Loop from the pop-up menu.

If you prefer to use a keyboard command, you can Control-click (Mac) or Start-click (Windows) the Play button in the Transport window or you can press Command-Shift-L (Mac) or Control-Shift-L (Windows). Or perhaps the simplest way is to press 4 on the numeric keypad when the Numeric Keypad mode is set to Transport.

> **NOTE**
> When 'Link Timeline and Edit Selection' is enabled and you select a clip in the Edit window, the start and end points are automatically entered into the Transport window's Play Selection fields and the toolbar display's Edit Selection Start and End locations. These are the locations around which playback will loop when you are in Loop Playback mode.

> **TIP**
> Pro Tools versions before 7.3 compelled you to stop playback if you wished to change the length of the loop selection, which made this feature slow to work with. With more recent versions, including Pro Tools 8, 9, 10, and 11, you can use Dynamic Transport mode, which lets you update your looping selection while the transport is running.

Using Separate Play and Stop Keys

If you are a Cubase user, you will probably have become used to having separate keys on the numeric keypad for Play and Stop.

Pro Tools lets you do this as well, but not by default. You have to specifically enable the 'Use Separate Play and Stop Keys' option in the Transport section of the Operation Preferences window—see Figure 2.31.

This option lets you start playback using the Enter key and stop playback using the Zero '0' key on the numeric keypad.

These keyboard commands are particularly useful for quickly starting and stopping playback when you are auditioning loop transitions.

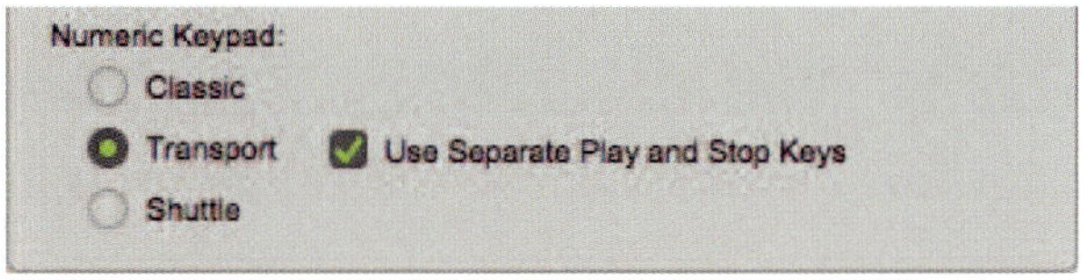

Figure 2.31
Preference for
separate Play and
Stop keys

> **NOTE**
> By default, the Enter key on the numeric keypad opens the New Memory Location window, so if you set Pro Tools to use this key to start the transport by enabling the 'Use Separate Play and Stop Keys' option you need some other way of opening the New Memory Location window. The software designers realized this, so all you need to do is to press and hold the Period (.) key on the numeric keypad while you press the Enter key and this will open the New Memory Location window instead.

Duplicate and Repeat

The basic 'Duplicate and Repeat' commands in the Edit menu could not be much simpler to use. Just select the clip that you want to repeat and use the 'Duplicate' command if you want to repeat this once, or use the 'Repeat' command to open the Repeat dialog, which lets you specify the number of repeats. If you prefer to use keyboard commands, you can press Command-D (Mac) or Control-D (Windows) for Duplicate or Option-R (Mac) or Alt-R (Windows) for Repeat.

Looping Clips

Instead of duplicating or repeating clips, you can loop audio or MIDI clips, or clip groups, using the Clip menu's 'Loop . . .' command. Looped clips offer much more flexibility than using the 'Duplicate' or 'Repeat' commands. For example, selecting and moving a looped clip selects and moves the source clip and all of its repeated clip loops as one.

When you select a clip or clip group and choose 'Loop . . .' from the Clip menu, the Clip Looping dialog opens—see Figure 2.32.

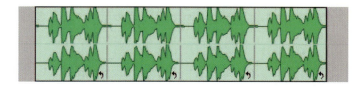

Figure 2.32
Clip looping dialog

Here you can enter the number of times to loop the clip, enabling a crossfade at the loop point if necessary. If you prefer, you can enter an exact duration for the loop, or you can loop until the next clip on the track or until the end of the session. When you OK the dialog, the looped clips are created in the Edit window. The looped clips show a loop icon in the lower right corner of each clip loop—see Figure 2.33.

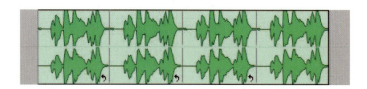

Figure 2.33
Looped clips with the Loop icon visible in the lower right corner

NOTE
Looping a clip does not loop any automation associated with the source clip, but you can always use the 'Copy Special' and 'Paste Special' 'Repeat to Fill Selection' commands to copy automation from the source loop to all the loop repeats.

The Clip menu also has an 'Unloop . . .' command that lets you unloop looped clips. You can either remove all the looped clips to revert back to the original clip or convert the looped clips into normal clips that contain copies of the original data—a process referred to as 'flattening' the loop—see Figure 2.34.

Figure 2.34
The Unloop Clips
dialog

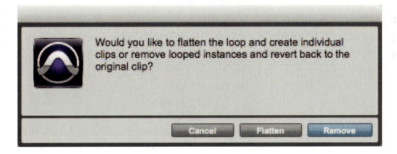

Trimming Looped Clips

Trimming looped clips can be done using the trimmer tool or the loop trimmer tool. The trimmer tool only trims the length of the individual loop iteration, which increases or decreases the number of loop iterations that are contained within the total length of the looped clip. If you make the individual loop iterations longer, for example, fewer of them will fit within the looped clip. The loop trimmer tool, on the other hand, changes the length of the entire looped clip, allowing you to increase or decrease the number of loop iterations contained within this clip.

Figure 2.35
Selecting the loop
trim tool

Creating and Editing Looped Clips

To use the loop trimmer tool to create looped clips, click the trimmer tool icon in the Edit window to access the Trim Tool Selector pop-up menu and select Loop—see Figure 2.35.

When you position the cursor over the top half of an unlooped audio or MIDI clip or clip group, the cursor changes to indicate that you can turn this into a loop—see Figure 2.36.

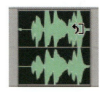

Figure 2.36
The cursor changes to the loop trim cursor when positioned in the top half of a clip

Click on the clip and adjust the length if you wish, and a loop icon will appear in the lower right corner to indicate that this is now a looped clip—see Figure 2.37.

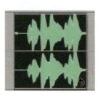

Figure 2.37
When you release the loop trim cursor after selecting an unlooped clip, this becomes a looped clip

Now you can just click at the end of the clip, and drag left or right to the point at which you want the loop to stop, or click at the beginning of the clip, and drag left or right to the point at which you want the loop to start—see Figure 2.38.

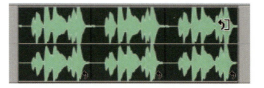

Figure 2.38
Three loop 'iterations' created using the loop trimmer tool and dragging the end of the clip to the right

TIP

If you hold down the Control key (Mac) or Start key (Windows) while using the trim tool, this will constrain the trim tool to removing or revealing individual loop iterations from the looped clip. For example, with the grid set to quarter notes and a loop length of one bar, holding the modifier key while using the trim tool prevents you from shortening or lengthening the looped clip by amounts less than the original one-bar loop length. This is useful if you have set a grid that is less than one bar but you want to simply remove or reveal individual loop iterations, as opposed to quarter-note parts of the loop.

If you position the cursor over the bottom half of the clip, you get the standard trimmer tool cursor instead. You can use this to adjust the length of the original clip, making it shorter—see Figure 2.39. Take a look at both of the accompanying screenshots to see how this works.

Figure 2.39
Using the standard
loop trimmer to
reduce the duration
of the original
looped clip

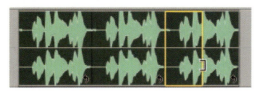

In the example shown in Figure 2.39, the loop is one bar in length, Pro Tools is in Grid mode and the grid is set to quarter notes, and I am in the process of shortening the loop length by two quarter notes to make it last for just half a bar. Take a look at Figure 2.40 to see how the total length of the looped part is preserved, while the original looped clip has been shortened.

Figure 2.40
The total length
of the looped part
remains the same
after trimming the
original clip length

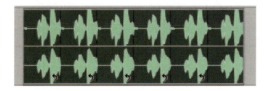

Trim Clip Commands

In addition to the trimmer tools, Pro Tools provides several menu options and equivalent keyboard commands for editing clip and clip group boundaries—see Figure 2.41.

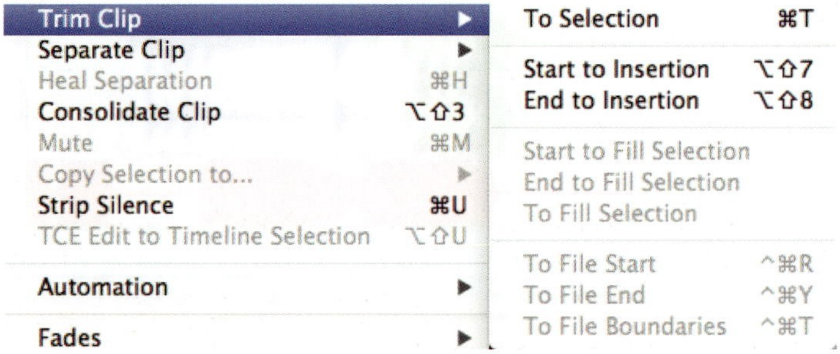

Figure 2.41
Trim clip commands
in the Edit menu

Trim Clip	▶	To Selection	⌘T
Separate Clip	▶		
Heal Separation	⌘H	**Start to Insertion**	⌥⇧7
Consolidate Clip	⌥⇧3	**End to Insertion**	⌥⇧8
Mute	⌘M		
Copy Selection to...	▶	Start to Fill Selection	
Strip Silence	⌘U	End to Fill Selection	
TCE Edit to Timeline Selection	⌥⇧U	To Fill Selection	
Automation	▶	To File Start	^⌘R
		To File End	^⌘Y
Fades	▶	To File Boundaries	^⌘T

'Trim to Selection' Command

If you select part of a clip (as in Figure 2.42), you can use the 'Trim to Selection' command to create a new clip that includes only the selection you have chosen—the rest is removed.

Figure 2.42
Before using the 'Trim to Selection' command

After selecting part of a clip, choose Trim to Selection from the Edit menu, or press Command-T (Mac) or Control-T (Windows) if you prefer to use a keyboard command. The rest of the clip will be trimmed, leaving just the selected part—see Figure 2.43.

Figure 2.43
After using the 'Trim to Selection' command

'Trim to Insertion' Commands

Sometimes, you will want to remove part of a clip from the beginning of the clip up to a certain point within the clip. In this case, you can use the selector tool to place the insertion cursor inside the clip or MIDI note you want to trim, and use the 'Trim Start to Insertion' command. Alternatively, enable the Keyboard Command Focus mode and simply press the 'a' key. The clip's start point will be automatically trimmed from the beginning of the clip up to the insertion point. The keyboard command for this is Option-Shift-7 (Mac) or Alt-Shift-7 (Windows). See Figures 2.44 and 2.45.

Figure 2.44
Before trimming to the insertion point

Figure 2.45
After trimming to the insertion point

Similarly, you can trim a clip or MIDI note by automatically removing the material between the edit insertion point and the end point of the clip. Place the insertion cursor where you want the clip to end and use the 'Trim End to Insertion' command or press Option-Shift-8 (Mac) or Alt-Shift-8 (Windows). Alternatively, enable the Keyboard Command Focus mode and simply press the 's' key. The clip's end point will be automatically trimmed to the insertion point.

TIP

You can also trim the start or end points of a clip by nudging them. First, set the nudge value that you want to use, then use the time grabber tool to select the clip you want to trim. To trim the start point, hold the Option (Mac) or Alt (Windows) key and press Plus (+) or Minus (–) on the numeric keypad. To trim the end point, hold the Command (Mac) or Control (Windows) key and press Plus (+) or Minus (–) on the numeric keypad.

'Trim Clip to Fill Selection' Commands

The 'Trim Clip to Fill Selection' commands let you automatically reveal underlying material in the gaps between clips or before and after individual clips. You can use these commands to fill any gaps between clips that emerge after quantizing clips, for example.

The 'Trim Clip Start to Fill Selection' command expands the selected clip backward until it reaches the end of the previous clip—or as far as possible if there is not enough underlying material to cover the gap. To trim from a start point to fill gaps you must use the selector tool to select across at least one gap between clips before using the command.

You can also trim from the end points to fill the gaps using the 'Trim Clip End to Fill Selection' command. In this case, the end point of each selected clip that has a gap following this is expanded forward until it meets the beginning of the next clip, or as far as possible if there is not enough underlying material to cover the gap.

'Trim Clip to Fill Selection' command trims both the start and end points of a selected clip to fill a selection that extends before and after this clip. For this command to work, you must make a selection that includes the clip that you want to extend and gaps that exist both before and after this clip. The 'Trim Clip to Fill Selection' command then becomes active. When you invoke the command, the start point of the clip is trimmed backward to the edit selection start, or as far as possible if there is not enough underlying material to cover the selection, and the end point of the clip is trimmed forward to the edit selection end, or as far as possible if there is not enough underlying material to cover the selection.

NOTE

The purpose of these commands is to save you time when trimming clips, compared with the time it would take to make these trim edits using the trim tool.

Quantizing Clips to Grid

One of the most useful editing commands when you are working with clips is the 'Quantize to Grid' command. This adjusts the placement of selected audio and MIDI clips so that their start points (or sync points, if present) align precisely to the nearest grid boundary.

TIP

If you have recorded a rhythmic part that is not as accurate as you would like, you may be able to use the 'Strip Silence' or the 'Separate Clip on Grid' or 'At Transients' commands in the Edit menu to chop this into individual 'hits' that should fall on Bar|Beat boundaries—then use 'Quantize to Grid' to make these fall in time.

To quantize one or more clips, select the grid value, then, using either the selector or time grabber tool, select the clip or clips you want to quantize. The clips can be on multiple tracks, but beware that only clips that are entirely selected will be quantized. When you choose 'Quantize to Grid' from the Clip menu, all the clip start times (or sync points) within your selection will be aligned to the nearest boundaries for the defined grid. If you prefer to use the keyboard, Command-0 (Mac) or Control-0 (Windows) will invoke the 'Quantize to Grid' command.

For MIDI clips, only the clips are quantized and all MIDI data contained within the clips (such as notes) are moved equally, thereby retaining their rhythmic relationships.

For Elastic Audio clips, only the clips are quantized and all Elastic Audio events contained within the clips (such as transient events) are moved equally, thereby retaining their rhythmic relationships.

> **NOTE**
>
> To quantize individual MIDI notes or Elastic Audio events, use the Quantize Event Operation. You can also use the Quantize Event Operation to quantize audio clips.

> **TIP**
>
> To fill any gaps that may be created when quantizing clips to a new tempo, you can use the 'Trim Clip to Fill Selection' commands. With drum or percussion loops, you will normally use Trim Clip Start to Fill Selection, which is the same as using the trim tool to take the start of the clip and move this to the left until it reaches the previous clip boundary. If you use the Trim Clip End to Fill Selection, this is the same as using the trim tool to take the end of a clip and moving this to the right until it meets the next clip boundary. In the case of a drum loop, this would likely expose the next 'hit' in the drum pattern, causing a double hit or a glitch at the clip boundaries.

Loop Playback Mode

If Loop Playback mode is enabled, when you make a selection in the Edit window and start playback, Pro Tools will play the range you have selected, then instantly loop back to the beginning of the range you have selected and play this again and continue like this—until you stop playback.

If you haven't made any range selection, playback will operate normally—playing back from the current cursor location.

TIP

Looping playback is a useful way to check the rhythmic continuity of a selection when working with musical material. If you're working with one-bar selections, you can loop playback to see if the material loops cleanly. If it seems to skip, you should then adjust the length of the selection until it works 'musically' within the context of the playlist and the other tracks.

Dynamic Transport Mode

Dynamic Transport mode lets you decouple the playback location from the timeline selection. It is similar in this respect to using the 'Unlink Timeline from Edit Selection' feature, which is very useful for MIDI editing. But Dynamic Transport mode offers a lot more than unlinking these selections: the timeline selection can be freely moved around by clicking and dragging the grey area that defines the selection in the Main Timebase ruler, for example, and you can start playback from anywhere on the timeline without losing your timeline or edit selections. So, for example, you can use Dynamic Transport mode in conjunction with Loop Playback mode to quickly audition loop transitions.

You can enable Dynamic Transport mode by selecting this from the Options menu or, if you prefer to use a keyboard command, you can press Command-Control-P (Mac) or Control-Start-P (Windows) to toggle Dynamic Transport mode on or off.

NOTE

When you enable Dynamic Transport mode, it disables the Link Timeline to Selection function. This allows you to play back either the timeline selection or an edit selection—whichever you prefer. You can press the left bracket key to audition the edit selection and use the Enter key to play the timeline selection. This lets you carry on making your edits, very flexibly, while Pro Tools is playing back. And if you want the timeline selection to follow the edit selection, you can always reenable the Link Timeline to Selection button whenever you like. Then the Dynamic Transport will follow each time you make a new selection in the Edit window.

Using Dynamic Transport Mode

When Dynamic Transport mode is enabled, the Main Timebase ruler expands to double height and reveals the play start marker, which marks the location from which playback will begin when you start playback.

> **NOTE**
> The upper part of the main Timebase ruler is referred to as the Timeline Selection Move Strip and the lower part is referred to as the Play Start Marker Strip.

If you move the play start marker to within the timeline selection that you are cycling around, then start playback, it starts from this location, plays to the end of the timeline selection, then subsequently loops around the whole of the timeline selection. You can always move the play start marker to a different location on the timeline—inside or outside of the timeline selection—any time you like. So you can cycle around a section of your song in Dynamic Transport mode while you do some edits on the fly, then simply grab and drag the timeline selection to another part of the song to do more edits there.

> **TIP**
> If you want to record, say, a MIDI part in 4-bar sections, each with a 1- or 2-bar Pre-Roll, you can set the timeline selection to cover four bars, set the play start marker a bar or two before this, record this section, then grab and drag the timeline selection to the next four bars, record this section, and so on. If you want to change the length of the timeline selection, just drag the start or end markers for the selection back or forth along the timeline so that it encompasses however many bars you want.

The Play/Start Marker

The accompanying screenshot (Figure 2.46) shows the play start marker located approximately halfway between bars 6 and 8 in this example, which has a time signature of 6/4.

The edit selection, as shown by the darkened clip in the Edit window and by the edit selection start and end locations displayed in the toolbar, does not play back when you press Play—the timeline selection, shown by the darkened area in the Main Timebase ruler (the Bars|Beats ruler in this example) plays back instead.

Figure 2.46
When Dynamic Transport is enabled, the Main Timebase ruler expands to reveal the play start marker

Here's an example of how you might use Dynamic Transport mode in Loop Playback mode:

If you move the play start to a location within the timeline selection that you are cycling around (see Figure 2.47), then start playback with Loop Playback mode enabled in the Options menu, it will start from this location, play to the end of the timeline selection, then subsequently loop around the timeline selection.

Figure 2.47
Play start marker located within a timeline selection

NOTE
To reposition the play start marker, you can point to any location within the play start marker strip in the Main Timebase ruler and click once, or you can simply click on the play start marker and drag this to wherever you like along the timeline.

A more practical scenario is to move the play start marker to a bar or two before the timeline selection, so that it acts as a Pre-Roll to the cycle around the timeline selection.

In the accompanying screenshot (Figure 2.48), the play start marker is positioned one beat before the edit selection. When you press Play, playback will start from the play start marker, play through the edit selection that follows this, play through the timeline selection that follows this, and then loop around the timeline selection until you stop playback.

Figure 2.48
Play start marker positioned independently of the edit selection and of the timeline selection

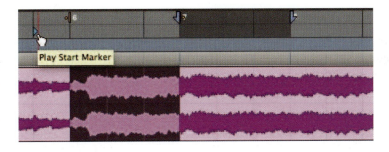

You can even reposition the play start marker *during playback* and playback will continue from the new location.

So, if you want to judge how well the transition into the looped timeline selection works, but you don't need to start from quite so far back next time, you can simply drag the play start marker, during playback, to a new location (see Figure 2.49) and when you let go of the mouse, playback will jump to that location and continue from there.

Figure 2.49
Dragging the play start marker to a new location: this even works during playback

Moving the Timeline Selection

The timeline selection can also be freely moved around by pointing at the timeline selection with the mouse, then clicking and dragging the grey area that defines the selection in the Main Timebase ruler while holding down the mouse. Take a look at the two accompanying screenshots, Figures 2.50 and 2.51, to see how this looks—the mouse cursor changes to a 'hand grabber' icon when you point the mouse at the Timeline Selection Move Strip.

Figure 2.50
The timeline selection about to be dragged from bar 6

Figure 2.51
The timeline selection after being dragged from bar 6 to bar 7

TIP

Remember also that you can always change the timeline selection by clicking in the Main Timebase ruler using the selector tool and dragging along the timeline to encompass a new selection. Alternatively, with any edit tool selected, you can just click and drag the timeline selection start or end markers to move them to new locations along the timeline.

Keyboard Shortcuts for Dynamic Transport Mode

Because it is very probable that you will want to locate the play start marker to the beginning or end of the timeline or edit selections, either during playback or when the transport is stopped, several keyboard shortcuts are provided for these and similar actions.

Pressing Period (.) on the numeric keypad then the left arrow moves the play start marker to the start of the timeline selection: pressing the right arrow moves it to the end of the timeline selection instead.

Pressing Period (.) on the numeric keypad then the down arrow moves the play start marker to the start of the edit selection.

You can nudge the play start marker backward along the timeline, a bar at a time, by pressing 1 on the numeric keypad. To go forward, press 2.

Assuming that you have Bars|Beats selected as the Main Timebase ruler, you can move the play start to a specific bar using the numeric keypad: hold the Asterisk (*) key while you type the bar number, then let go of these keys and press the Enter key.

Temporarily Linking the Timeline and Edit Selections

Whenever you have an edit tool selected, you can move the timeline selection backward or forward along the timeline by clicking and dragging. When the timeline and edit selections are not linked, only the timeline selection will move. If you want to temporarily link the timeline and edit selections while you move the timeline selection, press the Option key (Mac) or Alt key (Windows) while you drag the timeline selection. The edit selection will be made to correspond with the timeline selection, and they will move together to the new location.

Recording in Dynamic Transport Mode

If you are recording in Dynamic Transport mode, you can start independently of the timeline selection, using the play start marker as a manual Pre-Roll before the timeline selection, and Pro Tools will punch in to record at the start of the selection and punch out at the end of the selection.

Setting Selection Start and End in Dynamic Transport Mode

When you enable Dynamic Transport mode, Pro Tools automatically enables Loop Playback mode and automatically disables 'Link Timeline and Edit Selection'. As a consequence, when you select a clip in the Edit window, the start and end points are *not* automatically entered into the Transport window's play selection fields or the toolbar display's edit selection start and end locations. However, a special keyboard command becomes available in Dynamic Transport mode to let you do this manually: just press the letter 'O' on your computer keyboard.

Auditioning Loops in Dynamic Transport Mode

You can audition a series of loops without stopping and restarting playback when Dynamic Transport is enabled. Simply select a new clip that you wish to loop around during playback and press the letter 'O' to update the loop start and end locations into the Transport window and Pro Tools will immediately jump to the start of the new loop clip.

Preferences for Dynamic Transport

If you want the play start marker to always snap to the timeline selection start marker when you move the timeline selection, or when you draw a new selection or adjust the timeline selection start, you can set a preference for this—'Play Start Marker Follows Timeline Selection'—in the Operation Preferences window—see Figure 2.52.

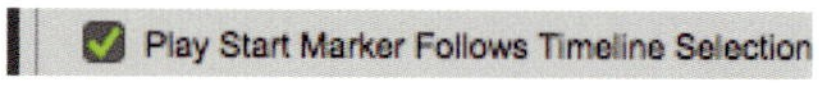

✅ Play Start Marker Follows Timeline Selection

Figure 2.52
Selecting Play Start Marker Follows Timeline Selection in the Operation Preferences

The way this works when you are playing back is that if you click anywhere in the Edit window, playback will immediately jump to that position—great for quickly checking through the session to see how everything is sounding while comparing sections. Similarly, if you drag in the Edit window to make a selection, playback will immediately jump to the beginning of that edit selection.

Another important setting to choose in the Operation Preferences window is the playback behavior—that is, what happens when you stop playback. When the option 'Timeline Insertion/Play Start Marker Follows Playback' is selected in the Operation Preferences window (see Figure 2.53), the timeline insertion and the play start marker both move to the point in the timeline at which playback is stopped. So when you start playback again, playback continues from the point at which you stopped.

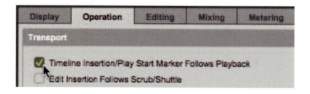

Figure 2.53
Selecting Timeline Insertion/Play Start Marker Follows Playback in the Operation Preferences

Figure 2.54
Setting 'Timeline
Insertion/Play Start
Marker Follows
Playback' using the
button in the Edit
window toolbar

This is such a regularly used preference that a dedicated button is available to let you toggle this on and off from the Edit window toolbar—see Figure 2.54.

Figure 2.54
Setting 'Timeline Insertion/Play Start Marker Follows Playback' using the button in the Edit window toolbar

If you want the timeline insertion and the play start marker to stay at the location from which you originally commenced playback, you need to make sure that this option is deselected. Then, when you start playback again, it starts from this original starting point—not from the place where you stopped.

> **TIP**
> You can toggle the 'Timeline Insertion/Play Start Marker Follows Playback' option on and off using the 'n' key on your computer keyboard.

> **NOTE**
> Be careful not to accidentally hit the 'n' key on your computer keyboard, or you will find that your session is behaving differently from the way you expect!

Summary

Whether you are working with MIDI or with audio in Pro Tools, you need to have a good understanding of how to work with clips (formerly called regions) in the Edit window. The way many people develop arrangements using MIDI—or using audio 'samples' of drum grooves, bass riffs, keyboard patterns, guitar licks, or whatever—involves selecting, copying, pasting, moving, or 'looping' MIDI or audio clips to build these arrangements.

It is important to understand how to link and unlink the timeline and edit selections, for example, and to have a clear understanding of how the playback and edit markers work.

The time compression and expansion features are extremely powerful, especially when used with the highest-quality algorithms provided by X-Form or other optional plug-ins.

Duplicating or repeating clips in the Pro Tools Edit window is the most basic way to create 'loops', but you should also make sure that you are familiar with the clip loop features, which can play a very important role if this is your style of working. You can also use the Dynamic Transport mode in conjunction with the Loop Playback mode to quickly audition loop transitions when you are setting up looped clips.

in this chapter

CHAPTER 3

Beat Detective

Using the 'Identify Beat' Command

The 'Identify Beat' command lets you create a tempo map for audio that has been recorded without a click. This command is also very useful when you don't know the tempo of imported audio.

The way this works is that by careful listening you identify a range of beats with a particular time signature, such as one bar of 4/4 or four bars of 11/8, then you select this range in the Edit window and use the 'Identify Beat' command to open the Add Bar|Beat Markers dialog. In this dialog, you specify the start location and the end location (in terms of bars and beats) of the selection that you have made, and specify the time signature that you consider is correct. When you OK this dialog, Pro Tools works out what the tempo of this selection range must be if the information that you have provided is correct. Bar|Beat markers for the calculated tempo are then inserted and appear in the Tempo ruler at the beginning and end of the selection. A meter event indicating the time signature will also be inserted into the Meter ruler if necessary.

Pro Tools does not work out where the beginning of the bar is and which bar within the music this is, or the Bar|Beat location where your selection ends, or what the time signature of this selection of music is: you make these decisions and select the range in the Edit window accordingly. Obviously you need to make an extremely accurate selection for this to work out properly, so you will need to zoom the display, look at the waveform, choose zero waveform crossings for your start and end points, and so forth.

> **TIP**
> It always helps to first set the tempo of your Pro Tools session to somewhere close to the actual tempo of the audio for which you wish to identify the beats. You can do this by trial and error, trying plausible BPM settings until you get close. Then the bars and beats in Pro Tools will not be too far away from those in your audio.

For example, in the accompanying screenshot, Figure 3.1, I have selected one bar of drums where what the drummer has played doesn't correspond exactly to the bars and beats in Pro Tools—but it is not too far off.

Figure 3.1
Using Identify Beat

This selection starts on bar 2 in the timeline and finishes just a little after the beginning of bar 3. If the drummer had played exactly in time with the click, the tempo of the Pro Tools session would have matched the tempo of this audio exactly, and the selection would start exactly at the beginning of bar 2 and finish exactly at the beginning of bar 3.

When you select the 'Identify Beat' command from the Event menu or by pressing Command-I (Mac) or Control-I (Windows) on your computer keyboard, the Add Bar|Beat Markers dialog opens—see Figure 3.2.

In this example, you know that the audio should start at 2|1|000 and end at 3|1|000, so type the correct start and end Bar|Beat locations and click OK. If the time signature is anything other than 4/4, you should also set the time signature here. See Figure 3.3.

Pro Tools works out what the tempo would have to be for this selection to represent this many beats with this time signature. Then it inserts blue-colored Bar|Beat markers for the calculated tempo into the Tempo ruler at the beginning of the selection—see Figure 3.4. It was not necessary for Pro Tools to insert a meter event into the Meter ruler in this example because the music was in 4/4 so the session was left at the default meter setting of 4/4.

Figure 3.2
The Add Bar|Beat Markers dialog

Figure 3.3
Corrected Bar|Beat markers entries

Figure 3.4
Bar|Beat markers at the beginning and end of the selection

Creating a Tempo Map using Identify Beat

Going through a long piece of music to identify every beat in this way can take time, but you can get quick results if the music only changes tempo occasionally throughout the piece. If the tempo is drifting very slowly, you may only need to identify the beat every 16, 32, or more bars.

> **TIP**
> If you find that the tempo is changing within a bar on individual beats or sub-beats, you may find that Beat Detective will generate accurate Bar|Beat markers more quickly than using this manual 'Identify Beat' command.

Beat Detective Overview

Beat Detective identifies the individual beats in your audio selection by looking for the peaks in the waveform. You can adjust the settings until you have identified most of these and then edit manually to fine-tune the choices. The points identified are referred to as 'beat triggers' in Pro Tools, and these can be converted to Bar|Beat markers. Once the beats have been identified correctly, Beat Detective can extract the tempo from the audio—creating a tempo map. Beat Detective can also be used to create Bar|Beat markers from MIDI recordings. Other audio clips and MIDI tracks can then be quantized to these markers.

First define a selection of audio material—either on a single mono-channel or multichannel track or across multiple tracks. Then adjust the detection parameters so that vertical beat triggers appear in the Edit window based on the peak transients detected in the selection. For example, with a 'boom-chick' bass drum and snare drum beat you would see a vertical line in the display immediately before each bass drum and snare drum beat. Examine these triggers visually to make sure that there are none in the wrong places for any reason. When you are satisfied, go ahead and generate Bar|Beat markers, based on these beat triggers, to form a tempo map that you can use for your session.

So, for example, you can use Beat Detective to extract tempo from audio that was recorded without listening to a click—even if the audio contains varying tempos or material that is swung—and you can then quantize other audio clips or MIDI tracks to this 'groove'. You can also do the opposite of this: if your session already has the right tempo, you can use Beat Detective to 'conform' any audio with a different tempo (or with varying tempos) to the session's tempo or tempo map or to a groove template. You can choose to keep a percentage of the original feel if you like, to 'tighten up' the performance, and you can even increase or decrease the amount of swing in the conformed material.

One of the most useful applications is aligning loops with different tempos or feels. For example, you could take a drum or percussion loop, separate each beat in the loop, then make the loop run at the session's tempo—without using time compression or expansion, which can affect the quality of the audio. Then if one loop has a subtly different feel or groove you can use Beat Detective to impose that groove onto another loop. This is great for remixes, where you often need to extract tempo from the original drum tracks or even from the original stereo mix. New audio or MIDI tracks can then be matched timing-wise to the original material, or, alternatively, the original material can be matched to the new tracks.

Beat Detective Features and Operation

First, you must accurately select the audio clip that you want to work with in the Edit window. Beat Detective cannot generate beat triggers accurately unless the length of the selection has been correctly defined and the time signature correctly specified.

Always make sure that the selected passage starts exactly on the attack of the first beat and finishes at the end of the last beat. Zero in on the audio clip's start and end points and zoom in to sample level so that you can choose zero-crossing points on the waveform. The 'Tab to Transients' feature can be helpful here.

> **TIP**
> It is often a good idea to play your selection back using Loop Playback mode to make sure that you have selected exactly the correct number of bars and beats, in which case it should loop with no glitches.

It is also very important to capture the selection in Beat Detective as soon as you open its window after making a new selection. If the tempo of the audio matches the session tempo and the audio is correctly aligned with the bars and beats, the bars and beats corresponding to the range that you have selected will be automatically entered for you when you click the Capture button. However, if the tempo of the audio does not match that of the session, the bars and beats will not be aligned correctly with this audio, and the bars and beats that are automatically entered will not be correct (unless the discrepancy is very small). So you must tell Pro Tools how many bars and beats you have selected by manually entering the time signature and the start and end Bar|Beat locations. For example, if you have a four-bar selection that starts on bar 1, beat 1 of the session, you would enter 1|1 and 5|1.

> **TIP**
>
> If you are unsure of the length of the material, enter the time signature and the start Bar|Beat location; then start playback and click the Tap End B|B button repeatedly. Beat Detective will automatically calculate the end Bar|Beat number for you. This may take Bar|Beat a little while with a long selection, so keep tapping until the end Bar|Beat settles to a particular number.

Once you have made your selection accurately, you need to analyze it to produce beat triggers. Three analysis algorithms are available to suit different types of audio material. Enhanced resolution works well with the broadest range of material, such as full mixes and loops, and is the default analysis option. High emphasis may work better with high-frequency, inharmonic material, such as cymbals and hi-hats, while low emphasis works well with low-frequency material such as bass guitar and kick drum, as well as with other harmonic material, such as piano or rhythm guitar.

When you have finished analyzing the audio, you can use Beat Detective to separate your selection into separate clips corresponding to bars, beats, or sub-beats, based on the beat triggers, then conform these new clips to the session's tempo map or to a groove template.

The Beat Detective Window

The Beat Detective window is divided into three sections: Operation, Selection, and Detection. The Operation section lets you choose which of the five operation modes to work with. The Selection section lets you define and capture the selection, while the Detection section changes according to the operation mode that you have selected: it contains the detection parameters when you are using the first three operation modes and swaps these for the conform parameters and the smoothing parameters in the last two operation modes.

Bar|Beat Marker Generation

Beat Detective's first operation mode, Bar|Beat Marker Generation (see Figure 3.5), automatically generates triggers corresponding to transients detected in the audio selection.

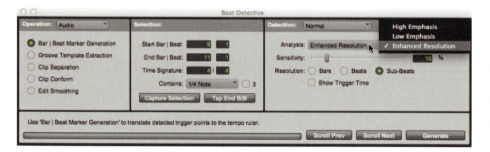

Figure 3.5
Beat Detective window showing Bar|Beat Marker Generation with the analysis algorithms pop-up in use

> **NOTE**
> Bar triggers appear as thick lines, beat triggers as medium-thick lines, and sub-beat triggers as thin lines.

By selecting the relevant resolution button in the Beat Detective window, you can show just the bars, or bars and beats, or bars, beats, and sub-beats. You should view these in turn to make sure that they have been correctly identified, then remove or move any that are in the wrong places, such as the extra bar trigger just before the end of the second bar in the example shown in Figure 3.6. To remove a trigger, just Option-click (Mac) or Alt-click (Windows) on the trigger using the grabber tool.

Figure 3.6
Bar triggers in the Edit selection

Once Beat Detective has detected the transient peaks in the audio selection to find where to place the bar and beat triggers, and you have corrected, added, or removed these as necessary, these triggers can be converted to Bar|Beat markers by pressing the Generate button. These Bar|Beat markers are then added to the Tempo ruler, along with their associated tempo values, to act as a 'tempo map' for the session—see Figure 3.7.

Figure 3.7
Bar|Beat markers generated by Beat Detective are added to the Tempo ruler.

To examine these beat triggers closely to make sure they appear in sensible locations, you can zoom the waveform display, to sample level if necessary, and use Beat Detective's Scroll Next and Scroll Previous buttons to move to the next or previous trigger within your selection.

If beat triggers are missing in any places where you believe they should be, you can manually insert these. Select the grabber tool in the Edit window, then click in the audio selection where you want to insert the new trigger. To move a beat trigger that is too late or too early, drag it to the left or right until it is exactly where you want it to be.

> **NOTE**
> In practice, I have found that sometimes it is impossible to insert a trigger at a particular location. Similarly, it can sometimes be impossible to move the trigger to exactly the correct location, and, occasionally, the beat trigger disappears altogether when moved.

If 'false triggers' appear where there are no beats, you can either delete them or reduce the sensitivity setting.

> **TIP**
> To help you to figure out whether the beat triggers are in the correct locations, you can enable Beat Detective's 'Show Trigger Time' option. This shows the Bar|Beat locations next to the beat triggers in the waveform display.

Fixing Problems with Beat Triggers

Sometimes, when you raise the sensitivity high enough to detect all the important triggers that you need to find, 'false triggers' appear at locations where you do not want to create beat triggers. You can delete these individually, but this can take time.

A faster way to deal with this situation is to 'promote' the beat triggers that you want to retain by Command-clicking (Mac) or Control-clicking (Windows) on the ones that you want, then lower the sensitivity until the others disappear. Promoted beat triggers will not disappear until you reduce the sensitivity to zero, but all the others should be long gone when you get down to just a few percent.

> **TIP**
>
> If you mess up and want to return the beat triggers to their original state, just click Analyze again to demote all the beat triggers in your selection.

Sometimes a transient is detected that is slightly off the beat, located a little earlier or later. If a beat trigger is not assigned to the correct metric location during the analysis, you can relocate it using the Identify Trigger dialog. You will know the start Bar|Beat of your selection, so should be able to work out where the wrongly placed beat should be located by listening to your audio selection, observing where the beat trigger falls, and zooming the display to look at this more carefully. If you think it is not exactly where it should be, open the Identify Trigger dialog by double-clicking on the beat trigger and type in the correct location—see Figure 3.8.

Figure 3.8
Identifying a trigger

Groove Template Extraction

When a 'master' musician plays with a rhythmic 'groove' or 'feel' that includes notes played on subdivisions of the main beats that are deliberately placed outside the strict tempo subdivisions as represented by the quantization grid, Beat Detective can be used to analyze and extract the timing of the sub-beats that define this 'groove' or 'feel'. So, if you have a recording of some drum beats (or a bass line or a percussion 'groove') that the musician has played with a great 'feel', Beat Detective will let you capture this and apply it when you quantize MIDI notes (or audio clips on tick-based tracks).

Beat Detective's Groove Template Extraction mode extracts the timing information from selected audio clips (or MIDI notes) and puts this onto the Groove Clipboard so you can use it to immediately apply this 'feel' when quantizing another track. Alternatively, you can save this timing information as a groove template so that you can apply it to other audio clips or MIDI tracks.

To extract the groove from a selection, first make sure that when you capture the selection, you specify that this contains small enough beat subdivisions, such as 16th or 32nd notes, click the Analyze button, then set the detection resolution to sub-beats so that beat triggers will be produced corresponding to the inner rhythms within each bar, then raise the sensitivity slider to reveal the sub-beat triggers—see Figure 3.9.

Figure 3.9
Beat Detective
groove extraction

When I tried this on a particular drum track, even using 100 percent sensitivity and enhanced resolution, Beat Detective was unable to find all of the sub-beats. Nevertheless, it found most of them—and the missing triggers were easy to spot. Take a look at Figure 3.10 and you will easily see which were missing.

Figure 3.10
Missing triggers

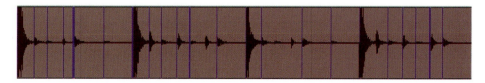

I clicked in the waveform display to insert the missing beat triggers and, with the grabber tool selected, I moved a couple of beat triggers to position them more accurately—see Figure 3.11.

Figure 3.11
Clicking to a beat trigger using the grabber tool

When you click the Extract button to create the groove template, you are presented with a dialog window that lets you specify whether to save this to the Groove Clipboard or to disk—see Figure 3.12. You should also check to make sure that the correct number of bars that you have analyzed is entered here, and put in the correct figure if necessary.

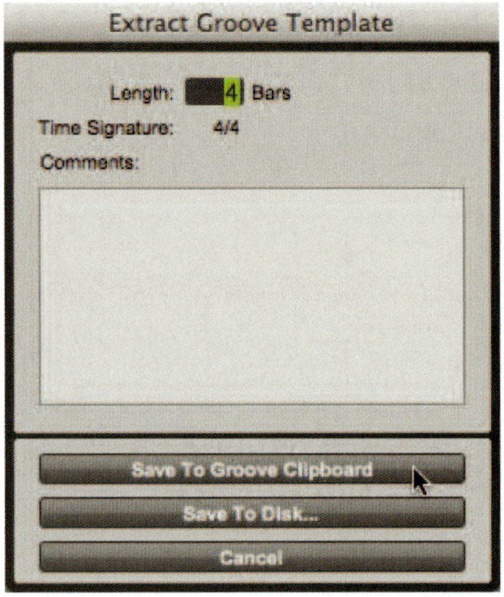

If you choose 'Save To Disk . . .', this brings up the Save dialog. Here you can choose the folder in which to save your groove templates and, by default, it will save them into the Grooves folder within the Pro Tools folder in the User: Documents folder—see Figure 3.13.

Extract Groove Template

Length: 4 Bars
Time Signature: 4/4
Comments:

Save To Groove Clipboard
Save To Disk...
Cancel

Figure 3.12
Extract Groove Template dialog

Grooves saved into the Grooves folder will appear in the Event menu's Event Operations sub-menu's Quantize dialog's Quantize Grid pop-up selector. (Whew, what a mouthful! Don't worry—it's a lot easier than it sounds!)

Groove templates can be used to transfer the feel of a particular performance to selected audio clips using Beat Detective's Groove Conform feature.

You can also apply groove templates or the contents of the Groove Clipboard to selected MIDI data using the Event Operations Quantize dialog that you can access from the Event menu—see Figure 3.14.

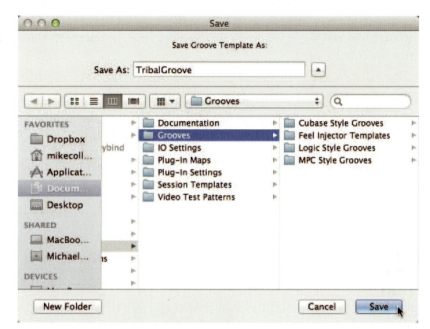

Figure 3.13
Save to disk

Figure 3.14
Applying a groove
quantize using the
Event Operations
Quantize dialog

Using the Groove Template Extraction feature, it is very easy to build up your own library of groove templates to use alongside the preset groove templates supplied with Pro Tools.

> **NOTE**
> Beat Detective uses a very powerful 'groove extrapolation' logic that can extract tempo data from a wide range of material when creating groove templates. This groove extrapolation logic can extrapolate from peak transients that it does find to produce beat triggers for beats that logically should exist—even if it does not find any peak transients at these locations. If a drum loop did not have a hit on beat 3, for example, there would be no peak transient on that beat and no beat trigger would be generated for that beat. However, when you extract the groove template from your selection, Beat Detective will extrapolate from the other beat triggers in the selection and create a trigger for beat 3 in the groove template. Any such extrapolated triggers will also preserve the timing of triggers generated from detected peak transients. For example, if three beat triggers were detected within a particular bar, all of which were 20 ticks ahead of the beat, then any extrapolated beat trigger would also be generated 20 ticks ahead of the beat.

Clip Separation

Clip Separation mode lets you automatically separate and create new clips based on transients detected in the audio selection—see Figure 3.15. This allows you to separate audio clips into individual sections that can then be quantized to the grid to conform to the session tempo or to follow a tempo map or groove template. Typically, you will go on to use the Clip Conform mode to quantize the separated clips; then use the Edit Smoothing mode to smooth over the edit transitions.

> **NOTE**
> To make sure that the separation process does not cut off the beginnings of any transients at the edit points, you can enter a trigger pad value of up to 50 ms (although typically you would set this to 5 or 10 ms). This will ensure that the clip separation points are created that distance away from the beat trigger sync points, while the beat trigger sync points will still be used to determine where the clip is placed.

Figure 3.15
Clip Separation
mode

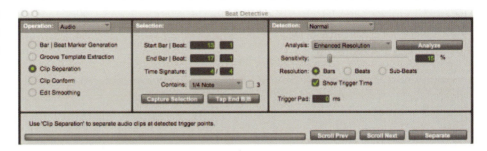

To test this, I worked on a four-bar drum groove, played at approximately 112 BPM, recorded with mono tracks for the bass drum and snare drum, a stereo track for a pair of overhead mics, and another stereo track for a front pair of microphones positioned in front of the kit to capture the 'audience perspective'. My goal was to separate this into individual beats so that I could make sure that each beat was accurately placed.

First I set the beat sub-division to quarter notes and the resolution to bars and analyzed the selection. I set the sensitivity to 15 percent and beat triggers appeared at the beginning of each bar. When I looked at these, I saw that an extra, incorrect, beat trigger had been created near the end of the selection. I removed this by Option-(Alt-)clicking on the trigger with the grabber tool selected—see Figure 3.16 (this shows only the bass drum track to conserve space).

Figure 3.16
Removing an
incorrectly
generated beat
trigger

Then I set the resolution to beats and examined each beat trigger, with the 'Show Trigger Time' option checked in the Beat Detective window, so that I could make sure that each beat trigger was in the right place. The bass drum was playing 'fours'—in other words, this was being played on every beat—so I knew that if the beat triggers were placed just before each bass drum, they would be in the right places. However, the off-beat snare drum was 'leaking' into the bass drum track, and this caused every other beat trigger to be placed, incorrectly, on the snare beats, so I had to use the grabber tool to drag each of these onto the correct beat before proceeding—see Figure 3.17.

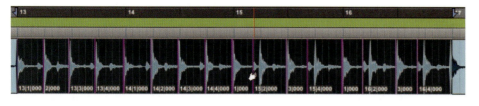

Figure 3.17
Beat triggers on each bass drum, with wrongly positioned triggers being moved using the grabber tool

Once I had all the beat triggers correctly in place, a click on the Separate button caused the selected clip to be separated at the beginning of each beat—see Figure 3.18.

Figure 3.18
Four bars of 4/4 music separated into individual beats

Clip Conform

Clip Conform mode lets you conform all the separated clips within the selection to your session's current tempo map. The standard conform settings let you keep some of the original feel of the material, depending on how you set the 'Strength', 'Exclude Within', and 'Swing' parameters—see Figure 3.19.

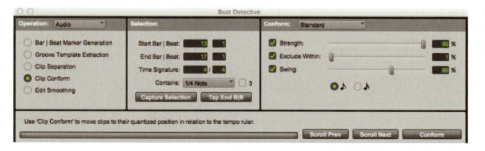

Figure 3.19
Clip conform options

Setting the strength to 100 percent will conform the separated clips exactly to the grid, while a setting of zero percent will not move the clips at all. The Exclude setting will only move clips that are sufficiently far away from the theoretically correct values to be regarded as wrong. Any clips that are very close to the grid will not be moved unless you set a very high value. This preserves some of the natural variation that can make a performance sound more 'human' and less mechanically precise. The Swing setting affects either 8h note or 16th note grid positions, depending on which note value you choose, and determines how far toward the corresponding triplet note positions selected clips will be moved. If you set this to 100 percent, for example, it would move selected clips exactly to their corresponding triplet value positions. Typically, you might set this to 30 percent or 60 percent.

Beat Detective can also conform audio clips to groove templates—see Figure 3.20. Cubase, Feel Injector, Logic, and MPC-style groove templates are provided as standard, and you can add your own grooves to the list whenever you create these.

Figure 3.20
Beat Detective
'Conform to Groove'
showing the groove
templates pop-up
selector

If you select the 'Timing' option and enter a percentage value, this will affect how strongly the clips conform to the groove template. Lower percentage values keep more of the original feel while higher percentage values will align the clips more tightly to the groove template's grid.

> **NOTE**
> If you have enabled the 'Timing' option, you are also given a further option to enable 'Pre-Process using Standard Conform'. If you choose this option, the clips will first be conformed using the Standard Conform settings before the groove template is applied. This gives better results with sloppy

(Continued)

performances, by making sure that the performance is accurately mapped to the correct bars, beats, and sub-beats before the groove template is applied. Using this option first is particularly recommended if you will subsequently be applying a groove template that has a heavy 'swing' feel.

To make my four-bar drum groove conform to a steady click after slicing it into separate beats, I selected Clip Conform in the Operation section of Beat Detective, checked again that the selection was captured correctly and that the time signature and beat subdivision were set correctly, then clicked the Conform button.

I had set the tempo of the session to exactly 112 BPM and the drummer had played to the click, but he had played some beats slightly faster or slightly slower than this. Conforming these separate one-beat clips to 112 BPM caused any one-beat clip slices originally played slower than 112 BPM to overlap the following clip, while any originally played faster than 112 BPM would finish before the end of the beat, leaving a gap before the next clip slice. As a consequence, there were gaps between some of the clips after conforming these—see Figure 3.21—and I could hear these as 'glitches' or short interruptions in the sound.

Figure 3.21
Two bars of the selection showing individual one-beat clip 'slices' conformed to the grid with some gaps visible

Edit Smoothing

After conforming clips, if there are gaps left between these, you can use the Edit Smoothing feature to fill any gaps between clips, automatically trimming them and inserting crossfades as necessary. This can save you an awful lot of detailed editing work that would otherwise be necessary to avoid pops and clicks at the clip boundaries that would normally need to be trimmed and crossfaded. It also has the advantage of preserving the ambience throughout the track to keep this constant despite the edits.

The 'Fill Gaps' option trims the clip end points so that the gaps between the clips are filled. The 'Fill and Crossfade' option not only trims the clip end points but also automatically adds a pre-fade directly before each clip start point. You can specify this crossfade length in milliseconds and a typical value for a short crossfade would be the default value of 5 ms. See Figure 3.22.

Figure 3.22
Beat Detective edit smoothing

TIP

You can also use the Edit Smoothing feature with any audio track that contains lots of clips that need to be trimmed and crossfaded, such as sound effects for postproduction—it's a great time saver when used for this purpose as well!

To get rid of the 'glitches' at the edit points in my four-bar drum groove, I selected 'Edit Smoothing' in the Operation section of Beat Detective, chose the 'Fill and Crossfade' option with a 5 ms crossfade length, and clicked the Smooth button to automatically trim and crossfade all the selected clip slices. I've zoomed in to show one of these edit points in detail in Figure 3.23.

Figure 3.23
An edit that has been filled and crossfaded—with a clip sync point visible just after the crossfade

The final separated, conformed, and smoothed four-bar drum groove now played back 'steady as a rock' exactly at the 112 BPM tempo—and with no glitches at the edit points! See Figure 3.24, which shows all four bars split into

16 separate beats, conformed and smoothed, for all four drum tracks: mono bass drum, mono snare drum, stereo front pair, and stereo overheads.

Figure 3.24
The final separated, conformed, and smoothed four-bar drum groove

I found it convenient to group these four bars together using the 'Clip Group' command so that I could move them around in the Edit window more easily—see Figure 3.25.

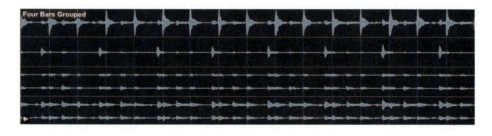

Figure 3.25
Sixteen separate one-beat slices grouped to form one four-bar multitrack clip group for ease of editing

When I had finalized my arrangement, I selected this clip group and used the 'Consolidate Clip' command from the Edit menu to create new 'edit-free' audio files on disk—see Figure 3.26.

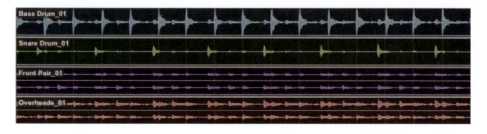

Figure 3.26
Four separate consolidated files—one for each of the four drum tracks: Bass Drum, Snare Drum, Front Pair, and Overheads

> **NOTE**
>
> The process of separating, conforming, and smoothing with Beat Detective can leave tracks with many clips and many crossfades. If you are working with multiple tracks, the density of these edits may lead to system performance problems. So, once you are satisfied with the results from Beat Detective, it is recommended that you 'flatten' the tracks using the 'Consolidate' command. This procedure produces a single audio file for each track that you select (in this case for each track within my clip group), each containing just the selected audio. Using these consolidated audio files, Pro Tools has fewer edits to process during playback.

MIDI Mode

Beat Detective also works with MIDI recordings. You can use Beat Detective to generate Bar|Beat markers corresponding to MIDI notes by analyzing MIDI clips, then other MIDI (or tick-based audio) clips can be quantized using the Bar|Beat positions derived from these.

To generate beat triggers from a MIDI selection, set the MIDI track to Notes view and select a range of MIDI notes, starting and finishing at sensible positions—such as on bar lines. When you choose MIDI from Beat Detective's Operation pop-up menu, just two of the operational modes can be selected: either Bar|Beat Marker Generation or Groove Template Extraction—see Figure 3.27.

Everything works very similarly to the way you work with audio clips. So, for example, you have to carefully define your selection or capture this in Beat Detective's Selection section, and set the time signature and beat subdivisions appropriately.

The main difference is that the Analysis pop-up menu offers various MIDI chord recognition algorithms instead of the audio analysis algorithms. These include Last Note, First Note, Loudest Note, Average Location, Highest Note, and Lowest Note—all of which are self-explanatory. Beat Detective uses these algorithms to decide where to insert the beat triggers whenever you play a chord containing two or more notes at the same time.

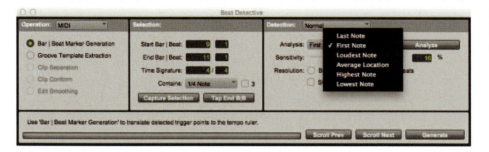

Figure 3.27
Beat Detective's
MIDI mode

Beat Detection across Multiple Tracks

Beat Detective will let you select several tracks and analyze these at the same time. Typically, you would do this with a group of drum tracks.

Beat Detective will still analyze each track independently, and transients that are detected in any track will appear as beat triggers across all the selected tracks—even if the detected transient only exists in some of or even just one of these tracks. The snare drum might be picked up strongly by the overheads as well as by the snare microphone, for example, but not by the bass drum microphone.

The clever thing here is that when events, such as a snare or bass drum hit, are closely aligned on multiple tracks, Beat Detective will only insert a beat trigger for the first transient that is detected on any of these tracks.

I tried this with a bass drum, a snare drum, a stereo pair in front of the kit, and a stereo pair of overheads. Sound travels about 1,100 feet every second, so a distance of two or three feet represents a time delay of about two or three milliseconds. So with a setup like this, transients in the tracks recorded using the front pair and the overhead microphones, which might be two or three feet away, will be captured later in time than those from the snare and bass drum mics, which are normally placed just a few inches away from the drums.

Take a look at Figure 3.28 to see how this works out in practice. Here you will see four selected drum tracks that have been analyzed to find the beats. The transient in the bass drum track has been identified and a beat trigger inserted across all the tracks. The transients in the snare, front pair, and overhead tracks have been recognized as belonging to this same bass drum hit, even though they occur a few milliseconds later, so no beat triggers have been inserted corresponding to these.

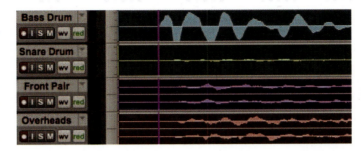

Figure 3.28
Working with
multiple tracks:
The transient is
detected for the
bass drum track
because this occurs
before the other
tracks

NOTE
This ability to 'intelligently' analyze multiple tracks is why Beat Detective scores highly compared with using the 'Separate Clip At Transients/At Grid' command from the Edit menu—which works fine for separating single tracks, but cannot take account of multiple tracks of drums (or other instruments) in the same way.

Creating a Tempo Map Using Beat Detective

You can use Beat Detective to create a tempo map so that you can overdub other instruments to music that was recorded without playing to a click and have this line up with the bars and beats in Pro Tools.

Here's the scenario: I recorded a drummer playing various four-bar patterns so that I could choose one of these later as the basis for a new song. The drummer did not play to a click, but did play very accurately, with his own 'feel'.

After the recording session, I chose the four-bar pattern that I wanted to use, trimmed the start and end points of the audio to get rid of the unwanted material, and made sure that my selection started exactly at the beginning and finished exactly at the end of the four bars.

TIP
You can zoom the display in using Command-[on the Mac or using Control-[on Windows to check that the edit points are falling on zero waveform crossings; to zoom the display out, use Command-] on the Mac or Control-] on Windows.

I estimated the tempo as 137.5 BPM and set the Pro Tools session tempo to this. Then I moved the clip to start at the beginning of bar 2 (to leave room for a count-in to use when overdubbing more instruments).

93

Listening to the drums, I recognized that the smallest beat subdivisions that the drummer was playing were 16th notes and that the time signature was 4/4, so I set these parameters to match in the Selection section of the Beat Detective window.

> **NOTE**
>
> To be able to make informed decisions about time signatures and beat subdivisions, you need at least a basic knowledge of music theory: if you don't know about this stuff, either ask a musician or go and learn about music theory—it can be very useful!

Having made these preparations, I selected the four-bar clip and clicked once on Capture Selection. As the drums were not exactly in time with the estimated tempo, this came up with the start Bar|Beat correctly at 2|1 but with the end Bar|Beat incorrectly at 5|4. See Figure 3.29.

Figure 3.29
Capture selection

Knowing that I had selected exactly four bars, I corrected this by typing the correct end Bar|Beat as 6|1—see Figure 3.30.

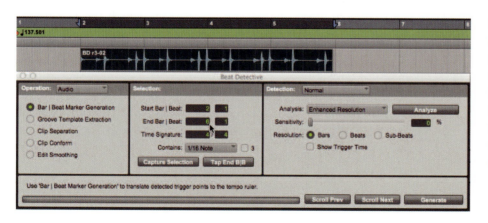

Figure 3.30
Edit window showing a bass drum track with the Beat Detective window being used to capture and define the selection range

NOTE

If you have placed the clip to start accurately on a bar line, the start Bar|Beat will always be correct—but the end Bar|Beat probably will not be. If you have counted the bars in your selection accurately and made your selection accurately, then you know what the end Bar|Beat should be, so you can go ahead and type this to correctly define the selection range.

TIP

If you have made your selection accurately, but are not quite sure what the end Bar|Beat is, you can either count the bars carefully while playing back, or you can simply play the selection and click on the Tap End B|B button in the Beat Detective window until Beat Detective works out what this should be. Don't assume that Beat Detective has got this right: always check by counting each bar while also listening to the Pro Tools session metronome click.

In the Detection section I chose enhanced resolution and clicked on the Analyze button. After a few seconds the analysis was complete and I selected bars as the resolution and raised the sensitivity slider until purplish-red beat trigger lines appeared at the beginning of every bar—see Figure 3.31.

Figure 3.31 Raising the sensitivity to detect bars

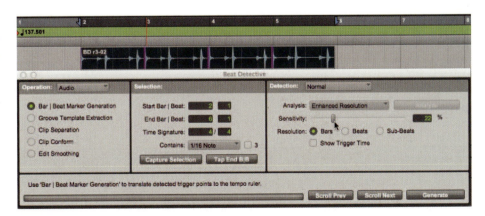

To create a tempo map for this selection now just required one click on the Generate button at the bottom right of the window. After clicking this button, the Realign Session dialog appears, asking whether to preserve the sample positions or the tick positions of clips on tick-based tracks—see Figure 3.32. In this case, the correct choice was to preserve the sample position so that my sample-based audio clip would not move from its original position.

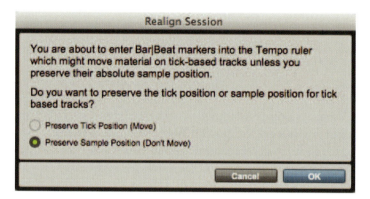

Figure 3.32
The Realign Session dialog

NOTE

When you keep the sample positions fixed, the Bar|Beat positions will change as the new Bar|Beat markers and tempo changes are created, so that they reflect the actual tempo changes within the audio and match the bars and beats in the Pro Tools session to these.

As can be seen in Figure 3.33, after approving the Realign Session dialog, a series of beat markers was inserted in the Tempo ruler, with tempo changes calculated to ensure that the start of each bar of the audio aligned itself exactly with each bar in Pro Tools.

Figure 3.33
Bar|Beat markers generated by Beat Detective are shown in the Tempo ruler

> **TIP**
> Depending on how fast you are at using the 'Identify Beat' command, and how unlucky you could get using Beat Detective, it is quite possible that you may find it quicker to create an accurate tempo map manually.

Beat Detective's Most Powerful Feature: Collection Mode

You can also use Collection mode when you are working with multiple audio tracks. In many ways, this is Beat Detective's most powerful feature. Instead of having to manually insert, move, or delete incorrect or false triggers, which is often the case in Normal Detection mode, Collection mode lets you set the detection parameters individually on any track or tracks within a group of tracks to generate the particular beat triggers that you need to suit your particular purpose.

Using Beat Detective's Bar|Beat Marker Generation, Groove Template Extraction, or Clip Separation modes, you can collect beat triggers generated separately using two or more tracks from a group of tracks, switch the Detection section from Normal to Collection mode and save these as a collection of triggers, then apply this collection to the group of tracks. As with Beat Detection across Multiple Tracks, this Collection mode is quite intelligent: if successive triggers in the collection are located closely together (for example, because of microphone leakage), Beat Detective keeps only the earlier triggers.

You can use different detection settings for each track that you analyze (or even use different detection settings on a single track), collect and save a set of beat triggers from these different tracks using Collection mode, then use the collection of triggers to generate Bar|Beat markers or groove templates or to separate your selection into new clips. This last step (saving triggers using Collection mode) must be carried out from the Collection Mode section, where the triggers from the different tracks will be displayed in different colors.

Drum Tracks and Collection Mode

Using Collection mode, you can analyze each drum track separately, one at a time, optimizing the Detection settings for each track until you get the triggers you want. All the triggers for each track can be added successively to the collection, or you can choose only the unique triggers from each track that don't exist on the other tracks. The collection of triggers that you have put together can then be used to generate Bar|Beat markers or a groove template, or to separate the selected clip into new clips.

Here's a typical scenario: you have recorded a drum kit using separate tracks for bass drum, snare drum, a front microphone pair, and an overhead microphone pair. If you analyze a selection across all of the tracks, you are likely to get lots of false triggers when you raise the sensitivity slider high enough to capture the sub-beats that you are interested in. The hi-hats, for example, would be mostly in the overhead mics, together with all the bass drum and snare drum beats. But if you were to only generate beat markers based on the overheads, the resulting beat triggers would be slightly later than the material on the other tracks (because it takes more time for the sounds from the drum kit to reach the overhead mics). And if you did decide to use the beat triggers from the overheads and you extended this selection to the other drum tracks so that you could separate these into separate clips, the clip slices from the bass drum and snare drum tracks would not be in the right places. The cuts would be made further along the timeline than they should be—in line with the beat triggers from the overheads track. The solution is to use Collection mode to take only the beat triggers you want from the different tracks, put these into your collection of beat triggers, then use these to slice the tracks.

Example Session

I tried this out with a recent recording. The group of drum tracks included a mono bass drum, a mono snare drum, a stereo front pair, and a stereo overhead pair. The front pair was about two feet away from the drum kit and the overhead pair was about three feet above the drum kit, so sounds from the bass drum and snare drum would be delayed by two or three milliseconds in the front and overhead pairs of microphones.

I wanted to have a set of beat triggers corresponding to each of the main bass drum and snare drum beats within the four-bar drum groove section that I selected. The main snare beats were on the 2 and the 4, while the main bass drum beat was on the 1. There was no main bass drum beat on the 3, although I wanted to split the pattern at that point.

I couldn't analyze the audio from the front or overhead microphone pairs to find triggers for the bass drum and snare drum because these beat triggers would be located slightly later in time than the bass drum and snare drum on the mono tracks. So I analyzed the bass drum track first, to find the main bar triggers, and added these to the collection. Then I analyzed the snare track to identify the backbeats on the 2 and the 4. Then I analyzed the bass drum track again, using a very low sensitivity that did not produce any beat triggers, manually inserted beat triggers on beat 3 of every bar, and added these to complete the collection.

Here's how it went: having carefully prepared my four-bar selection containing all four drum tracks in the Edit window, I selected the bass drum track on its own, opened Beat Detective, and chose Bar|Beat Marker Generation.

In the Selection area, I checked that the correct four bars were selected in 4/4 time signature, and I chose 8th notes as the beat subdivision. I set the detection resolution to bars, then hit Analyze, and raised the sensitivity slider until I found the triggers I needed.

You can see these triggers, shown in a purplish-red color, at the beginning of each bar in the bass drum's waveform display, as shown in Figure 3.34.

Figure 3.34
Analyzing the bass
drum track to find
bar triggers

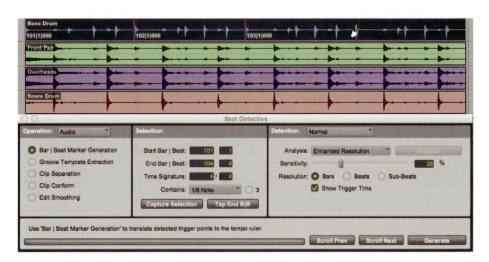

To add these to my new collection of beat triggers, I switched to Collection mode using the pop-up selector in the Detection section and clicked 'Add All' to add the bar triggers for the bass drum track—see Figure 3.35.

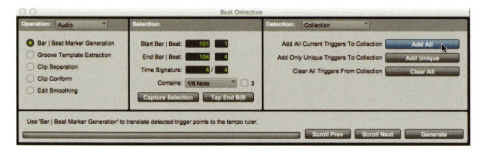

Figure 3.35
Adding the bass drum track's bar triggers for the collection.

To pick out the main snare beats, I moved the selection down to the snare track, switched back to Normal Detection mode, set the detection resolution to beat, clicked Analyze again, then raised the sensitivity slider until beat triggers appeared next to all of the main snare beats—see Figure 3.36.

> **TIP**
>
> To move the selection down, press Control-semicolon (Mac) or Start-semi-colon (Windows) or, if you have the Commands Keyboard Focus enabled, you can just press the Semicolon (;) key to move the selection down. [Press P to move up.]

Figure 3.36
Analyzing the snare drum track to find beat triggers for the backbeat

Switching back to Collection mode, I added these unique new beat triggers from the snare drum to my collection of beat triggers—see Figure 3.37. (These beat triggers are 'unique' in the sense that they are not in the same locations as any previously added to the collection.)

Figure 3.37
Adding the beat triggers from the snare drum track to the collection

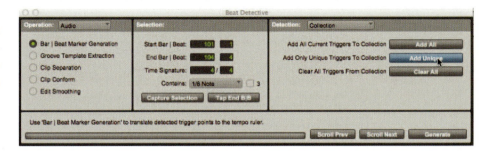

In Collection mode, I could see the red-colored bar triggers that I had added to the collection previously together with the snare drum beat triggers, colored light green in this example—see Figure 3.38.

Figure 3.38
After adding the snare drum track's beat triggers to the collection you see red triggers on the bar lines and light green triggers at the beginning of each snare backbeat.

At this point, I realized that I needed beat triggers for beat 3 of each bar. The bass drum was never played on beat 3, but I wanted to be able to separate the clip according to where the main beats actually fell, on beats 1, 2, and 4, and also to make a separation where beat 3 would fall in relation to these. So I moved the selection back up to the bass drum track so that I could insert a beat trigger on beat 3 in every bar.

> **TIP**
> To move the selection up, press Control-p (Mac) or Start-p (Windows) or, if you have the Commands Keyboard Focus enabled, you can just press 'p' to move the selection up. [Press ';' to move down.]

With the bass drum track selected, I switched back to Normal mode, chose Beat Resolution this time, clicked Analyze again, then set the sensitivity to a low percentage so that any beat triggers that I inserted manually would be visible—while none would be detected automatically. Then I manually inserted a beat trigger on beat 3 of every bar. These were displayed in a light green color—see Figure 3.39.

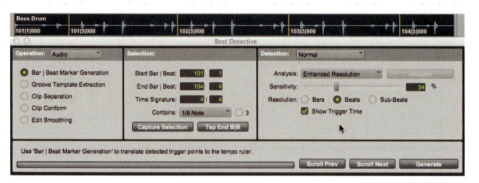

Figure 3.39
Four beat triggers manually inserted into the bass drum track on beat 3 of each bar

To make sure that these were accurately positioned, I zoomed the display so that I could finely adjust the position of each of these beat triggers—watching the cursor display as I moved each trigger until it read exactly beat 3|000—see Figure 3.40.

Once I had the exact triggers that I needed, I switched to Collection mode and added these unique new triggers to complete my collection—see Figure 3.41.

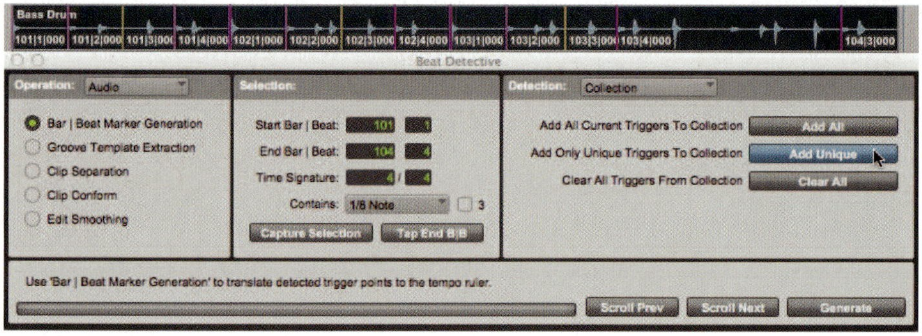

Figure 3.40
Adjusting a beat trigger to fall exactly on beat 3 of the bar

Figure 3.41
Adding beat triggers from the bass drum track to the collection

Now that I had all the beat triggers that I needed in my collection, with one for each beat of each bar, it was time to use these. I selected all the other drum tracks by Shift-clicking on each track so that the beat triggers would extend throughout all the tracks. Then I switched to Clip Separation mode (leaving the Detection section's Collection mode active) and clicked on the Separate button—see Figure 3.42.

Figure 3.42
Separating the clips

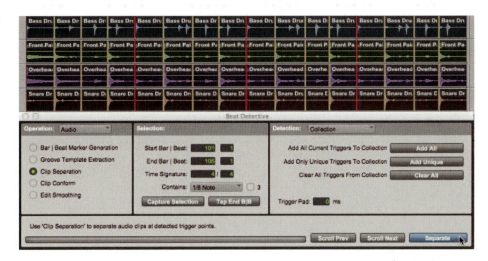

This worked perfectly—'just like it says on the tin'! Beat Detective created 16 clips on each track, each clip containing one beat from the original four-bar drum groove. The beat triggers that were detected for the bass drum and snare drum were also used to separate the overheads and the front pair, so that the cuts to these fell exactly in the correct places, lining up with the cuts in the bass drum and snare tracks—see Figure 3.43.

Figure 3.43
The 16 separated clips for each track

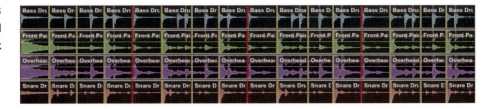

Figure 3.44
A single bass drum 'hit'

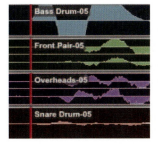

To check that the slices had been made correctly, I zoomed the display so that I could look at each of the main bass drum and snare drum beats in detail. Looking at a bass drum slice, I could see the bass drum 'kick' showing up a little later in the snare drum track at much lower level, then in the front pair and overheads tracks a little later again. Notice the thick, red bar marker. See Figure 3.44.

Looking at a snare drum slice, I could see the snare drum 'hit' showing up a little later in the bass drum track at much lower level, then in the front pair a little later and in the overheads track even later still. Notice the thinner, light green, beat marker. See Figure 3.45.

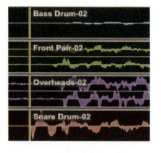

Figure 3.45
A single snare drum 'hit'

So, mission accomplished again: Beat Detective's Collection mode had helped me to produce the exact slices that I wanted from the four-bar drum groove that I chose to work with by automatically analyzing the bass drum and snare drum tracks and by allowing me to manually insert additional beat triggers in the bass drum track.

Tips for using Beat Detective

- Beat Detective works best with rhythmic tracks such as drums, bass, and guitars—so don't expect great results with more ambient material such as strings, vocals, or synthesizer pads. And if you have a rhythm track that is too complex or too far out of time, you may not get such good results either.

- Where Beat Detective works well is when analyzing short sections of audio— especially when you want to detect the beats and sub-beats within a bar.

- It is often best to split longer clips into shorter eight-bar or four-bar sections (or less) to make these easier to work with.

- It is crucially important to define selections accurately, zooming the display to sample level and editing at zero waveform crossings whenever possible.

- It is also crucially important to make sure that you have set the correct time signature.

- Be aware that beat triggers can often be placed wrongly or inaccurately during the analysis process, so you may need to move these, delete any that are not needed, and insert any that are missing.

- Be aware that Beat Detective is not perfect! You should check visually and aurally when you have created the Bar|Beat markers to make sure these are accurately placed and move them manually if they are not.

- Remember that it can be quicker to use the 'Identify Beat' command to build tempo maps, especially if you find that Beat Detective is not successfully placing beat triggers after analyzing a selection.

■ If you are creating a tempo map, you should be using sample-based audio tracks so that the audio clips stay fixed while the bars and beats change in relation to the audio when you generate Bar|Beat markers.

■ After analyzing one clip, if you want to analyze subsequent clips, make sure that you capture or define each new section accurately in the Beat Detective window and that you click the Analyze button again for each new clip that you work with. Otherwise the beat triggers that you see may be based on a previous analysis.

■ Try using the different Analysis modes—enhanced, low, or high—if you are not getting sensible results. These different modes use slightly different algorithms to analyze the audio material that may be more appropriate for the audio you are working with.

■ If you are creating a groove template, make sure that there are no Bar|Beat markers within your selection.

■ It is often a good idea to get the average tempo of your selection set as accurately as possible—either by using Identify Beat or by generating a Bar|Beat marker using Beat Detective for your whole selection by pressing Generate as soon as you have set the selection parameters correctly. Then when you subsequently analyze the selection to produce bar and beat triggers, you will immediately see if these are located in the correct positions—and Beat Detective will find it easier to place them in the correct positions.

■ Make sure that you fully understand how Beat Detective works with multiple tracks—and how to use Collection mode. If you do need to work with multiple tracks, it may be easier to work with them individually—or using Collection mode.

Summary

Beat Detective can produce excellent results, saving you a lot of time when you are editing audio clips. However, to get the best out of using this, you must always make sure that you have set the time signature correctly, that you have made your clip selections accurately, and that you have entered the correct number of bars and beats when defining your clip selection.

I have found that I get much more accurate results more quickly if I make sure that the session tempo is as close as possible to the actual tempo of the audio before I try to do anything ambitious with Beat Detective. This way, it is easy to see if the beat triggers are being produced in sensible locations. You can do this by trial and error—guessing the tempo and trying this to see if your guess is correct—or by using the 'Identify Beat' command.

Also, if Beat Detective puts any Bar|Beat markers in the wrong places when you generate these, you can manually move them to the correct positions by dragging the markers along the timeline. So it is always a good idea to play through your selection when you think you have all these markers in the right places, with the display zoomed so that you can see how accurately they are actually positioned, and move them as necessary. You can even use the 'Identify Beat' command for this purpose. So don't feel that you can only use Beat Detective—you can use a combination of methods to achieve the end result.

Ultimately, if you practice using Beat Detective on a range of different material, you will get used to using it and find the ways that work best for you. Good luck!

in this chapter

CHAPTER 4

Elastic Audio

Introduction

'Elastic Audio'—say what? The clue lies in the name. As you have probably guessed, it lets you stretch and squeeze audio as though it were elastic. According to Avid: 'Elastic Audio lets you apply Time Compression and Expansion (TCE) in real-time to individual audio tracks using high-quality transient detection algorithms, beat and tempo analysis, and TCE processing algorithms.'

Pro Tools analyzes the audio first, to identify events such as drum hits, individual sung notes, individual chords, or whatever. Each audio track can be switched to display the Analysis view in the Edit window to let you see the event markers generated by this analysis. These detected events, when viewed in Warp view, can be fixed as control points around which you can manually 'warp', in other words 'time stretch', the audio. In Warp view, you can grab and move individual event markers back and forth along the timeline to warp audio that lies before, after, or in between warp markers that fix chosen event markers on the timeline. You can use this feature to realign even a single note that was played just a little late!

Audio events can also be warped automatically, allowing you to quantize events or match the audio to the session tempo. Elastic Audio pretty much lets you treat audio like MIDI, quantizing and moving notes around more or less at will. You can use Elastic Audio to quickly beat match an entire song to the session tempo and Bar|Beat grid. Or you can quantize audio to a groove template so that other audio clips take on the same feel. For instance, you might take the feel from one drum loop that you like and apply this to another. And if you are a fan of Beatles-style recording effects, you can use the Varispeed algorithm to create tape-like effects with speed changes that cause pitch changes.

Elastic Audio is exactly what you need to use if you work with rhythmic loops and samples. You can preview these in Workspace browsers either at their original tempo or at the session tempo using Elastic Audio—even while the session is playing back. When you find a loop or sample you like, just drag it from the browser and drop it into the session. It will automatically be assigned to an Elastic Audio–enabled track and will automatically adjust its tempo to the session tempo map and Bar|Beat grid.

You can also use Elastic Audio in many other types of music production or postproduction work. If you are recording 'live' musicians, you will often encounter situations where the timing of a performance drifts more than it should, and Elastic Audio can be an incredibly effective tool to use when you need to encourage the piano player, the horn section, the rhythm guitar player, the drummer, or the percussionist to get back into the 'groove'. Or to creatively change the groove! And the potential for creating special effects is, as they say, only limited by your own imagination . . .

In the track controls area in the Edit window you can enable Elastic Audio on any audio track by choosing one of the four algorithms provided: Polyphonic, Rhythmic, Monophonic, and Varispeed. The first three are time stretch algorithms that let you make the session tempo faster or slower and cause the tempo of the audio on that track to change with the session tempo while the pitch of the audio stays the same. Or you can choose the Varispeed algorithm, in which case the pitch of the audio on the track will change as you change the session tempo—just like when you vary the speed of a tape recorder.

> **NOTE**
> Although the Polyphonic, Monophonic, and Rhythmic algorithms used for Elastic Audio processing are very good, you may still hear artifacts disturbing the audio quality when using these real-time algorithms. If this bothers you, you should use X-Form processing—which is only available as a rendered process. It can take a while to apply this rendered processing, but the results are much, much better.

Sample Based versus Tick Based

Audio is normally sample based and does not change in any way if you alter the session tempo. MIDI is normally tick based and plays back faster or slower if you change the session tempo. As with all Pro Tools tracks, Elastic Audio tracks can be either sample based or tick based.

Sample-based Elastic Audio–enabled tracks let you apply real-time or rendered Elastic Audio processing by editing in Warp view, applying Quantize, and using the TCE trimmer tool. However, sample-based Elastic Audio–enabled tracks *will not* automatically change the tempo of the audio when the session tempo changes.

You can also do all of these things with tick-based Elastic Audio–enabled tracks, but with one difference: tick-based Elastic Audio–enabled tracks

112

will automatically change the tempo of the audio when the session tempo changes—and without the pitch of the audio changing.

How does this happen? Any audio on tick-based Elastic Audio–enabled tracks is automatically warped to conform to the session tempo whenever this tempo changes. So when you change your tracks to tick based and change the session tempo, time compression or expansion is automatically applied to the audio so that its tempo will match the session's tempo—Elastic Audio is an incredibly powerful audio manipulation tool!

Elastic Audio Algorithms

Polyphonic

The Polyphonic plug-in uses a general-purpose algorithm that works effectively with most material. This is the one to use for complex loops and multi-instrument mixes and is the default plug-in for previewing and importing from DigiBase browsers or the clip list.

This plug-in has two control options: 'Follow' and 'Window'. Selecting the 'Follow' option enables an envelope follower that simulates the original acoustics of the audio being stretched. The Window control lets you adjust the size of the analysis window used for the TCE processing. It is worth experimenting with different settings to achieve best results (see Figure 4.1).

Figure 4.1
Elastic Audio
Polyphonic plug-in

Rhythmic

The Rhythmic plug-in works best with material that has clear attack transients, such as drums (see Figure 4.2). When you are using the Rhythmic plug-in, any gaps between transients resulting from time stretching are automatically filled in with audio.

A decay rate control lets you set a fade-out rate so that you can decide how much of the decay from the transients will be heard in the processed audio. If you adjust the decay rate up to 100 percent you will hear the audio that is filling the gaps created by the time stretching with only a slight fade. Adjusting down to 1 percent will allow the audio to completely fade out to silence in the gaps between the original transients.

For example, if you stretch a bass drum or snare drum track, the decay part of the sound can become a little messed up because of the time-stretching process. If this happens, you can reduce the decay rate until the messed up part of the decay is no longer heard—allowing you to achieve a better-sounding result.

Figure 4.2
Elastic Audio
Rhythmic plug-in

Monophonic

The Monophonic plug-in works best with monophonic material. It preserves formant relationships with material such as vocals, and also works well with monophonic instrumental parts played on guitar, bass, or other instruments. It has no parameters that you can adjust.

The Monophonic plug-in provides higher-quality time compression and expansion with pitched material because the algorithm analyzes pitch as well as peak transients. Because of its extra complexity, this analysis does take a little longer than the analysis for the Polyphonic, Rhythmic, and Varispeed algorithms. Also, when you switch to this plug-in from any other Elastic Audio plug-in, the audio has to be reanalyzed so that the pitch analysis can be added to the peak transient analysis data.

Varispeed

Like the Monophonic plug-in, the Varispeed plug-in has no control parameters to tweak—it just works. With this enabled, when you change tempo in Pro Tools the audio on the track plays back faster or slower, changing pitch accordingly, just like a tape machine would do.

Because varispeed changes the pitch when you speed up or slow down the audio, this can be a very desirable effect creatively—think robots and chipmunks if you slow down or speed up vocals very obviously. The Beatles were one of the first groups to regularly use small amounts of varispeed on the vocals, and occasionally on other instruments, to make the tracks sound 'special' in more subtle ways. With vocals running just a little faster or a little slower, the effect is not always blatantly obvious to the listener, but it will always have at least a subliminal effect that can make the vocals sound larger than life—although not totally natural. Pop music is all about catching the ear of the listening public, of course, so artificially enhancing sounds is often the producer's intention . . .

NOTE

If you apply pitch transposition to an audio clip using the Polyphonic or Rhythmic algorithms, that transposition data is stored in the metadata for the clip. If you then switch to Varispeed, change the TCE factor, and switch back to the original Elastic Audio algorithm, the clip will revert to the original amount of pitch shifting while maintaining the amount of time compression/ expansion applied by the Varispeed algorithm.

TIP

Here's a neat trick that you can try: with the Varispeed algorithm selected and 'Conform to Tempo' enabled in the browser, drag a drum loop from the Workspace browser into a Pro Tools session to create a new Elastic Audio track. If you change the session tempo this will varispeed the loop up or down in pitch, causing a change in the timbre. When you are happy with the sound, disable Elastic Audio and commit to this change to render a new version of the loop. If you then bring this loop back to the original session tempo using the Elastic Audio Rhythmic algorithm, it will retain the altered timbre that you just created using the varispeed process. You can also play with the decay rate to create stuttering or other effects. This is a great way to create new sounds from old loops to blend in with your other tracks!

X-Form

The real-time Elastic Audio algorithms, although very good, do not produce the highest-quality results. You can use these while you are working out ideas and when achieving the highest audio quality is not a priority. When this is a priority, use X-Form rendered processing. This takes significantly longer—but the results are well worth the wait!

Two control parameters are provided. The Quality pop-up menu lets you choose either Maximum or Low (Faster) quality. Maximum is the slowest processing algorithm, but provides the highest-quality results. Low (Faster) produces relatively good results and is much faster than the Maximum setting. For audio material with clear formants, you can enable 'Formant' to preserve the formant shape of your audio when you are applying TCE processing (see Figure 4.3).

Figure 4.3
Elastic Audio
X-Form plug-in

The Elastic Properties Window

The Elastic Properties window displays information about the Elastic Audio processing, for one or more clips, which can be examined and changed as necessary from this one convenient place. If more than one clip is selected, the Clip property shows the number of selected clips. The Clip property displays the name of the clip (or states how many clips are selected) and also includes a small metronome icon if it is on a tick-based track.

The Input Gain property lets you attenuate the gain of the audio signal before it is processed to avoid clipping, and the Pitch Shift property lets you pitch shift clips using the Polyphonic, Rhythmic, or X-Form plug-ins.

The TCE Factor property shows how much time compression/expansion has been applied to the selected clips or clips as a percentage. You can change this displayed value to apply TCE to selected clips on sample-based tracks. On tick-based tracks, this property cannot be edited—it is for display only.

The Pro Tools Reference Guide explains the Event Sensitivity property in some detail that is worth quoting directly here: 'The Event Sensitivity property lets you filter Event markers based on the analysis confidence level. The confidence level for any detected transient event is based, in part, on the clarity of the transient. For example, if the file is a drum loop, loud accented hits will be analyzed with a higher degree of confidence than a soft, unaccented hit. The Event Sensitivity acts like a threshold for showing only the transient events that were detected with a high degree of confidence.'

The Reference Guide goes on to explain why it can be a good idea to adjust the event sensitivity when you are working with the event markers in Analysis view: 'In Warp or Analysis track views you will see the number of Event markers decrease or increase as you lower or raise the Event Sensitivity. Lowering the event sensitivity can help reduce the number of erroneously detected transients. In turn, this can result in better sounding Elastic Audio processing.' In other words, don't assume that the automatic analysis will always pick out all the correct transients to mark with event markers. In fact, on many occasions, it will not, as the Reference Guide also makes clear: 'Pro Tools preserves the detected transients when applying TCE in order to avoid flamming and granulation of the transients. Consequently, false transients are also preserved and the resulting sound quality can be less than desirable. If you are working with audio material that does not have clearly defined transients, you may want to lower the Event Sensitivity in the Elastic Properties, or you may want to even edit the Event markers in Analysis view.'

> **TIP**
> In practice, I have found that a combination of adjusting the event sensitivity and manually editing the event markers produces the best results with most material.

Elastic Properties for Sample-based Tracks

On sample-based tracks, a smaller set of Elastic Audio properties is displayed that includes the Clip property, the TCE Factor, Event Sensitivity, Input Gain, and Pitch Shift properties. On tick-based tracks, additional properties including Source Length, Source Tempo, and Meter are displayed.

> **NOTE**
> If you open the Elastic Properties window when clips on both tick-based and sample-based tracks are selected, only the smaller set of properties is displayed in the Elastic Properties window (see Figure 4.4).

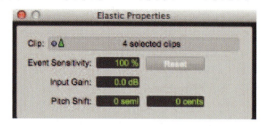

Figure 4.4
Elastic Properties window for a clip on a sample-based track

Elastic Properties for Tick-based Tracks

With tick-based tracks, the Elastic Properties window also allows you to change the length of the clip that the Elastic Audio analysis has calculated, which it refers to as the *source length*. Sometimes, the analysis will produce a source length that is actually half or twice the duration of the analyzed clip. If this happens, you can easily correct this by clicking the 1/2 or X2 buttons. Otherwise, simply type the correct bar, beat, and tick values into the Source Length field.

The Source Tempo property shows the average tempo of the analyzed clip. If the analysis was not successful, you may also need to change this. A pop-up selector lets you choose the correct beat value—you would choose 1/8 for music in a 12/8 time signature, for example. As with the source length, the 1/2 and X2 buttons can be used to make instant corrections if the tempo was incorrectly analyzed as half or twice the correct tempo. Otherwise, simply type the correct tempo into the Source Tempo field. There is also a field that lets you set the correct meter if this is other than 4/4 (see Figure 4.5).

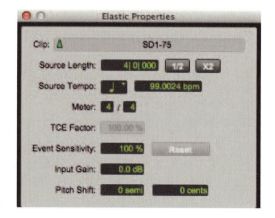

Figure 4.5
Elastic Properties window for a clip on a tick-based track

Conforming a Loop to the Session Tempo

Here is an example of how you can use the Elastic Properties window to conform a loop to the session tempo. Just drag and drop a two-bar loop with an original tempo of 120 BPM onto an Elastic Audio–enabled, tick-based track, in a session with a tempo of 60 BPM. This will play back in the time of one bar with the session set to run at half the loop's original tempo, that is, 60 BPM, because the audio within the loop will still play back at 120 BPM, that is, twice as fast. If you select this clip and open the Elastic Properties window, it displays the source length as one bar and the source tempo as 60 BPM—having picked up this information from the Pro Tools session. See Figure 4.6.

Figure 4.6
A two-bar loop containing audio that plays at 120 BPM occupies one bar with the session tempo set to 60 BPM.

But we know that the source length is two bars and the source tempo is 120 BPM, so we can enter this information into the Elastic Properties window so that Pro Tools can use the correct information about the audio. In this example, because the correct length and tempo are twice what is displayed in the Elastic Properties window, all that is necessary is to click on the X2 button to enter the correct values. The audio clip in the Edit window is immediately time stretched to run at the session tempo of 60 BPM, and now lasts for two bars on the timeline, as can be seen in Figure 4.7.

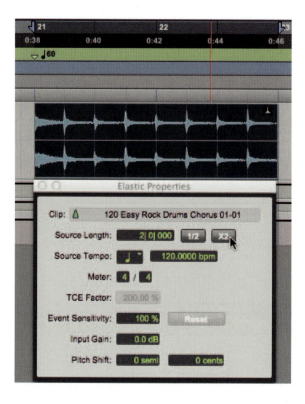

Figure 4.7
After entering the correct source length and tempo, the selected audio clip is immediately time stretched to last two bars, running at the session tempo of 60 BPM

> **NOTE**
> Another way to achieve the same result is to select the clip and use the 'Conform to Tempo' command from the Clip menu. The Elastic Properties window will be updated accordingly. The advantage of using the Elastic Properties window is that you are presented with more detailed information and control possibilities. You can, for example, adjust the beat resolution and the meter from this window.

'Conform to Tempo' Command

A very quick way to make the tempo of an audio clip conform to the session's tempo is to use the 'Conform to Tempo' command. Assuming that a tempo is already set in Pro Tools, all you need to do is to enable one of Elastic Audio's analysis modes. When you select an audio clip within the track, the 'Conform to Tempo' command becomes available in the Clip menu.

When you choose the 'Conform to Tempo' command, it analyzes the selected clip to determine its tempo and duration in bars and beats, then applies Elastic Audio processing to match the tempo of the clip to the tempo of the session.

> **NOTE**
>
> Pro Tools actually analyzes the clip's entire file to work out where the Elastic Audio events should fall and what the tempo should be. If a tempo is detected and a duration in bars and beats is determined, the clip conforms automatically to the session tempo. If a tempo and duration is not detected (such as with a clip that contains a single drum hit or with a long clip with no clear tempo), the clip will not be conformed to the session tempo.

I tried this with a two-bar loop with a tempo of 120 BPM. I set the tempo of the session to 120 BPM and dropped the loop into a track. As you would expect, the loop occupied two bars along the timeline (see Figure 4.8).

Figure 4.8
Two-bar loop, running at 120 BPM, placed into a session running at 120 BPM. So the loop occupies exactly two bars along the timeline at this tempo

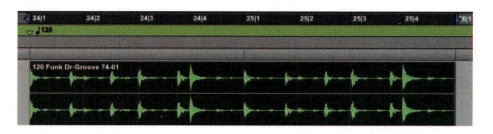

When I set the session tempo to 80 BPM, the two-bar loop, still running at 120 BPM, occupied just a little more than one bar along the timeline (see Figure 4.9).

Figure 4.9
With the session tempo at 80 BPM and the loop tempo at 120 BPM, the two-bar loop is out of step with the bars and beats in Pro Tools

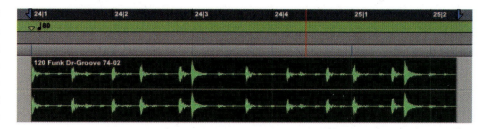

I enabled Elastic Audio's Rhythmic algorithm on the track, selected the clip, and applied the 'Conform to Tempo' command from the Clip menu. The audio was immediately time compressed to run at the session tempo of 80 BPM. Mission accomplished! (See Figure 4.10.)

Figure 4.10
After applying the 'Conform to Tempo' command, the loop is adjusted to play back at the tempo of the session, so it plays for two bars at 80BPM

NOTE

'Conform to Tempo' can only be applied to clips and cannot be applied to clip groups. To conform clip groups to the tempo you must first ungroup the clip group, then apply 'Conform to Tempo' to the underlying clips, and then regroup those clips.

TIP

'Conform to Tempo' does not always work as expected. It worked fine for me with straightforward percussive loops from sample libraries, but I could not get it to work properly with longer audio clips that I had recorded using 'live' musicians. The Pro Tools Reference Guide suggests that you use Elastic Audio's telescoping warp if you want to conform long clips to the session tempo—and my experience confirms that this is the best way to handle these.

Out-of-range Processing

While I was trying out the 'Conform to Tempo' command, I put in a tempo of 30 BPM just to see what would happen. The first thing I noticed was that the processed clip turned pink! It turned out that it does this to warn that the processing is out of the range that can be successfully handled. And, sure enough, when I played it back, the audio quality did not sound too good.

According to the manual, the valid range for Elastic Audio TCE processing is from a quarter of the tempo up to four times the tempo for all Elastic Audio plug-ins except X-Form, which supports a wider range—from 1/8 x to 8 x. Time compression or expansion less than or greater than these ranges, that is, less than 25 percent or greater than 400 percent, is out of range—so Pro Tools warns you about this by turning the clip pink.

In this case, the original tempo was 120 BPM, so the lowest tempo within this range is 120/4 = 30 BPM. When I tried it, 31 BPM worked fine, but at 30 BPM the display turned pink; 479 BPM also worked fine, but at 480 BPM the display turned pink (see Figure 4.11).

Figure 4.11
Audio clip warped from 120 BPM to 30 BPM. This is out of range, so the display turns pink to warn you about this

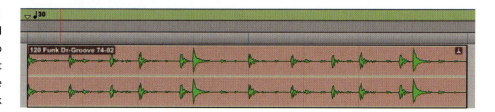

As the manual points out, when you exceed the valid range, it is possible that not every sample will be processed correctly and you may encounter audio dropouts. In truth, such extreme shifts are only likely to produce sounds that fall into the realm of sound effects rather than pristine audio in which dropouts could not be tolerated. By the time you have shifted to such extremes, the resulting sound will already be full of artifacts, so any such dropouts may not be noticeable or may actually help the effect!

Audio Tempo follows Session Tempo

One of the most basic things you can do with Elastic Audio is to have the tempo of your audio tracks speed up or slow down when you change the tempo in your Pro Tools session.

You can try this out with anything you like, so why not pick one of your favorite mixes or a track off one of your favorite CDs; put this onto a track in a new Pro Tools session and see how well Elastic Audio lets you do this!

Click on the pop-up Elastic Audio Plug-in selector in the Track Controls area and choose the Elastic Audio plug-in that best suits your track material and the effect you are trying to achieve (see Figure 4.12). The options include Polyphonic, Rhythmic, Monophonic, Varispeed, and X-Form. For example, with a stereo mix that you want to speed up or slow down without changing pitch, you will normally get best results using the Polyphonic algorithm.

Real-Time Processing will be selected by default. The audio clip in the track will temporarily go offline and the waveform will be greyed-out for a short time until the analysis is completed (see Figure 4.13).

Figure 4.12
Enabling Elastic
Audio polyphonic
real-time processing

Figure 4.13
Changing the track
timebase to ticks

Make sure that the track's timebase is set to Ticks instead of Samples. Then all you need to do is to move the manual tempo up or down and the tempo will follow immediately: it couldn't get much easier than this!

> **TIP**
> To change the tempo manually, first make sure that the Conductor icon is deselected in the Transport window. Use the mouse to point, click and hold, then drag up or down on the tempo field in the Transport window—see Figure 4.14.

Figure 4.14
Changing the
tempo of an Elastic
Audio–enabled
track

NOTE

Although the tempo of the audio will change when you change the session tempo, the audio will not necessarily match the tempo of the session (so the bars in the audio won't correspond with the bars in the Pro Tools session)—unless the tempo of the audio is constant and you had originally set the tempo of the session to this tempo (in which case the bars in the audio will correspond with the bars in the Pro Tools session).

TIP

To match a recording with changing tempos to the session tempo, you need to use the more detailed warp matching features explained in the next section.

Analyzing and Warping Audio

Using Elastic Audio is all about analyzing audio to discover where transient events occur that coincide with beats within the audio that you want to lock to particular locations along the timeline so that you can expand or compress the audio before, after, or in between these locked beats to suit your purpose . . .

Analyzing Audio

After enabling Elastic Audio, you can view the results of the analysis by switching the track or tracks you are working with to Analysis view. Here, you will see the event markers that have automatically been generated to correspond to the transient events that were identified during the analysis. In Analysis view, event markers appear as black, vertical lines that do not fully extend to the top and bottom of the track—see Figure 4.15.

Figure 4.15
Analysis view
showing event
markers

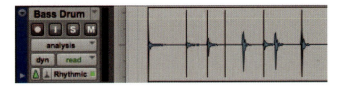

125

In simple situations, where you are analyzing just a few bars of audio containing a basic drum loop in 4/4, for example, you may not need to view these event markers. If the analysis worked properly and picked out all the important events correctly, then everything will work the way you want it to and expect it to—'just like it says on the tin'. But this will not always be the case. As the manual explains, 'Pro Tools may not accurately detect every audio event or may erroneously detect audio events with some types of audio material. This is especially true for audio without clear transients, such as legato strings, melismatic vocals, or soft synth pads.'

If you have recorded a funky or jazzy drummer, for example, and he doesn't play the beats in simple, obvious places, and especially if he plays very dynamically, using a mix of soft, medium, and strongly played beats, then Pro Tools is not really going to be able to automatically figure out which of these beats should be identified for use when subsequently warping the audio. That is something that you will have to do manually—using your human analytical skills!

This is where Analysis view becomes important. Here, you can add, move, promote, and delete event markers. To add an event marker, you can use the pencil tool to click at the location or use the grabber tool and double-click or Control-click (Mac) or Start-click (Windows) at the location where you want to add the event marker. You can use the pencil tool or the grabber tool to drag an event marker to a new location. To delete an event marker using the pencil tool or grabber tool, simply Option-click (Mac) or Alt-click (Windows) on the event marker that you want to delete. You can also use the selector tool to make an edit selection that includes any event markers you want to delete, then press Delete or Backspace on your computer keyboard—or Right-click the selection and choose 'Remove Event Marker' from the pop-up menu.

> **TIP**
> If you edit event markers (or warp markers) on an Elastic Audio–enabled track that is part of an edit group, the edits you make will be applied to all the other tracks in the group.

Fine-tuning the Event Markers

One of the best ways of making sure that you have the most appropriate set of event markers is to examine these in detail in Analysis view while adjusting the event sensitivity in the Elastic Properties window.

You can adjust the event sensitivity percentage to filter out markers that are detected with lower percentages of certainty (or 'confidence' as the manual puts it): event markers with the least confidence are filtered out first as you reduce the percentage value from 100 percent—see Figure 4.16.

Figure 4.16
Adjusting event
sensitivity in the
Elastic Properties
window allows you
to selectively reveal
event markers (the
vertical lines that
can be seen in
Analysis view)

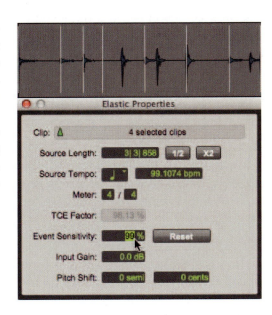

When you reduce the event sensitivity, hoping that the less relevant event markers will be removed, some event markers that you would like to retain may disappear.

If this happens, just single-click on any event markers that you want to retain, using the grabber or pencil tool, and these will be 'promoted'. Event markers promoted in this way are regarded as having 100 percent confidence (they are definitely marking events that you want to mark)—as do manually created or moved event markers—so they will not disappear when you lower the event sensitivity.

What you are aiming for here is to make sure that you keep only the event markers that you are sure that you will want to use when you subsequently warp the audio. You could delete all the event markers and manually insert just the ones that you are 100 percent confident are in the right places and that you wish to use. With an audio clip that only last for a few bars, this would be a practical thing to do. With longer clips, you will appreciate the time savings that automatic analysis brings.

TIP

Don't forget that you can always manually add an event marker wherever you believe there should be one, delete any that will not serve your purpose, and move any that are not in the right places. You will, of course, have to carefully audition the audio while looking at the clip in Analysis view to figure all this out—and it becomes progressively more difficult to do this with longer selections.

Warping Audio

When you are satisfied that the event markers that you have left visible in Analysis view are just those that you want, and that they are all in the right places, it is time to switch to Warp view. Warp view is where you will manually 'time warp' the audio, or, in other words, apply time compression and expansion to correct or adjust the timing of a musical performance or to create special effects.

In Warp view, you will add warp markers to fix ('lock' or 'anchor') event markers that identify specific locations within the audio to specific locations on the timeline. When you have fixed or 'anchored' one or more locations within the audio to the timeline, placing warp markers on top of event markers to 'lock' these down, you are then free to drag other event markers along the timeline to 'warp', that is, time stretch, the audio in various ways.

There are many ways to add warp markers using the pencil or grabber tools. You can click anywhere in a clip with the pencil tool to add a warp marker at that location. Typically, you will click on an event marker to create a warp marker that 'locks' this event to its position on the timeline.

Alternatively, using the grabber tool, you can Control-click (Mac) or Start-click (Windows) at any location in the clip to add a warp marker. Also, with the grabber tool selected, if no warp markers are present in the clip, or if warp markers are only present prior to the location at which you wish to add the new warp marker, you can simply double-click an event marker to add a warp marker on top of this. And if there are no event markers in a clip, you can double-click at any location to add a warp marker. Another way to add a warp marker on top of an event marker is to single-click, using the grabber tool, on any event marker that is located before another existing warp marker in the clip. Finally, with any edit tool, you can Right-click anywhere in the clip and select 'Add Warp Marker' from the pop-up menu to add a warp marker at that location: and if there is an edit selection, warp markers are added both at the start and end of the selection.

To delete a warp marker using the grabber tool, just double-click on it. Or, with either the grabber or the pencil tool selected, you can Option-click (Mac) or Alt-click (Windows) on the warp marker to remove it. You can also Right-click any warp marker and select 'Remove Warp Marker' from the pop-up menu. To delete all the warp markers in a selection, make an edit selection that includes only the warp markers you want to delete (and none of the warp markers you want to keep). Then press Delete or Backspace on your computer keyboard or, with any edit tool, Right-click the edit selection and select 'Remove Warp Marker' from the pop-up menu.

> **TIP**
> You can move a warp marker to a new position any time you like with the pencil or grabber tool selected: simply Control-click (Mac) or Start-click (Windows) and drag the warp marker to any new location.

Warp markers appear as thick black vertical lines with a triangle at the base of each to help you to distinguish these markers. Event markers are visible in Warp view, but as grey, vertical lines that do not fully extend to the top and bottom of the track (see Figure 4.17). Note that you can't add, relocate, or delete event markers in Warp view—you have to go back to Analysis view to make these edits. However, with the grabber tool selected, you can drag event markers in Warp view—in fact, this is exactly what you are supposed to do to warp the audio, as we shall see shortly.

Figure 4.17
Warp view showing event markers in grey with a single warp marker inserted at the beginning of the clip

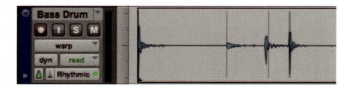

When a clip has been warped, it displays a warp indicator in the upper right-hand corner to make it clear that Elastic Audio processing has been applied to the clip (see Figure 4.18). This warp indicator is visible in any track view, but it will only be shown when a clip on an Elastic Audio–enabled track has actually been warped either manually or automatically by tempo change, quantization, or pitch shifting. Un-warped clips on Elastic Audio–enabled tracks do not display the warp indicator.

If you are unhappy with the results of your attempts at warping your audio, you can use the 'Remove Warp' command in the Clip menu to undo the warping effect from any uncommitted clip whenever you like. The warp markers are not deleted, so you can have another go at the warping using your existing warp markers— or remove these and start again.

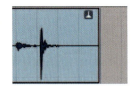

Figure 4.18
Warp indicator visible in the top right-hand corner of the warped clip

> **NOTE**
> 'Remove Warp' can only be applied to clips and cannot be applied to clip groups. To un-warp clip groups you must first ungroup the clip, apply 'Remove Warp' to the underlying clips, and then regroup those clips.

Manual Warping

Elastic Audio offers three different techniques for manually warping audio on sample- or tick-based tracks: Telescoping Warp, Accordion Warp, and Range Warp.

Telescoping Warp lets you warp the audio in the clip that occurs before the first warp marker that you have inserted or after the last warp marker that you have inserted. The important thing here is that no other warp marker should be constraining the section of the audio clip that you would like to warp.

Typically, you will apply Telescoping Warp after you have anchored the start of the clip to the timeline using a warp marker. Then, when you drag any later-occurring event marker in the clip to the left or right, all of the clip will be compressed or expanded in time from that fixed clip start point.

On the other hand, you could place a warp marker at the end of the clip. To warp audio before this marker, you would have to Option-click (Mac) or Alt-click (Windows) instead.

Accordion Warp lets you expand or compress the audio equally on both sides of a single warp marker in a clip. This is particularly useful when the most important event of interest, such as the first beat of a bar, occurs in the middle of the clip.

To use Accordion Warp, just add a single warp marker at the point in the clip that you want to remain fixed on the timeline and use the grabber tool to drag any event marker that occurs before or after the single warp marker to the left or the right. The audio will compress or expand, rather like the bellows of an accordion.

With Range Warp, the idea is that you fix two points within the clip to the timeline using warp markers, then warp the audio that lies in the range between these using the grabber tool to drag any event marker that is included within this range. As soon as you move the event marker to the left or the right, a warp marker is added on top of the event marker and the audio is compressed or expanded on either side of this marker. The audio that lies outside the two bounding markers is not affected. Range Warp is particularly useful when you need to make corrections or adjustments to sections within a selected clip without disturbing the audio elsewhere in the clip.

Matching Beats Using Telescoping Warp

One of the most useful applications of Telescoping Warp is to match the beats within a piece of music to the session tempo and Bar|Beat grid. Let's go through an example here. First you would set the Main Timebase ruler to Bars|Beats,

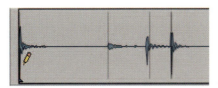

Figure 4.19
Adding a warp marker using the pencil tool

enable Grid mode, and set the meter and initial tempo appropriately. Then you enable Elastic Audio on the track, select Warp view, and use the pencil tool to add a warp marker at the first beat of the first bar within the clip—see Figure 4.19.

If the clip does not already start at the location on the timeline that it should start at, use the grabber tool to drag this first warp marker to the exact Bar|Beat location where you want the downbeat of the audio file to start—at the beginning of Bar 106 in the example shown in Figure 4.20. It is best to use Grid mode for this so that the clip will snap exactly to the bar.

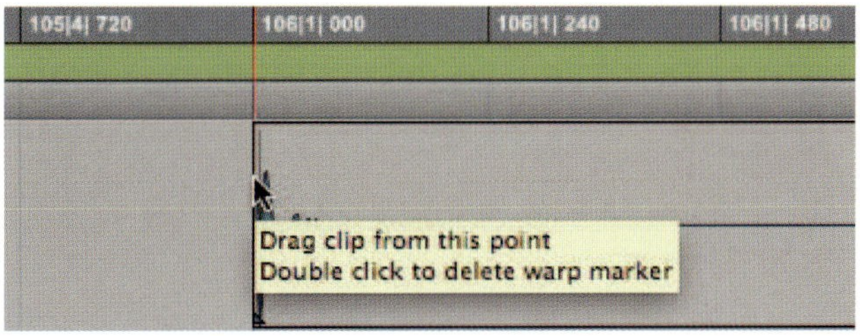

Figure 4.20
Dragging the clip by the first warp marker using the grabber tool

| 105|4| 720 | 106|1| 000 | 106|1| 240 | 106|1| 480 |

Drag clip from this point
Double click to delete warp marker

Once you have the first warp marker in place and positioned correctly on the timeline, click on the event marker on the first beat of the *second* bar of the clip (with the grabber tool selected). The cursor will change to a left-right arrow to indicate that you can move the marker to left or right to warp the audio—see Figure 4.21.

Figure 4.21
Using the Telescope
Warp feature

The warp marker at the beginning of the clip locks this to the timeline at this location (bar 106 in this example) so that when you warp the audio to reposition the start of the second bar (bar 107 in this example), the position of the start of the first bar does not move.

When you drag an event marker to line up with a bar line in the timeline (see Figure 4.22), this will warp the audio (a bit like stretching out or folding back a telescope with the eye piece fixed in front of your eye) until the event marker matches the corresponding bar number in the Bars|Beats ruler. This is why this is called 'telescope' warping.

> **TIP**
> If you use grid mode, all you need to do is to move an event marker toward a bar line and it will snap the marker to the bar line precisely.

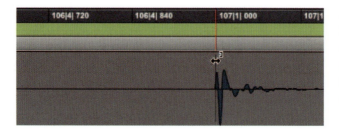

Figure 4.22
Notice that the event marker has been moved to line up with the beginning of the second bar of music, bar 107

If there are no tempo variations within the audio, all the first beats of each bar should now line up with the bar lines in Pro Tools. If there are tempo variations within the audio file or clip that you are working with, you will have to add a warp marker to the first beat of the next bar and repeat this process for each bar where the tempo changes.

In my example clip, there were slight tempo variations throughout the audio, so I added a warp marker at the beginning of the second bar of the audio clip to lock this to its correct location on the timeline—see Figure 4.23.

Figure 4.23
Add a warp marker to lock the beginning of the second bar to its correct location on the timeline

Then I examined the event marker at the start of the third bar of the audio— see Figure 4.24. This did not line up exactly with the correct bar in the timeline either, so I moved this to the correct position—see Figure 4.25.

Figure 4.24
The event marker in the third bar of the audio does not line up with the correct bar (bar 108 in this example) on the Pro Tools timeline

Figure 4.25
The third event marker is lined up with the correct bar line in the timeline as shown in the Bars|Beats ruler

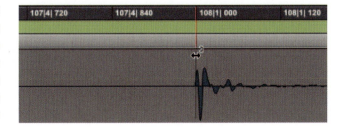

By this time I was beginning to understand how all this works, but realized that I was losing time by switching from the grabber tool to the pencil tool. So after I had moved the third event marker into position, I looked for a shortcut. As the manual explains, instead of changing to the pencil tool to add the warp marker, you can just Control-click (Mac) or Start-click (Windows) on the event marker using the grabber tool—which saves you the trouble of switching tools. See Figure 4.26.

Figure 4.26
Control-clicking on an event marker to add a warp marker

I went through the whole eight-bar clip, carefully positioning each event marker at the beginning of each bar to correspond with the correct bar line in the timeline, adding a warp marker at the start of each bar each time to lock down these positions so that the preceding audio would not be affected when I warped audio that followed. See the eight completed bars in Figure 4.27.

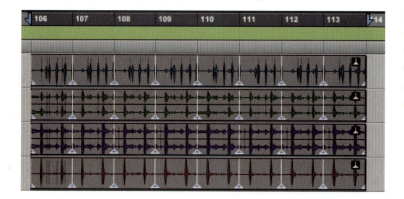

Figure 4.27
Eight bars of drums, with each bar warped to exactly match the bar lines in the timeline

> **NOTE**
> A small icon, the warp indicator, is added to the clip in the top right-hand corner of each channel to indicate that this clip has been warped—that is, that Elastic Audio processing has been applied to the clip.

You can also 'telescope' audio that occurs before the first warp marker in the clip. You might want to do this if you are working on a sound effect that preceded a visual event such as a creak before a door slam, for example. In this case, you would insert a warp marker to lock the position within the audio clip containing the sound of the door slam to the position on the timeline where this occurred, then telescope warp the sound of the creak that preceded this to get it to sound right. To telescope warp audio that is before the first warp marker, select the grabber tool, then Option-click (Mac) or Alt-click (Windows) and drag any event marker that is before the first warp marker in the clip to the left or right.

Accordion Warp

The way Accordion Warp works is that you lock down a particular position in the audio clip that you are warping, such as the central point, then drag the leftmost or rightmost event marker toward or away from this position to compress or expand the audio like an accordion's bellows.

To try out Accordion Warp, I took one bar of music in 6/4 time signature containing six quarter note drum hits and adjusted the position of this on the timeline in Analysis view so that the fourth beat lay exactly on a quarter note grid line. See Figure 4.28.

Figure 4.28
One bar of drums played in 6/4 meter, correctly lined up on the timeline in Analysis view

Then I switched from Analysis view to Warp view and added a single warp marker on top of the event marker at the fourth beat, right in the middle of the clip, using the grabber tool: ensuring that this beat would stay locked to the timeline. See Figure 4.29.

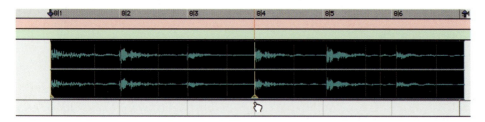

Figure 4.29
A warp marker is added in Warp view to lock down the fourth beat, in the middle of the clip

With the grabber tool still selected in Warp view, and using Grid mode with the grid set to quarter note resolution, I dragged the leftmost event marker in the clip to the right—see Figure 4.30. This was a bit like squeezing the bellows of an accordion: the audio either side of the warp marker was brought in toward the warp marker that was fixed in position on the timeline—allowing me to hear the effect of the six beats playing in the time of four.

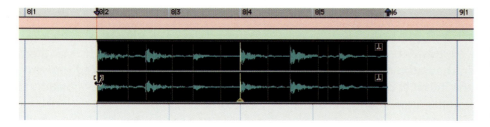

Figure 4.30
Dragging the leftmost event marker to the right using the Accordion Warp method causes both the start and end points of the clip to move equal distances toward the middle

When I dragged the event marker to the right again, this snapped to the next quarter note grid line—allowing me to hear the effect of these six beats playing in the time of two. See Figure 4.31.

Figure 4.31
Dragging the leftmost event marker more to the right compresses the audio on the timeline so that the six original beats now play in the time of two of the original beats.

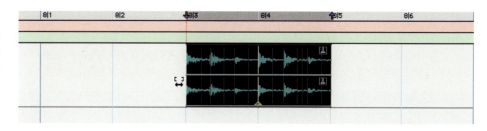

This Accordion Warp technique works fine in Slip mode as well, which would be the more appropriate mode to use in many circumstances where you want to continuously adjust the duration of the clip until it sounds the way you like it.

Range Warp

So far we have been moving event markers, warping the audio to the right or left of these, then locking these event markers into place by inserting warp markers on top of them. Elastic Audio also lets you warp the audio between two fixed points within a clip using its Range Warp feature.

Applying Range Warp between Two Warp Markers

In Warp view, simply add a warp marker at the first point that you want to fix to the timeline then add another at the end point that you want to fix to the timeline—see Figure 4.32.

Figure 4.32
A pair of warp markers is inserted at points to fix either side of the range of audio that is to be warped.

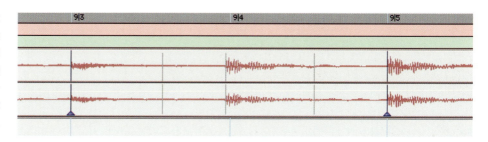

Using the grabber tool, drag an event marker located between the two warp markers left or right. As soon as you start this move, a warp marker is automatically added on top of the event marker and the audio between the two outer warp markers and the inner warp marker is compressed or expanded (depending on the direction in which you move the inner marker and which side of the marker you are looking at) as the inner marker is dragged to a new location. The audio in the rest of this clip outside the two bounding markers remains unaffected—protected by the outer warp markers. See Figure 4.33.

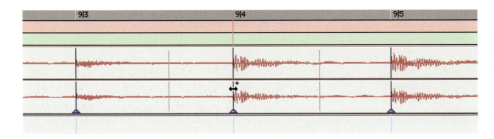

Figure 4.33
Audio between the two outer warp markers and the inner warp marker is compressed or expanded as the inner marker is dragged to a new location that corresponds with a grid line.

Applying Range Warp to a Selection

When you apply a Range Warp to an edit selection that lies between bounding warp markers, Elastic Audio keeps the timing intact between any other warp markers within the edit selection. To see how this works, create a pair of warp markers to encompass part of a clip on an Elastic Audio–enabled track. See Figure 4.34.

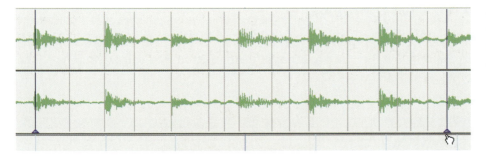

Figure 4.34
Create a pair of bounding warp markers within a clip.

Add a number of warp markers in between the bounding markers that surround an area of interest in the clip. In the example I used, a drummer had strayed from the beat within a couple of bars that started in the right place and finished in the right place. The timing between the three beats in the middle of the selection was fine, but the three beats were all played a little earlier than they should have been. See Figure 4.35.

Figure 4.35
Adding three more warp markers around the events of interest in this example

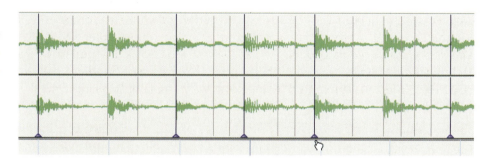

Having locked the relative positions of the three beats in question (the third, fourth, and fifth beats in the range), I used the selector tool to make an edit selection encompassing these three beats, including the three warp markers that lay in between the pair of bounding warp markers. See Figure 4.36.

Figure 4.36
Making an edit selection that encompasses a range between two bounding warp markers

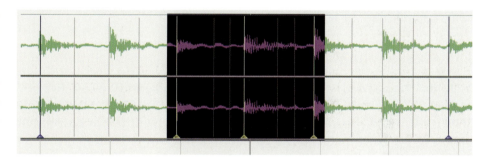

To line the fourth beat up with the grid line that can be seen just below the waveform display, I used the grabber tool to drag one of the warp markers within the edit selection a little to the right. I dragged the middle of the three warp markers, but I could have chosen to move any of these. See Figure 4.37.

Figure 4.37
The edit selection is moved to the correct beat. Audio to either side of this, in between the bounding warp markers, is warped to allow this to happen

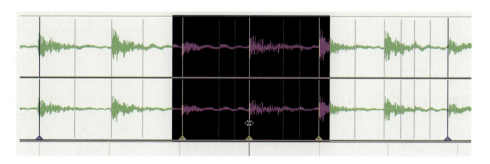

139

The audio between the first warp marker in the edit selection and the first bounding warp marker is warped; the audio between the last warp marker in the edit selection and the second bounding warp marker is warped; the audio within the edit selection is simply moved onto the correct beat, without being warped.

In other words, as the Reference Guide explains it, 'Elastic Audio processing is only applied to the audio between the bounding Warp markers and the first and last Warp markers in the Edit Selection. All audio events between the first and last Warp markers in the Edit Selection maintain their relative timing.'

The reason you would use Range Warp with an edit selection in this way is to fine-tune the timing of events within larger clips that you have already warped using one of the other methods. It allows you to focus on particular areas within a clip and move these around while leaving timings of events within the edit selection unaffected and timings outside the bounding warp markers unaffected. The assumption here is that everything else is okay: you just want to warp the audio between the bounding markers and the edit selection, while not affecting the audio within the edit selection.

Individual Range Warp

Individual Range Warp lets you apply warping to a single audio event. To apply Individual Range Warp, make sure that the grabber tool is selected, then simply Shift-click on a warp marker or an event marker. A warp marker will be inserted on top of the event marker and a pair of warp markers will automatically be created on the two adjacent event markers. You can then drag the central warp marker to the left or right to warp the clicked event.

> **NOTE**
> The event that you wish to move may already lie in between two existing warp markers. You can always Shift-click on any warp marker that lies between two already existing adjacent warp markers to add a third warp marker between these.

I tried this out with a snare drum track where one of the snare drum hits was played a little late. I made sure that the track was Elastic Audio enabled and switched to Warp view. No warp markers were present, so I simply Shift-clicked on the event marker at the start of the snare hit. See Figure 4.38.

Figure 4.38
Shift-clicking an
event marker
to prepare for
Individual Warp

A warp marker was inserted on top of this and two other warp markers were automatically created on either side of this. See Figure 4.39.

Figure 4.39
After Shift-clicking
an event marker
on a snare drum
track, warp markers
are automatically
created on top of
this and on the
adjacent event
markers.

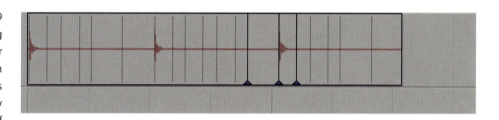

Dragging the central warp marker in between the outer warp markers to the left or right only affected the audio in the clip between the two outer warp markers. See Figure 4.40.

Figure 4.40
Dragging the
central warp marker
onto the correct
position in line with
the Bar|Beat grid

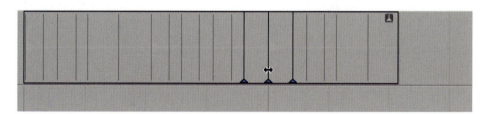

So it was easy to line the snare hit up with a Bar|Beat grid line without disturbing the rest of the audio clip. See Figure 4.41.

Figure 4.41
The individually
corrected snare drum
hit now lines up
correctly with the grid
and the previous and
next event markers are
not moved.

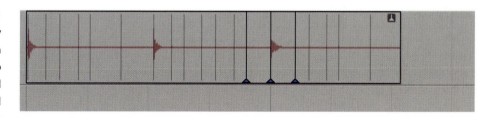

Elastic Audio Warping Summary

The basic principle to observe is that you should lock any audio that is already at the correct Bar|Beat position into place using warp markers, then drag other transient 'events' until they do line up correctly with Bar|Beat positions in Pro Tools. This will always work best if the bars and beats in Pro Tools already correspond fairly closely to the bars and beats in the audio that you are warping, so you need to take appropriate steps to ensure that this is the case.

For example, if you have a complete four-minute recording of drums for a song that you are producing and you realize that the drummer was just a little inaccurate in places, Elastic Audio can help you to quickly and seamlessly fix this. It might be that the drummer simply played the downbeat a little too late (or too early) after a fill at the end of a bar. In this case, you could place a warp marker at the beginning of the previous bar and another at the beginning of the next bar (assuming the drums line up correctly with these bar lines), then drag the offending beat into the correct position. If the inaccuracy is confined to just a couple of beats before and after the problematic beat, then you would place these locking warp markers just two beats before and two beats after the problematic downbeat.

Once you get used to doing this, you will realize how incredibly easy it is to use Elastic Audio to correct small timing errors in any performance.

Referring to this as *warping* audio and naming the techniques as *Telescope Warp*, *Range Warp*, and *Accordion Warp* could actually be counterproductive in my view.

Keep in mind that what is really going on here is that you are time stretching or compressing audio clips, with one or more locations within these clips 'locked' using markers so that the audio before and/or after these markers remains unaffected.

So you can think of Elastic Audio as giving you the ability to stretch or compress audio clips using the mouse to drag marked transient events to the right or the left along the timeline while particular locations within the clip are locked to prevent audio outside these locations from being affected.

Quantizing Elastic Audio Events

You can also quantize Elastic Audio events using the Event operations Quantize dialog. I tried this with four bars of drums that were a little unsteady. The drums had been played to a click running at 98 BPM while recording, but the drummer's time had wavered a little, so the bars in the audio did not line up exactly with the grid—see Figure 4.42.

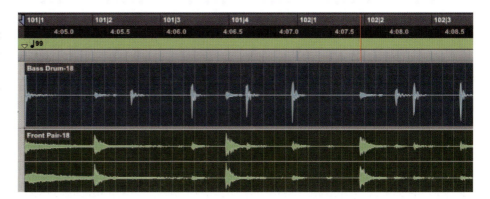

Figure 4.42
The drums not accurately lined up with the bar grid lines in Pro Tools

I decided that the best place for the event markers was on every quarter note. Unfortunately, the automatic analysis inserted far too many event markers on unimportant beats or in the wrong places. To counteract this, I reduced the sensitivity in the Elastic Properties window, which did reduce the number of event markers, although not enough—see Figure 4.43.

Figure 4.43
Using Analysis view

Figure 4.44
Adding an event transient marker using the pencil tool

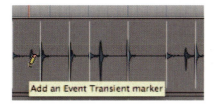

I still had to manually remove several event markers falling in between the main beats. You can remove markers by Option-clicking (Mac) or Alt-Clicking (Windows) on the marker, with the grabber tool selected. I also had to insert additional event markers manually using the pencil tool—see Figure 4.44.

I also had to reposition others until there was just one marker for every quarter note—see Figure 4.45.

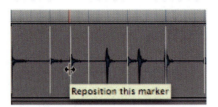

Figure 4.45
Using Analysis view to create event markers on every quarter note

Then I switched to Warp view and opened the Event Operations Quantize dialog—see Figure 4.46. In this window, I set the Quantize grid to quarter notes, then clicked Apply.

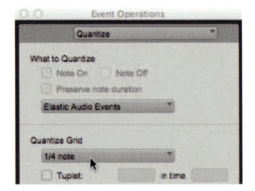

Figure 4.46
Quantizing Elastic Audio events using the Event Operations Quantize dialog

The event markers were immediately quantized to the quarter note grid and a warp marker was automatically inserted on top of each event marker to lock this into position on the timeline—see Figure 4.47. Voila! The drums were now exactly in time with the click!

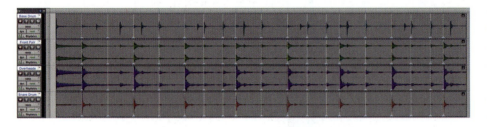

Figure 4.47
Warp view showing warp markers on every quarter note, with each quarter note snapped to the quarter note grid

144

Figure 4.48
Changing the editing preferences to switch the Track view to Warp view when using the Zoom Toggle feature

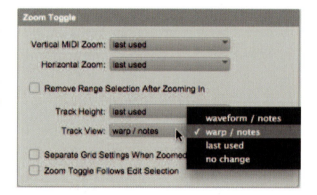

Rules for Elastic Audio

Elastic Audio analysis detects *transient events* in audio clips on sample- or tick-based Elastic Audio–enabled audio tracks and indicates these using event markers that are displayed in both Analysis and Warp track views. Elastic Audio analysis also calculates the tempo and the duration of the analyzed audio in bars and beats and stores this data, together with the transient analysis data, within the file.

In Workspace browsers, analyzed audio files are indicated by a check mark to the left of the file name, and these files display their duration in Bars|Beats, their timebase as ticks, and their native tempo in BPM.

Sample-based Elastic Audio–enabled tracks let you apply real-time or rendered Elastic Audio processing by editing in Warp view. Tick-based Elastic Audio tracks additionally let you automatically apply Elastic Audio processing based on tempo changes in the session.

If you would like new tracks that you create to be tick based by default, there is an option in the Timebase section of the editing preferences, 'New Tracks Default to Tick Timebase', that lets you enable this behavior. See Figure 4.49.

Figure 4.49
Enabling the 'New Tracks Default to Tick Timebase' option

When you create new tracks by dragging and dropping audio from a Workspace browser, if the 'Audio Files Conform to Session Tempo' option is enabled, Pro Tools creates tick-based Elastic Audio–enabled tracks regardless of whether the Elastic Audio–analyzed file is tick based or sample based. If this option is not enabled, Pro Tools creates sample-based tracks.

Committing and Disabling Elastic Audio Processing

If you use real-time Elastic Audio processing, this inevitably makes demands on your system's processing resources. There will be times when you are working your ideas out when it will be important to have the flexibility to keep the audio 'elastic' so that you can change tempo (or pitch) as you please. However, as soon as you have made your decisions about tempo and pitch, it makes a lot of sense to commit to those choices by writing the Elastic Audio–processed files to disk so that you can then disable the real-time processing to conserve precious system resources. To disable Elastic Audio, use the pop-up Elastic Audio plug-in selector that you will find among the track controls for each audio track in the Edit window—see Figure 4.50.

Figure 4.50
Using the Elastic Audio Plug-in selector to disable Elastic Audio

Pro Tools will always ask you to commit warped audio clips to disk whenever you select 'Disable Elastic Audio' using the Elastic Audio plug-in selector, and you should always choose to do this unless you decide that the Elastic Audio processing is not needed. Disabling Elastic Audio on a track affects all playlists on the track and all the playlists on the track are committed or reverted depending on which option you choose—see Figure 4.51.

Figure 4.51
Committing Elastic
Audio clips to disk

Also, if you move a clip from a track with Elastic Audio enabled to a track without Elastic Audio enabled, the Commit Elastic Audio dialog will open to ask you whether you want to commit the audio to disk or revert to unprocessed audio.

NOTE

Clips that are committed, either by disabling Elastic Audio on a track or by moving a clip to a track without Elastic Audio enabled, are written to disk as new audio files. These new audio files include the audio in the clip, plus any fades, and also an additional five seconds of audio before and after the clip (assuming that this additional audio actually exists in the original audio file).

Real-Time and Rendered Elastic Audio Processing

Elastic Audio processing can be real time or rendered. By default, it is set to 'Real-Time' so that any changes you make will take place immediately. Normally, this is what you will prefer to use, because rendering takes longer, so you lose the immediacy of real-time processing. However, non-real-time rendered processing is less demanding of system resources, so this is a better choice if you are working on a system with limited resources.

To use rendered processing on a track, click on the Elastic Audio plug-in button in the Track Controls area in the Edit window and select an appropriate Elastic Audio algorithm first—see Figure 4.52.

Then open the pop-up selector a second time and choose 'Rendered Processing'. When you subsequently make any changes to the processing, choose a different algorithm, or change tempo, any affected audio clips will temporarily go offline, new temporary 'rendered' audio files will be generated, then the affected audio clips will come back online.

NOTE
Temporary rendered files are automatically put into a Rendered Files folder in the session folder. Whenever you commit the processing to a track (by disabling Elastic Audio on a track and clicking Commit in the Commit Elastic Audio dialog), a new file is written to disk in the Audio Files folder and the temporary rendered file from which it was created is deleted from the Rendered Files folder.

TIP
If you Command-Control-click (Mac) or Start-Control-click (Windows) on the Elastic Audio Plug-in button, this will switch back and forth between real-time and rendered Elastic Audio processing.

Figure 4.52
Choose 'Rendered Processing' from the pop-up selector.

Clipping during Rendered Processing

It is possible that clipping of the waveform may occur during the rendering process. If this happens, an Elastic Audio Processing Clipping indicator, located just to the left of the Warp indicator in the clip's waveform display, will show red. See Figure 4.53.

To fix this, undo everything then lower the input gain for the clip in the Elastic Properties window (which you will find in the Clip menu) and go through the rendering process again.

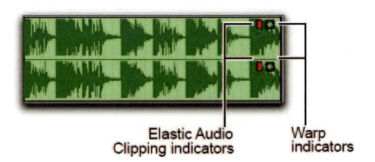

Figure 4.53
Elastic Audio
Clipping Indicators

Elastic Audio
Clipping indicators

Warp
indicators

Clipping during Real-Time Processing

Clipping of the waveform may also occur during real-time Elastic Audio processing when time compressing audio. To indicate that clipping has occurred, the Elastic Audio Plug-in button will turn red.

If clipping occurs, there are various ways to counteract this. For example, you could lower the input gain in the Elastic Properties window to attenuate the gain of the audio signal before it is processed. Or you could attenuate the default input gain in the Processing Preferences page—see Figure 4.54.

Figure 4.54
Adjusting the
default input
gain in the Elastic
Audio section of
the Processing
Preferences window

A third way is to undo any Elastic Audio processing, apply an appropriate amount of audio suite gain reduction, then try the Elastic Audio processing again.

Summary

Elastic Audio makes it possible to massage the timing of audio events in as finely detailed a way as your needs dictate. In simple and straightforward situations, where you are dealing with short loops of rhythmic material, you may not need to use the Analysis and Warp views with their powerful editing features: just go ahead and use the 'Conform to Tempo' command on sample-based tracks or simply change tempo whenever you like with tick-based tracks and the audio on Elastic Audio–enabled tracks will conform to the session tempo. However, if you take the trouble to learn how to use the Analysis and Warp view features, you will have ultimate control over the timing of your audio—however long the recordings that you are working with may be.

Once you get used to working with event markers and warp markers, you will find it very easy to correct any timing problems on your drum tracks, or other instrument or even vocal tracks, using a combination of the different warp methods. It is just so easy to go into Warp view, add a couple of warp markers around the audio you want to correct, and drag a warp marker in between these markers to correct inner events without disturbing the audio outside the bounding warp markers. You can do all this on sample-based tracks, but if you switch to tick-based tracks, additional possibilities come into play because the audio will now automatically stretch to follow whenever you change the tempo.

Real-time Elastic Audio processing can make significant demands on your system's processing resources, so it is always best to disable the processing and commit the processed audio to disk as soon as possible during your production workflow. And don't forget that you will always get much better results if you use rendered processing, so if quality is important, use this.

If you are working with short loops, the Elastic Audio features will usually give you instant gratification, letting you immediately conform loops to the session tempo or to a groove template and allowing you to quantize audio or change the tempo with tick-based audio tracks just like you can with MIDI tracks. It is important to realize that with longer audio recordings, or with more complicated beats, you will still have to make many decisions yourself about where to position event markers and which event markers are needed. Then you have to manually move warp markers to time stretch the audio that you want to affect.

Once you have mastered these powerful features you will almost certainly find that the new levels of control that you have gained will quickly become useful for almost any music production.

PART 2

Working with Virtual Instruments

Chapter 5 covers three plug-ins from AIR Music Technology—Strike 2.0 Ultimate Virtual Drummer Instrument, Transfuser 2.0 Interactive Melodic & Rhythmic Groove Creation Instrument, and Boom drum machine, which comes with every Pro Tools system.

Chapter 6 covers the rest of the AIR Instrument Expansion Pack virtual instruments, including Structure 2.0 Sampler Instrument Workstation; Hybrid 2.0 High Definition Analog Synth; Velvet 2.0 Vintage Electric Piano Instrument; Vacuum Pro 1.0.1 Polyphonic Analogue Tube Synthesizer; and Loom 1.0.1 Modular Additive Synthesizer. Also covered are Spectrasonics Stylus RMX, Trilian and Omnisphere, and Arturia's V-Collection of classic synthesizer and keyboard emulations.

Chapter 7 covers ReWire applications including Propellerhead Reason and Ableton Live. Lots of musicians use Reason, which features a whole rack of its own synthesizers, samplers, and drum machines, or Live, which is much loved by hordes of young DJs and 'beat-cutters' and also includes several virtual instruments. These can both be perfectly integrated with Pro Tools using ReWire technology.

in this chapter

CHAPTER 5

Creating and Working with Beats

Pro Tools includes AIR Music Technology's Boom drum machine, which features the ever-popular TR808 and other electronic drum sounds, but if you plan to work extensively with sampled drum kits, patterns, and loops, AIR Music Technology offers two much more advanced virtual instruments—Strike and Transfuser.

AIR Music Technology Strike

About Strike

AIR Music Technology's Strike is a virtual instrument that can be used to add very realistic drum tracks to your Pro Tools session. Similar in concept to the first drum machines that appeared back in the 1970s with preset rhythms such as rock, samba, waltz, and so forth, Strike gives you Dance+Electronic, Funk+RnB+Soul, Reggae+Dub, Indie+Alternative, Jazz, Pop, Rock, Ballad, Urban, Retro, and World categories to choose from. The big difference is that instead of extremely unrealistic synthesized analog sounds you get extremely realistic sampled sounds.

By default, Strike synchronizes to the software's tempo and to the current bar position—automatically locking its bar position to that of the host software. So, for example, a pattern that is triggered in the last quarter of the software's timeline will not play from its beginning but, instead, from the corresponding position in the pattern—and it will start the next bar at the same time as the software does.

Strike's preset patterns will automatically follow the tempo and meter of your Pro Tools session. So Strike will automatically follow a tempo map and will follow any time signature changes in the host software—automatically recalculating every pattern to match the new time signature.

You can also use Strike as a sampled drum instrument on which to play and record your own drum patterns using a MIDI keyboard or other MIDI controller—just like you can play sounds from Native Instruments Battery, for example. So whether you prefer to use the preset patterns, or modify those patterns and their variations, or whether you prefer to create your own MIDI tracks to play the drum sounds in Strike—the choice is yours.

DEFINITIONS

A *style* is a musical playing style, like samba, rock, or ballad. Strike comes with a range of preset styles that cover the most common musical genres. Each style contains 35 different patterns (see below) that are played and create a performance when combined.

Patterns are complex drum performances that have been captured as MIDI data. There are six types of patterns: verse, bridge, chorus, intro, fills, and outro. Each style contains 35 patterns. When Strike is playing, the patterns determine when and how any individual instrument (see below) is played.

A *part* is what one single instrument plays within a pattern—in other words, each pattern consists of several parts, each played by one instrument. Parts can be easily muted to create more dynamic and musically interesting performances.

An *instrument* is a single drum contained in a kit. There are all kinds of drum instruments in Strike, including such standards as kicks, snares, and crash cymbals. There are also exotic instruments, such as the darbuka or trash-ride.

A *kit* is a complete collection of instruments all stored in one file. Strike comes with many preset kits, but you can create your own custom kits using any of the instruments.

A *mix* is a file that contains all of the settings of the drum-mixing console within Strike, including the Level, Equalizer (EQ), and Insert Effect settings.

A *setting* is a single 'master' file that stores the complete state of Strike, including all style, kit, and mix settings.

TIP

You can mix and match any style with any kit and any mix settings. For example, you could have a reggae pattern play a jazz kit with a pop-style mix.

User Interface

The user interface window is divided into five sections with the control section occupying the largest part of the window, the browser to the left of this, a navigator section below the browser, a keyboard section underneath the control section, and a recorder section to the right of this.

The Browser

The browser section in the upper left part of the window (see Figure 5.1) lets you access all the instruments, kits, styles, mixes, and settings (which contain styles, kits and mixes in one file). At the top of the browser there are four 'folder' buttons for settings, styles, kits, and mixes. The currently selected folder is highlighted in blue. Each of the four main folders includes two sub-folders—Presets and User. The Presets folder contains factory files that cannot be modified; the User folder contains any custom files that you have created and saved.

Figure 5.1
The browser

TIP

You can quickly preview preset settings without having to load them: clicking and holding the Preview icon of a setting will play a short audio example of that setting.

If you click on the BPM legend at the bottom of the browser, this will display items ordered according to their BPM values. Click on the List legend to the right of this to display all the items as a scrollable list (not in folders). Click on the wrench icon in the bottom right-hand corner of the browser to open the Configuration page.

The Navigator

The collection of buttons at the bottom left of the screen is the navigator section. You can use these buttons to view the five 'pages' (Main, Editor, Style, Kit, and Mix) that can be displayed in the control section at the upper right of the screen. Whichever of these is currently selected is highlighted in blue. When you have chosen the settings you want for your Strike session, you can save all or specific parts of the session using the Save Setting button located at the bottom of the group of buttons.

The Keyboard

The keyboard section (see Figure 5.2) provides 72 keys for playing Strike, Key mode switches (to select Style or Kit), and a latch switch.

You can control Strike by clicking the on-screen keys or using a MIDI controller, or by sending MIDI data from an instrument or MIDI track in your music software.

Figure 5.2
The keyboard

The 72 keys in the keyboard section represent the keys of a MIDI keyboard and correspond to MIDI notes starting from C0 on the left to B5 on the right.

Style Mode

The Key mode is normally set to Style and lets you work with patterns.

- The first 12 trigger keys are used as pause keys to stop individual instruments from playing while the rest of the instruments continue to play within a pattern. For instance, key 1 can be used to pause the kick drum from playing.

- The next key, which is red, will stop the pattern from playing back.

- The white keys start playback of the various patterns that you might use for intro, verse, fill, bridge, chorus, or outro.

You can replace the pattern assigned to any white key on the keyboard. Say you have loaded a setting that you like but you would prefer to hear a pattern from a different style (for example, Bridge B from an unrelated style). Choose any style pattern from the Presets folder, then drag and drop this from the browser to any white key on the keyboard. Alternatively, you can double-click any pattern in the browser to automatically replace the white key corresponding to your selection. For example, double-clicking Chorus B in the browser will replace the pattern on the second chorus key on the keyboard.

■ The blue keys play individual hits, rolls, and articulations that you can choose by Right-clicking the key and selecting from a pop-up list.

■ When Latch mode is enabled using the button to the left of the keys, Strike continues playing even when you release a trigger key. The pattern continues to play until you hit the red Stop key or disable Latch mode.

Kit Mode

To play a General-MIDI-compatible file through Strike, or to trigger individual instruments using the General MIDI formatting, you can use Kit mode. This allows you to use the left four octaves of Strike's keys as a General MIDI kit that you can play using a MIDI keyboard or other controller. The blue keys can still be used to play individual hits, rolls, and articulations.

> ### TIP
> Assigning your MIDI keyboard to Strike on MIDI Channel 2 lets you trigger Strike in Kit mode, even if the Strike keyboard is set to Style mode. This way, you can trigger patterns (using MIDI Channel 1) and individual hits (using MIDI Channel 2) simultaneously.

The Recorder

Strike has a recorder (see Figure 5.3) that can capture drum loops as either digital audio or MIDI data. These data can then be dragged and dropped from the recorder's central display into your session timeline or onto your computer's desktop so that it can be used later.

Figure 5.3
The recorder

To start recording audio, make sure that the Audio button is selected (highlighted in blue) then click the red record button to record one, two, or four bars of the pattern that is playing.

A graphic representing the audio will appear in the display area—see Figure 5.4. Just click the record button once more to stop recording. If you want to replace this with another pattern, simply repeat this process. When you have finished recording, you can drag this either into the Pro Tools Edit window or onto the desktop where it will appear as a WAVE file.

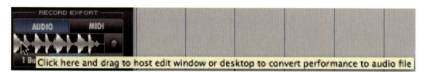

Figure 5.4
Audio in the
recorder

> **NOTE**
> The recorder intelligently synchronizes itself so that loops always start 'on the one' and are in sync with the session. In other words, you do not need to start and stop recording with perfect precision to capture a loop—the system handles this automatically.

To record MIDI, make sure that the MIDI button is selected (highlighted in blue), click the record button (see Figure 5.5), and play the patterns.

Figure 5.5
Recording MIDI

When you have finished recording, you can drag this into the Pro Tools Edit window or drag and drop onto the desktop where it will appear as a MIDI file (see Figure 5.6).

Figure 5.6
Exporting a MIDI
recording

> **NOTE**
> When in MIDI mode, there is no limit to the length of a performance that can be captured. You can even perform a full-length song and capture it all as MIDI data.

> **TIP**
> MIDI data can be captured either in Full mode (which makes maximum use of Strike) or as General-MIDI-compatible data if you select the GM button.

The Main Controls

From the Main window, shown in Figure 5.7, you can globally adjust the timing, the feel, the groove, the dynamics, the complexity, and much more—all in real time. You can use the automation features in Pro Tools to perfect your moves or use a MIDI controller to tweak your performance on the fly using knobs, sliders, drum pads, or anything else that sends MIDI data.

- The play display at the top shows which pattern is currently playing.

- The kit display in the middle shows a graphic representation of a kit. Instruments in this kit animate to indicate hits, including the intensity (velocity) of each hit.

- The style section parameters to the left of the kit display can be used to adjust the overall Strike playing dynamics and timing behavior.

- The kit section parameters to the right of the kit display influence the sound of the current kit and its instruments, allowing you to change the tuning of the drums, the playing style of the snare drum, and so forth.

- The mix controls underneath the kit display provide global controls over the mix.

Figure 5.7
Strike Main window

The Kit Controls

The Kit page, shown in Figure 5.8, lets you adjust the sound of each drum by tuning it, shifting the timbre, or changing the attack or decay, for example. The mute, solo, and volume fader controls from the Mix page are conveniently provided below the kit controls.

Figure 5.8
Strike kit controls

Strike lets you import your own WAV or AIFF sample sounds to create new instruments in a kit. You can do this by clicking the folder icon at the top of any instrument channel strip and locating your desired sound file in the prompt that appears. Once a custom sample is loaded, the instrument channel strip updates to show an audio file icon and the name of the file at the bottom of the channel strip. In this mode, the instrument load size and timbre shift controls are greyed out. You can mix the direct signal from the loaded sample using the close mics control in the Mix page. Strike emulates the room, overhead, and talkback mics for imported samples as well.

Figure 5.9
Instrument load size selector

You can use the instrument load size selector to adjust the amount of sample data that is loaded into your computer's RAM for each instrument, choosing from the three options available—see Figure 5.9.

- XXL is the largest possible instrument load size. XXL provides the maximum range of expression available, but also places the greatest demands on system resources.

- Mid is the Strike default instrument load size. Mid provides a good balance between system load and range of expression available.

- Eco is the smallest possible instrument load size. Eco uses fewer system resources for the instrument, but also limits the range of expression available.

The Style Controls

The style controls, shown in Figure 5.10, let you adjust the intensity and complexity, playing dynamics, and hit variations for each drum. The offset controls let you set timing offsets for the hits on each individual instrument to make these play earlier or later. Timing controls let you adjust the accuracy of the playing to be 'tighter' or 'looser'. Again, for your convenience, the basic controls from the Mix page are provided below the style controls.

Figure 5.10
Strike style controls

The range of styles, patterns, and variations can give you virtually instant gratification: just try these in turn until you find something that suits your purpose. And the good thing about Strike is that you can easily edit the patterns if they are not exactly what you are looking for!

The Style Editor

A click on the Editor button at the bottom left of the window switches the display to the style editor—see Figure 5.11. This is laid out as a grid, showing where the drum hits fall within the pattern.

You can choose the pattern you want to edit using the Strike keyboard, located below the pattern display, to select the various patterns, fills, or individual drums sounds. The pattern that you select will be identified by a red LED.

To edit patterns, select an appropriate item from the palette of tools provided at the top right of the Style Editor window—pencil, eraser, and others—then click in the edit section (located above the pattern display in the main part of the Style Editor window).

The style editor's edit section shows the currently selected pattern superimposed on a Bar|Beat grid. Each of the white vertical bars in the blue area represents an instrument hit. You can copy and paste parts in the edit section and, when you have finished, save the results of your editing.

The Style Editor's pattern display shows the rhythmic structure of the pattern with diamonds in vertically stacked lines—one for each instrument. To select a particular instrument's part for editing, just click on the horizontal line representing that instrument.

Figure 5.11
Strike style editor

Using the group of buttons to the left of the keyboard, you can copy, paste, or clear patterns; revert to the last saved version of the pattern; and enable or disable the Key Follow function. With Key Follow enabled, the display automatically changes to show the pattern that is currently playing as you change patterns. If you don't want the display to change, disable this.

The Mix Page

A click on the Mix button at the bottom left of the Main window reveals the Mix page—see Figure 5.12. Here you can change the levels from the close microphones, the overheads, the 'talkback' microphone, how much leakage, and so forth. Just being able to tweak the amount of leakage between the microphones, for example, makes an enormous difference to the sound of the kit.

Each instrument has a channel with a level fader, Mute and Solo buttons to the left of this, and a pan control above the fader. The section above contains horizontal faders that can be used to adjust the amount each instrument contributes to the overhead channel and the amount each instrument contributes to the room channel.

Figure 5.12
Strike Mix page

Each channel in the Strike mixer contains an EQ and two user-selectable insert effects. The controls for the EQ and insert effects are positioned along the top of the Mix page—see Figure 5.13. When you select a mixer channel by clicking on this, the controls for its equalizer and effects inserts are displayed here.

163

To modify the EQ settings, just click and drag the red, green, or blue dots in the Equalizer display. Drag left or right to adjust the cutoff frequency; drag up or down to adjust the gain. Control-click (Mac) or Right-click (Windows) and drag any of the dots up or down to change the Q value of the selected filter band.

Figure 5.13
EQ and insert effects

The two effects inserts are connected in series for each channel. These insert effects range from dynamics and limiting to reverb and delay—and you also get specialized tools such as a bit crusher and a filter—see Figure 5.14.

To select an effect insert, click the insert selector (just above the upper left of each insert slot) and select an effect from the drop-down menu. An Insert On button is also provided (just above the upper right of each insert slot). This can be used to enable or bypass the insert effect. When you click and drag the rotary controls to change the parameters of the insert effect, the value of the adjusted parameter is displayed in the information display below Strike's keyboard.

Eco Compressor
none
Dynamics 3
Opto Compressor
Eco Compressor
Brickwall Limiter
Gate
Envelope
Pumper
Dynamic EQ
Tube Saturation
Distortion
Bit Crusher
Enhancer
Mic Model
Vari Filter
Ring Modulator
Frequency Shifter
Delay
Dub Delay
Reverb
Phaser
Chorus
Oscillator

Figure 5.14
Available effects

The master section (see Figure 5.15) lets you apply final processing to the audio before this leaves Strike. This section has four channels—overheads, room, talkback, and master—each with its own specialized sound-sweetening controls.

Overheads

The overheads fader is a sub-master that lets you set the level of the sub-mix of overheads contributions from the individual instruments that will be routed to the master output.

Figure 5.15
Strike Mix page
master section

- The delay control lets you delay the overhead signal by up to 20 ms to simulate different overhead microphone distances.

- The cymbal width control lets you adjust the stereo width of all the cymbals in the kit.

- The width (overhead width) control lets you adjust the stereo width of the overhead microphones.

Room

The room microphones are stereo microphones similar to overheads, but are positioned further away in the room and capture a more diffuse sound.

The room fader acts as a sub-master that lets you set the level of the sub-mix of room contributions from the individual instruments that will be routed to the master output.

- The delay control lets you delay the room signal by up to 50 ms to simulate different room microphone distances.

- The size control lets you adjust the decay of the room signal to simulate smaller or larger recording spaces.

- The width control lets you adjust the stereo width of the room signal.

- The surround control sends an additional two channels of room ambience into the overhead channel, so a total of four available channels can be panned to the front and rear for a surround effect. The surround signal passes through the overhead EQ, inserts, and fader, and to the audio output selected for the overhead channel. The Surround LED is lit when activated.

Talkback

The talkback channel represents a mono microphone of the type that might have been originally placed in a recording room near to the drummer and heavily compressed so that the producer and engineers could hear what he said, but later used as an effect. This talkback microphone captures a sound similar to the overhead and room microphones, but has a very hard, compressed, and slightly distorted or 'dirty' quality.

- The drive control lets you adjust the gain of the talkback channel compression. Increasing the amount of drive makes the talkback signal sound more dense and distorted.

Master

The master channel is Strike's main output. All the individual channels are mixed down to this channel by default and then output to the instrument or auxiliary input track onto which Strike is inserted in Pro Tools.

- The close mics control lets you adjust the overall level of the close microphones routed to the master channel.

- The mic leakage control lets you adjust the level of 'bleed' across the different microphones. When recording drums in a studio environment, each microphone picks up some signal of all of the instruments in the drum kit. At the minimum setting, each microphone only captures the sound from one instrument.

- The snare buzz control lets you adjust the amount of sympathetic resonance of the snare drum when the kick drum and toms are played. In a real drum kit the snares of the snare drum rattle whenever another nearby drum is hit, particularly the kick drum. In Strike, the level is variable, so you can choose between an ultra-clean kick drum sound at a lower setting and a 'live'-sounding kit at a higher setting.

Setting the Outputs

When you insert Strike on an instrument track in Pro Tools, Strike's master output is automatically routed into Pro Tools to play back via this track.

Any of the Strike channels, or the overheads, room, or talkback sections, can be individually routed into Pro Tools using any of Strike's 15 additional two-channel outputs. A pop-up selector above each fader in the Mixer page allows you to select these outputs.

In the example shown in Figure 5.16, I have routed the kick drum and the snare drum to Strike outputs 1.L and 1.R, respectively, panning the kick drum all the way to the left and the snare drum all the way to the right to make sure these appear separately in the left and right channels of output pair 1. I also routed the overheads, room, and talkback to the two-channel output pairs 2, 3, and 4, respectively.

Figure 5.16
Assigning a Strike
channel to an
individual output

You also need to create auxiliary input tracks in Pro Tools to feed these additional outputs from Strike into your Pro Tools mix, selecting the appropriate outputs from Strike as inputs for these using the track input pop-up selectors in Pro Tools—see Figure 5.17.

Figure 5.18 shows a section of the Pro Tools mixer with the instrument track for Strike at the left, and three stereo and two mono auxiliary tracks to the right. With this setup, it is possible to set the levels of the bass drum and snare drum and the overheads, room, talkback, and main output mixes from Strike separately in Pro Tools. In this case, it is best to set the gain faders in Strike to zero boost/cut and control these levels in Pro Tools.

Figure 5.17
Routing Strike outputs to track inputs in Pro Tools

Figure 5.18
Pro Tools inputs set up to receive Strike outputs

MIDI Control

Strike lets you assign standard MIDI controllers to virtually any parameter so that you can control Strike from any MIDI controller, such as a MIDI keyboard, in real time.

To assign a MIDI controller to a parameter, Right-click (Windows or Mac) or Control-click (Mac) on a control, then select the desired MIDI controller from the Assign sub-menu. To un-assign a MIDI controller, select Forget.

If you select Learn, then move any control on your MIDI controller, the parameter is automatically assigned to that control.

For example, by default, the Latch button is assigned to MIDI Continuous Controller 64, which is normally used for a sustain pedal—see Figure 5.19. Sustain pedals are foot-operated on/off switches, so they are ideal for controlling this function.

Figure 5.19
Assigning MIDI
controller 64 (the
sustain pedal) to
the Latch button

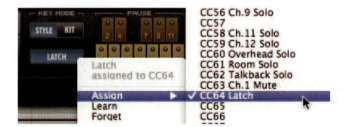

Configuring the Sync Settings

Strike has a Configuration page (see Figure 5.20) that contains several global parameters affecting its operation. To open or close this Configuration page, click on the wrench icon in the bottom right-hand corner of the browser.

- The 'Content Size' configuration option defines how much waveform content Strike loads into RAM by default.

- Using oversampling can help avoid aliasing noise when adjusting the tuning of instruments like crash cymbals, which can be rich in overtone content. Oversampling increases Strike's internal sampling rate to achieve higher quality at the expense of slightly higher CPU load.

- You can configure the maximum number of voices (instrument hits played at a time) that you want to reserve for Strike, using the Polyphony setting. Strike intelligently manages the number of voices to suit the current setting.

- The Tool Tips setting activates or deactivates 'Tool Tips for Strike'. A tool tip is a small window that displays descriptive text for a specific control. It appears next to the cursor when holding the cursor over a control.

- When 'Load Mix with Kit' is set to On, new mix settings are automatically loaded when a new kit is loaded. In the (default) Off setting, the previous mix settings are retained when new kits are loaded.

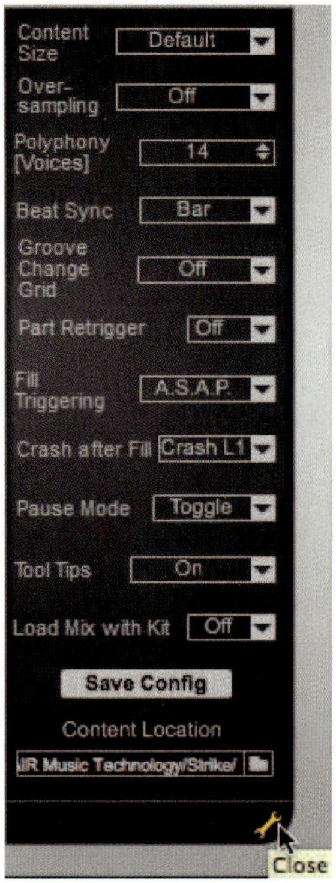

Beat Sync

The Beat Synchronization setting defines how Strike synchronizes to your music software. There are three settings:

Figure 5.20
Strike Configuration Page

- '**Bar**' synchronizes Strike to the software's tempo and to the current bar position.

This means a pattern that is triggered in the last quarter of the software's timeline will not play from its beginning but from the corresponding position in the pattern and will start the next bar at the same time the software does.

This means Strike automatically locks its bar position to the software. In this mode, it is not possible to play the first beat of a strike pattern on the third beat of the timeline—Strike will always play the third beat of its pattern on the third beat.

- '**Beat**' synchronizes Strike to the software's tempo and to the nearest beat.

This means a pattern that is triggered will start from its beginning at the next available quarter note beat in the software's timeline. Strike locks to beats but not to their position in the bar—for example, a pattern can be started on the third beat of the timeline but not between two quarter note beats.

- '**Off**' synchronizes Strike to the software's tempo, but not to the bar or beat position.

When a pattern is triggered, it synchronizes to the currently playing pattern; that means, for example, when you play Strike in Latch mode, every successively played pattern will synchronize to the previously played pattern. If no pattern is already playing, Strike immediately starts playing the newly triggered pattern from its beginning. In this mode, Strike beats will not automatically lock to the bars and beats—for example, if you trigger a pattern in the middle between two quarter note beats, Strike will play synchronized but with a steady 8th note time offset.

Groove Change Grid

The 'Groove Change Grid' option sets the position in the bar where Strike changes from the current pattern to the next. For example, if this parameter is set to half notes (1/2), the pattern change only occurs when the next half note in the bar is reached, regardless of when the pattern is actually triggered.

Pattern Retrigger

If Pattern Retrigger (labeled Part Retrigger in Strike's Configuration page) is set to On, each time a new pattern is triggered, it immediately replaces the pattern that is currently playing. If Pattern Retrigger is set to Off, the new pattern is not played immediately but only after the current pattern finishes playing, based on the time set in the Groove Change Grid.

Fill Triggering

The Fill Triggering setting defines how fills are triggered. There are three options:

- ■ '**Next**' triggers the fill at the next allowed position defined by the Groove Change Grid setting, plays the whole fill, and then changes back to the previously selected pattern.

- ■ '**A.S.A.P.**' triggers the fill at the next musically sensible position for the time the key is held and changes back to the previously selected pattern after the key is released.

- ■ '**While Held**' triggers the fill immediately and plays it for the entire time the key is held. It changes back to the previously selected pattern immediately after releasing the key.

Crash after Fill

When the 'Crash after Fill' setting is On, Strike plays a crash cymbal hit at the first beat of the next bar after a fill is played.

Pause Mode

This configuration defines how the pause keys work. There are three options:

- ■ '**Toggle**' pauses instrument playback when a pause key is triggered. Playback resumes only when the pause key is triggered again. This is the default setting.

- ■ '**Held**' pauses instrument playback as long as the pause key is held.

- ■ '**Released**' pauses all instruments by default. Single instruments' playback resumes as long as their pause keys are held.

Quickstart Guide to Using Strike

From the browser, choose a suitable style that runs at a tempo close to the one you will be using, then audition the various verse, chorus, and bridge patterns to find one that suits your song.

By default, Strike is set up so that the patterns will play back at the tempo your Pro Tools session is set to. So you can play your session, click on any pattern and it will start to play in sync. Just click a new pattern to try the alternatives while the session is still playing.

It makes sense to start your session playing back in your session's chorus while you look for a chorus pattern in Strike—see Figure 5.21—then play the verse, the bridge, and so forth, choosing patterns for these.

Figure 5.21
Choosing Strike patterns

You may also want to choose intro and outro figures and fills that you can use to lead into new sections or to provide interest during long sections—see Figure 5.22.

Figure 5.22
Choosing Strike
intro pattern

While you are choosing the patterns, make a list of the most likely contenders to use for each section together with the MIDI note number that will trigger each pattern.

When you have chosen your patterns, go to the instrument track in the Pro Tools Edit window, put this into Notes view, choose Grid mode set to bars, and then insert the appropriate notes to trigger your chosen patterns at the start of the relevant sections—see Figure 5.23.

TIP

If you make each note last as long as the section throughout which you want the associated pattern to play, then you can stop and start playback anywhere within the section and the pattern will play.

Figure 5.23
MIDI notes inserted
to trigger patterns
for each song
section

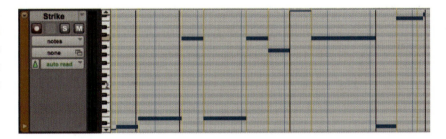

If any of the instruments are too loud or too quiet in relation to the others, you can adjust the levels using the Mixer page—see Figure 5.24. Here, you can adjust the relative levels of direct and room sound and the level of the overheads or the bottom and top mics on the snare, for example.

You can also adjust the complexity of, say, the bass drum pattern or the intensity (how hard it is being hit) of the snare drum, or introduce variations

Figure 5.24
Adjusting levels on
the Mixer page

Figure 5.25
Adjusting relative
intensity of a Strike
instrument

into the playing dynamics and instrument 'hits', or introduce timing offsets and variations to emulate a more human 'feel'. See Figure 5.25.

You can also adjust the kit parameters, such as the tuning of the snare—see Figure 5.26.

Figure 5.26
Adjusting kit
parameters

> **TIP**
>
> When you have finished, record the audio to disk so that you can deactivate the instrument track to conserve CPU power.

AIR Music Technology Transfuser 2.0

If you like to work with loops and samples, AIR Music Technology's Transfuser lets you slice, dice, shuffle, trigger, rephrase, and resequence audio loops and phrases—a lot like Ableton Live. It's an ideal tool for making electronic dance music (EDM), and can be very useful in all sorts of other musical genres as well.

Transfuser comes with its own comprehensive 2 GB library of loops, phrases, and samples, with everything from beats, breaks, and basses to acoustic and ethnic instruments, synth sounds, and more. You can also drag and drop your own ACID, REX, Apple Loops, AIFF, or WAV files directly into Transfuser.

> **NOTE**
>
> It is no longer possible to drag and drop audio clips from the Pro Tools timeline into Transformer or to relink samples.

When you drag an audio file from the browser into the Tracks pane, Transfuser slices up the audio into a slice sequence track; or time stretches it into a beat-matched, trigger sequence track; or converts the audio into a drum sequence track, with individual drum samples mapped to its 12-pad drum sampler—depending on the conversion type you have chosen.

Each track has sequencer, synth, and effects modules, with editors for these. You can use the sequencer modules to change around the beats and the 'feel' of a drum loop, varying notes in phrases or altering the sequence speed (in relation to the session tempo), and more—all in real time. The synthesizer modules can be used to modify individual sounds and manipulate the audio frequency, velocity, filter, attack, and decay and you can use the effects modules to insert up to four of the 20 available effects onto each track. These effects include tape delay, pumper, beatcutter, compressor, vinyl, filter, spring reverb, distortion, and lo-fi.

Transfuser 2.0 adds two new polyphonic synthesizers, Analoge and Electric, and two new polyphonic sequencers, Poly-Seq and Chord-Seq. All Transformer's sequencers now have new multitool and randomization features; the Browser pane now includes a Preset tab and a Save Preset button; and the Configuration page has two new features: a sequencer scale link and an auxiliary output routing control.

Then there is M.A.R.I.O.—the Musically Advanced Random Intelligence Operations randomization algorithm. This feature lets you create variations of your sequencer patterns by clicking a single button and is available in all of the Transfuser sequencers, with algorithms specifically tailored for each sequencer. M.A.R.I.O. lets you explore musically inspired variations of a loop at the click of a button, with some human input into the processes if you like. For example, you can tell M.A.R.I.O. how much variation to apply or which parameters— such as pitch, rhythm, level, or filter—to affect.

Performance Control

Transfuser has excellent live performance-oriented features that allow you to play Transfuser from Pro Tools MIDI tracks or in real time using an external MIDI controller.

You can easily assign MIDI notes and MIDI continuous controllers to control virtually any pattern trigger and its interface, including its module and parameter controls, smart knobs, trigger pads, four-octave keyboard, and crossfader.

To assign a control, Right-click the control you want to assign and select 'Learn CC', then move the knob, fader, button, mod wheel, foot switch, or other controller on your hardware MIDI controller. Several standard MIDI continuous controllers

(CC) are set as default assignments, for example: CC 1, the Mod Wheel, controls Transfuser's X-Fade between Busses 1 and 2; CC 7, Volume, controls Transfuser's Master Level; CC 24 controls Play All; and CC 23 controls Stop All.

> **NOTE**
>
> MIDI continuous controller assignments in Transfuser work in MIDI's Omni mode, so they are received by Transfuser tracks on all MIDI input channels. This lets you use a single controller for common controls on different Transfuser tracks—even if the tracks are assigned to different MIDI input channels. Of course, if the tracks are assigned to different MIDI input channels, you can always use these different MIDI input channels to trigger and play the various sequencer and synth modules independently using MIDI notes.

The User Interface

Let's take a look at the all-important user interface. When you insert Transfuser on a stereo instrument track in Pro Tools and open the plug-in's window, you will immediately notice that Transfuser lets you access everything from this one main window—see Figure 5.27.

Figure 5.27
Transfuser user interface

There are four main 'panes' or sections within the Transfuser window, with the Browser and Tracks panes in the upper half and the Info and Editor panes in the lower half.

The Browser pane is located at the upper left of the window. There is a Save Preset button at the top, with three tabs below this—one for Transfuser tracks, one for Transfuser presets, and one for audio files. When you find an audio file or track that you want to use, you can just drag and drop this into the tracks pane. And to open a preset in the Tracks pane is even easier—just double-click on this in the browser.

> **NOTE**
>
> A preset typically contains a set of two or more Transfuser tracks that can be used together.

To hear a sequence play back, just click the Play button in the Track's sequencer section. Alternatively, you can click a key on the keyboard in the controller section, or play a note on an external MIDI keyboard that you have routed through to Transfuser via Pro Tools. In this case, any Transfuser track set to the corresponding MIDI input channel will play back.

> **NOTE**
>
> If the Latch button is lit (enabled), the sequencer keeps playing until you click Stop. If Latch is disabled, playback stops as soon as you release the mouse.

If you have loaded a preset or have manually added more than one track module, you can play all of these together by clicking the Play button in Transfuser's master transport section. This starts and latches playback for all the tracks you have loaded into Transfuser.

For example, if you load the 'R 'n' B Ballad' factory preset, which has five tracks including pizzicato, bass, and electric piano sliced musical phrase sequences, together with echo perc and drums sequences, you can resize the Transfuser window (by dragging the bottom edge of the window upward to minimize this) so that just one track is visible with the controller and master sections below this, as in Figure 5.28.

Figure 5.28
Transfuser
minimized

Figure 5.29
Hide Editor Pane
button

Notice that the Info pane and the Editor pane have also been hidden from view by clicking on the small icon at the bottom right of the Tracks pane—see Figure 5.29.

Figure 5.30
Browsing disks to
find audio files

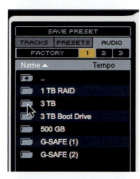

Using the Browser

The browser's Audio tab lets you search through the factory audio files or any audio files on disk drives attached to your computer—see Figure 5.30.

Double-click on any disk drive icon to open this in any of the three available viewing panes—1, 2, or 3. Double-click on any disk drive icon to display its contents in the viewing pane—see Figure 5.31. To hide the contents in the viewing pane, double-click on the top folder, which has the '+' sign.

Figure 5.31
Browsing disks

You can also browse through the factory-supplied tracks—see Figure 5.32—or browse through any Transfuser tracks that you have created and saved as user presets—which can be accessed by clicking on the User tab in the browser.

TIP

Transfuser lets you filter items in the browser list by alphanumeric characters. This way you can quickly find the track or sample you want. To filter the browser list, type the alphanumeric characters in the Filter field to show all items in the list containing those characters.

NOTE

You can click the Preview button at the bottom of the browser to enable the Preview feature. With this enabled, you can listen to any track or audio file by clicking its name in the list. Preview lasts as long as you hold down the mouse (or, if Latch is enabled, for the duration of the audio file).

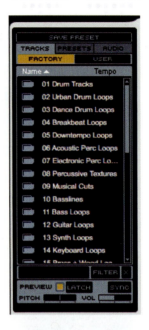

Figure 5.32
Browsing the factory tracks in the Browser pane

The Info Pane

Point at anything in the Transfuser window with your mouse, and the Info pane at the bottom left, just below the browser, will explain what this is—see Figure 5.33. It tells you what the sections are for and what the controls do and it also gives you information about audio files that you have selected in the browser. Audio file info includes the bit depth, number of channels, sample rate, duration, and the BPM (if this is available).

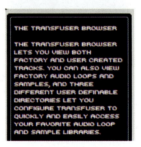

Figure 5.33
The Info pane

Transfuser Tracks

The larger section to the right of the Browser pane is called the Tracks pane. This is a bit like a virtual rack of instruments—similar to Propellerhead's Reason or Native Instrument's Kontakt sampler in some ways. Transfuser 'tracks' actually contain a set of modules for automation, sequencing and triggering, generating sound, adding effects, then mixing the results. Each track can be played back and mixed individually and you can crossfade between tracks while playing back, similarly to the way you can with a DJ mixer.

When you first insert Transfuser in a Pro Tools track, the Transfuser Tracks pane is empty and a reminder text is visible that says 'Drop Track or Audio Files here'—see Figure 5.34. As it suggests, you can simply drag and drop existing Transfuser tracks or audio files from the Transfuser browser to add or create Transfuser tracks.

Figure 5.34
An empty
Transfuser tracks
pane

When you drag and drop a factory track into the Tracks pane, this appears as a 'rack' unit subdivided into its various sections—see Figure 5.35.

At the top of the Tracks pane you will see the sections within the pane labeled Track, Seq, Syn, EFX, and Mix. You can use the controls provided within these sections to set up the sounds you want for each track.

Figure 5.35
A Transfuser track
dragged and
dropped from the
factory tracks in the
Browser pane

Figure 5.36
Saving track
settings

Saving Tracks

Once you have set up your tracks the way you want them, you can save each customized track individually, with or without any imported samples, using the drop-down menu in the module's track section—see Figure 5.36.

TIP
If you want to save with samples, there is a faster way to do this: simply drag the track module from the Tracks pane into the browser.

Whichever method you have used to initiate your 'save', the Save dialog will appear so that you can name your customized track—see Figure 5.37. When you okay this dialog, the track will be saved into the User Presets folder in the browser.

Figure 5.37
The Save Tracks
dialog

Multiple Tracks

You can add up to 16 tracks of paired sequencer and synth modules and play these back simultaneously in a single instance of Transfuser. So you can create anything from simple grooves to complex arrangements that can then be triggered using MIDI data from tracks in Pro Tools. You can also play these sounds in real time using Transfuser's on-screen interface or using an external MIDI controller.

> **TIP**
> If you need more than 16 independently controllable Transfuser tracks, you can always insert additional instances of Transfuser on additional Pro Tools Instrument tracks.

The Editor Pane

The lowest section of the Transfuser window contains the Editor pane, which provides access to all of the controls for the selected module or effect—for an example, see Figure 5.38.

Figure 5.38
The Editor pane
for the Analoge
synthesizer

TIP

To see the controls for the sequencer or the synthesizer, click on one or other of these sections in the track to highlight it first. To see the controls for any of the inserted effects, click on the effect in the insert slot to highlight this.

The Track Module

The first module in each Transfuser track, called the track module, provides controls for managing tracks. Here you can assign a MIDI channel that can be used to control the track. You can also assign a low and a high key to define a MIDI note range for each track, so that different keyboard ranges can be used to control different tracks.

Track Automation

If you click the Automation button in the track module, this reveals the track's automation editor in the Editor pane lower down in the Transfuser window.

This automation editor provides 12 lanes of automation that can be assigned to any control in any module in the track by Control-clicking or Right-clicking on the control using your computer keyboard and mouse—see Figure 5.39.

Each automation lane provides 32-step sequencing for automation. Steps can be measured in 1/4-note to 1/32-note subdivisions of the beat (including

Figure 5.39
Assigning a control to an automation lane

triplets). Each sequence can play straight through, first to last and then repeat, or forward then backward. Also, each can be step automated or interpolated on a curve (curve automation is good for continuous controls).

Along the top of the window, you will find an Enable/Disable button for the automation, two drop-down menus, a pair of Undo/Redo arrows, a multitool that includes a pencil to let you draw and manipulate automation events and adjust their values, a selector tool to select automation events, and an eraser tool to erase automation events—see Figure 5.40.

Figure 5.40
Writing automation into an automation lane in the track automation editor

Sequencer Modules

Transfuser tracks can use any of six different types of sequencer modules with any of seven different types of synthesizer modules.

■ The drum sequencer provides a 12-note step sequencer for playing the drums synth module.

■ The phrase sequencer provides a monophonic MIDI sequencer for playing the phrase synth module.

- The slice sequencer provides a slice sequencer for playing the slicer synth module.

- Poly-Seq is a polyphonic sequencer that can be programmed to send MIDI data to a Transfuser synthesizer module.

- Chord-Seq is a polyphonic sequencer that makes it easy to send chords, strums, and other polyphonic 'blocks' of notes to a polyphonic Transfuser synthesizer.

- The Thru module simply lets any MIDI input play the track synth module directly.

Synthesizer Modules

- The slicer synth module automatically chops up your audio into individual events that can be played back in any order and with all kinds of processing.

- The phrase synth module converts imported audio to beat-matched audio (using time compression and expansion) that can be played back at different transpositions, times, and durations.

- The drums synth module converts imported audio to individual drum samples mapped to a 12-pad drum sampler.

- The bass synthesizer is a monophonic analog bass line synthesizer.

- Analoge is a polyphonic synthesizer that uses analog modeling technology to provide everything from classic 'old-school' analog sounds to cutting-edge modern sounds.

- Electric is a polyphonic synthesizer that emulates electric pianos with all kinds of tones ranging from mellow and laidback to 'forward' and aggressive.

- The audio input synthesizer passes audio from disk or from the audio input of the track on which Transfuser is inserted. This lets you directly process Pro Tools audio in Transfuser without having to import it. Three controls let you modify this audio: you can adjust the attack time and release time of the envelope, while Thru lets you adjust the level passing through.

Synthesizer Randomization

The Analoge, Electric, bass and drums synthesizer modules (but not the input, phrase, and slicer) all have an intelligent randomization system that lets you create new sounds quickly.

The way this works is that you just click the Randomize button and all the synthesizer's parameters will be randomly set to new values. If you don't like the result, just click again to try different settings—until you find some that you do like.

There is also a Target button next to the Randomize button—see Figure 5.41. Click on this to open a menu that lets you uncheck sets of parameters that you would like to prevent being randomized. For example, if you just want to try new filter settings, uncheck all the other sets using this Target menu before you click the Randomize button.

Figure 5.41
Randomize and Target features

TIP

Notice the small arrow buttons at either side of the Target and Randomize buttons that point to the left and the right, respectively. If you click Randomize once, you can undo the changes by clicking the left arrow button, then reapply the new settings by clicking the right arrow button. If you have clicked the Randomize button several times, then, using these left and right arrow buttons, you can switch through the different settings you have created so that you can audition these in turn.

Track Effects Section

Every Transfuser track has an effects section that allows you to insert up to four effects onto each track—see Figure 5.42. These effects can be arranged in serial or parallel configuration depending on whether the 'PAR' or 'SER' button is selected, and a bypass button is provided, located just to the right of these. Four buttons to the left of these can be used to change the display to show each of the four effects in turn.

Figure 5.42
Effects section

Under these buttons, four pop-up selectors appear that can be used to load your choice of the available effects into each of the four available 'slots'—see Figure 5.43.

Each effect has three rotary controls for the most important parameters and these change according to the effect type selected.

The Editor pane below the Tracks pane displays the full set of parameters for each effect when you have loaded and selected any of the four effects 'slots' in the track module.

Figure 5.43
Choosing the effects

Figure 5.44
Track mix section showing the output bus pop-up selector

Track Mix Section

Every Transfuser track also has a mix section with level meters at the top, a pair of buttons to the left of this that let you solo or mute the track, two sliders to control the volume and pan position of its audio output, two rotary FX sends that can be used to add more effects, and a pop-up output bus selector.

The output bus selector can be used to route the track's audio output to one of eight buses or to a cue output—see Figure 5.44.

Output Buses 1 and 2 can be used in conjunction with the X-Fade control in the controller section to crossfade between a pair of tracks set to Output Buses 1 and 2, respectively, to let you do DJ-style crossfading.

To allow more control of the audio, you can assign tracks from Transfuser to tracks in the Pro Tools mixer for further mixing and processing, or to record the output from Transfuser to disk.

If you want to audition Transfuser tracks in a headphone mix, you can assign these to the Cue Output Bus. You would then need to route this Cue Output Bus into an auxiliary input track in your Pro Tools mixer, and assign the output of this Aux track to whichever outputs you are using to feed your headphone-monitoring amplifier—see Figure 5.45.

Figure 5.45
Routing Transfuser's Cue Bus via a Pro Tools auxiliary input track to the Pro Tools output pair used to feed the headphone-monitoring amplifier

> **NOTE**
> By default, all of Transformer's eight output busses, and the Cue Output Bus, are routed to Transfuser's main outputs (which play through the Pro Tools track on which Transfuser is inserted) until they are selected as the input to a Pro Tools auxiliary input or audio track.

Importing Audio into Tracks

To import audio, simply drag and drop any WAVE or AIFF audio file from the Transfuser browser or from your computer's desktop into Transfuser's Tracks pane. A dialog window will appear to ask you how you would like Transfuser to handle this—see Figure 5.46.

Figure 5.46
Transfuser Import
dialog

The Slice Sequencer

If you choose 'Sliced Audio and Slice Sequence' the audio will be sliced up and dropped into a slice sequencer that plays back the sliced audio—see Figure 5.47.

Figure 5.47
Audio dragged
and dropped
into Transfuser's
Tracks pane and
automatically sliced
and sequenced to
create a Transfuser
track

The Phrase Sequencer

If you choose 'Time-Stretched Audio and Trigger Sequence', the dropped audio is converted to time-stretched, beat-matched audio that can be triggered by mapped MIDI notes—see Figure 5.48.

Figure 5.48
Audio dragged and dropped into Transfuser's Tracks pane and automatically converted into a time-stretched audio phrase with an associated phrase sequence within a Transfuser track

The Drum Sequencer

If you choose 'Drum Kit and Drum Sequence', the audio is converted to a drum pattern sequence and plays back a sampled drum kit—see Figure 5.49.

Figure 5.49
An audio region dragged from the Edit window is automatically converted to a drum pattern sequence and plays back a sampled drum kit within a Transfuser track

> **NOTE**
> Transfuser extracts drum samples from the dropped audio as effectively as possible for playback via the drum synthesizer, but how successfully will, of course, depend on the nature of the audio used.

Synthesizer Tracks

In addition to the sample replay tracks, Transfuser also offers three types of synthesizer tracks: Analoge, Electric, and bass.

The bass synthesizer is usually paired with a phrase sequencer, while the Analoge and Electric synthesizers are typically paired with either a chord or polyphonic sequencer—see Figure 5.50.

Figure 5.50
Transfuser
synthesizers

Figure 5.51
Bass synthesizer
module

The Bass Synthesizer

The monophonic bass synthesizer module has filter cutoff and resonance, envelope modulation, and distortion drive controls—see Figure 5.51.

In the Editor pane, you will find a single oscillator, with filter, envelope, accent, and distortion controls—see Figure 5.52.

Figure 5.52
Bass Synthesizer
Editor pane

The Analoge Synthesizer

Analoge is a polyphonic synthesizer module that has controls for complexity, tone, punch, and length—see Figure 5.53.

See its Editor pane in Figure 5.54.

Figure 5.53
Analoge synthesizer
module

Figure 5.54
Analoge Synthesizer
Editor pane

The Electric Synthesizer

The polyphonic Electric piano synthesizer has controls for pickup type, pickup distortion, attack, and decay—see Figure 5.55.

See its Editor pane in Figure 5.56.

Figure 5.55
Electric synthesizer
module

Figure 5.56
Electric Synthesizer
Editor pane

The Sequencers

All the sequencer modules (except the Thru-Seq) have play, latch, and stop controls for the sequence and a groove control that lets you apply the groove micro timing as configured in the Editor pane.

> **NOTE**
> Pressing the sequencer's Play button engages a temporary latching (indicated by a green latch light). This is so that you don't have to hold the Play button to keep hearing a sequence.

You can choose preset sequences from the drop-down menu at the top of each module or you can create your own using the module's Editor pane.

The Poly-Seq, Phrase-Seq, and Drum-Seq sequencer modules all let you record short sequences of MIDI notes, one to four bars in length, from an external keyboard or controller into any or all of the sequencer module's 12 patterns. These sequencer patterns can then be used to play the track's synthesizer module.

> **TIP**
> Alternatively, using the Thru Sequencer, you can play any Transfuser track's synthesizer module directly from MIDI sequences on Pro Tools MIDI tracks or using an external MIDI keyboard or controller.

The Poly-Seq, Drum-Seq, Phrase-Seq, and Slice-Seq all have three playback modes or note range settings that can be applied when playing keys in the note range area of the on-screen keyboard. These modes also apply when playing notes within this range from an external MIDI keyboard or controller. You can choose the mode using the drop-down note range selector located beneath the transport controls in the sequencer's Editor pane.

> **NOTE**
> The Chord-Seq does not have note range settings and the note range settings for the Slice-Seq are a little different.

Sequencer Controls

Speed

The Speed setting lets you set the speed of the sequencer as a ratio to the Pro Tools tempo.

Sync

Sync selects the sequencer's playback position when triggered in relation to the host's timeline or other Transfuser tracks: use Bar for performing and playing around in Transfuser; use Beat or 1/16 for arranging songs in your host's timeline; select Off to turn off automatic synchronization.

Groove

All of the Transfuser sequencer modules let you apply varying degrees of 'groove' as configured in Transfuser's master groove. In Transfuser sequencers, each step is normally locked to a strict grid and the groove matches a single bar in the sequencer module. Applying groove lets you adjust the micro timing of each of 16 steps. You can also select one of the preset grooves or even import a Pro Tools groove template. Groove presets include 1/16 SW, a 16th note swing groove; 1/8 SW, an 8th note swing groove; 1/4 SW, a quarter note swing groove; 'Laid Back', a laid back groove that is just a little bit behind the beat; and 'Ahead', which applies a pushing groove pattern that is a little bit ahead of the beat. Each sequencer can also be set locally to use the master groove or local groove preset.

In the sequencer editor, the knob above the Groove menu can be used to set the amount of groove, from 0–100 percent, and the drop-down Groove menu lets you select the groove micro timing. Here, you can choose either Transfuser's master groove or another local groove option. Each sequencer can be individually set to different grooves. Choose 'Master' to apply the master groove locally or choose from any of the standard groove presets to use these locally. Additional choices that can be applied locally include 1/32 SW, a 32nd note swing groove, and 'Random'.

M.A.R.I.O.

All of the Transfuser sequencers feature M.A.R.I.O. (Musical Advanced Random Intelligent Operations)—see Figure 5.57. This musical randomization algorithm lets you create variations of your sequencer patterns, with algorithms specifically tailored for each sequencer type.

Figure 5.57
M.A.R.I.O.

To use M.A.R.I.O., simply click the Apply button on any sequence that you wish to randomize. If the sequencer is playing, you will immediately hear and see the changes that have been made. You can continue clicking the Apply button if you would like to hear multiple variations until you find one that you like. The left and right arrow keys next to the Apply button let you undo and redo the changes made by M.A.R.I.O. This makes it easy to audition multiple variations of a pattern. You can adjust the amount of randomization that will be applied by turning the 'wheel' around the Apply button clockwise from Min to Max.

The Target drop-down menu lets you select which parameters of a sequence you would like to randomize. This is useful in scenarios where you like certain parts of a sequence (the rhythm or timing, for example) but would like to hear variations in other areas of the sequence. To do this, simply check the parts you would like to randomize (or uncheck the parts you would like to retain) and click the Apply button to hear the results.

The Multitool

All the sequencers have a useful multitool in the Editor pane that allows you to perform a variety of tasks in the display area without having to change tools (such as switching from pencil to eraser). For example, you can move one or more events by clicking and dragging the note(s), or you can copy these if you hold the Option key (Mac) or Alt key (Windows) while you do this. You can also delete a note or event by double-clicking it or reset an event to its default value by Alt (Windows) or Option (Mac) clicking it.

The Polyphonic Sequencer

The polyphonic sequencer (see Figure 5.58) lets you choose from preset sequences or create new sequences to play a synthesizer module on the same Transfuser track.

Figure 5.58
The polyphonic sequencer

The drop-down Sequencer menu at the top of the module lets you load preset sequences and save, copy, and paste sequences. This menu also allows you to choose a different sequencer module to replace the current sequencer module.

To the right of this, there is a button that lets you enable or disable the module. Lower down, there are sequencer Transport controls for play, latch, and stop together with a groove control that lets you set the amount of 'groove' to apply to the sequence.

Poly-Seq Editor Pane

In the Poly-Seq Editor pane (see Figure 5.59), the transport controls include a Record button that you can use to record a sequence by playing the notes on a MIDI keyboard instead of entering the notes using the mouse in the editor's display section.

Figure 5.59
Poly-Seq editor display

Selecting any of the scales in the 'Force Notes to Scale' drop-down menu constrains the notes allowed in the phrase pattern to the notes of the selected diatonic scale.

The Note Range drop-down menu offers three choices:

- If you choose 'Trigger Phrase', playing keys within the note range will trigger and retrigger the phrase pattern.

- If you choose 'Transpose Phrase', playing keys within the note range will trigger and retrigger the phrase pattern and will transpose this chromatically when you play any key other than C3 (which plays the phrase pattern at its original pitch).

- If you choose 'Play Notes', keys that you play within the note range will be passed through so they play the module directly without triggering the pattern—allowing you to play your own melodies.

The Chord Sequencer

Chord-Seq is a polyphonic sequencer that makes it easy to send chords, strums, and other polyphonic 'blocks' of notes to a polyphonic Transfuser synthesizer—see Figure 5.60.

Figure 5.60
The Chord-Seq module

You can choose these from the drop-down menu at the top of the module (see Figure 5.61) or you can create your own using the Editor pane.

Figure 5.61
Choosing chord sequences from the Chord-Seq drop-down menu

Chord-Seq Editor Pane

The chord sequencer's Editor pane (see Figure 5.62) has the usual speed, sync, groove, M.A.R.I.O., and pattern controls—but does not have the note range feature.

Figure 5.62

The Chord Sequencer Editor pane

The chord sequencer has a unique 'Humanize' feature that lets you add random variations to velocity and micro timing to make for more dynamic-sounding sequences. The associated drop-down menu determines whether the 'Humanize' feature will apply to the velocity, timing—or both.

The Phrase Sequencer

The phrase sequencer (see Figure 5.63) lets you choose from the preset sequences or create new sequences to play a phrase, bass, Electric, or Analoge synthesizer module on the same Transfuser track.

The drop-down Sequencer menu at the top of the module lets you load preset sequences and save, copy, and paste sequences. This menu also allows you to choose a different sequencer module to replace the current sequencer module.

To the right of this, there is a button that lets you enable or disable the module. Lower down, there are sequencer transport controls for play, latch, and stop together with a groove control that lets you set the amount of 'groove' to apply to the sequence.

Figure 5.63

The phrase sequencer

Phrase Sequencer Editor

The Phrase Sequencer Editor pane (see Figure 5.64) has the usual speed, sync, groove, M.A.R.I.O., and pattern controls and also has a 'Force Notes to Scale' feature.

Selecting any of the scales in the 'Force Notes to Scale' drop-down menu constrains the notes allowed in the phrase pattern to the notes of the selected diatonic scale.

Figure 5.64
The Phrase
Sequencer Editor
pane

The Note Range drop-down menu offers three choices:

■ If you choose 'Trigger Phrase', playing keys within the note range will trigger and retrigger the phrase pattern.

■ If you choose 'Transpose Phrase', playing keys within the note range will trigger and retrigger the phrase pattern and will transpose this chromatically when you play any key other than C3 (which plays the phrase pattern at its original pitch).

■ If you choose 'Play Notes', keys that you play within the note range will be passed through so they play the module directly without triggering the pattern—allowing you to play your own melodies.

Drum Sequencer

The drum sequencer (see Figure 5.65) lets you choose from the preset sequences or create new sequences to play a drums synthesizer module on the same Transfuser track.

The drop-down Sequencer menu at the top of the module lets you load preset sequences and save, copy, and paste sequences. This menu also allows you to choose a different sequencer module to replace the current sequencer module.

Figure 5.65
The drum
sequencer

To the right of this, there is a button that lets you enable or disable the module. Lower down, there are sequencer transport controls for play, latch, and stop together with a groove control that lets you set the amount of 'groove' to apply to the sequence.

199

Drum Sequencer Editor

The Drum Sequencer Editor pane (see Figure 5.66) has the usual note range, speed, sync, groove, M.A.R.I.O., and pattern controls.

Figure 5.66
The Drum
Sequencer Editor
pane

The Note Range drop-down menu offers three choices:

- If you choose 'Trigger Pattern', playing keys within the note range will trigger patterns 1–12.

- If you choose 'Transpose Pattern', playing keys within the note range will trigger the selected pattern and transpose it according to the note played (or clicked). The original pitch transposition of the loop is mapped to C3. Notes below C3 transpose the loop down and notes above transpose it up.

- If you choose 'Play Pads', playing keys within the note range will play the individual pads 1–12 in the drums synth.

The simplify control lets you gradually simplify any drum pattern by muting notes to thin out the number of notes playing. In the sequencer, these muted notes are shown greyed out.

The drum pattern editor lets you create and edit a 12-note step sequencer. It has a Select menu with various commands for changing the note event selection; an Edit menu that lets you access the 'Undo History', 'Revert Pattern to Initial State', 'Create Rhythmic Template', and various selection edit commands. It also has the multitool, selector, and eraser tools.

Below these tools, the main editor display shows the note 'lanes' for each 'slice' (typically containing a drum or percussion sound). This display has Bar View and

Loop Range selectors at top right, with Parameter Page selectors for note velocity, note timing, pitch transposition, filter control values, and pan control values.

The sequencer pattern editor only has room to display one bar at a time, so the bar view selector lets you select the bar you want to view. The loop range selector lets you set the loop points for the sequencer pattern.

The Slice Sequencer

The slice sequencer (see Figure 5.67) lets you choose from the preset sequences or create new sequences to play a slicer synthesizer module on the same Transfuser track.

The drop-down Sequencer menu at the top of the module lets you load preset sequences and save, copy, and paste sequences. This menu also allows you to choose a different sequencer module to replace the current sequencer module.

Figure 5.67
The slice sequencer

To the right of this, there is a button that lets you enable or disable the module. Lower down, there are sequencer transport controls for play, latch, and stop together with a groove control that lets you set the amount of 'groove' to apply to the sequence.

Slice Sequencer Editor

The slice sequencer editor has the usual speed, sync, groove, M.A.R.I.O., and pattern controls—see Figure 5.68. It also has a quantize control that lets you apply a variable amount of quantization to the timing of slices in the slice pattern. The following quantize values are available: 1/32—32nd note; 1/16T—16th note triplet; 1/16—16th note; 1/8T—8th note triplet.

Figure 5.68
The slice
sequencer editor
pane

The Note Range drop-down menu offers three choices:

■ If you choose 'Trigger Loop', playing keys within the note range will trigger the sliced loop at its original pitch. All keys in the note range have exactly the same function.

■ If you choose 'Transpose Loop', playing keys within the note range will trigger the loop and also transpose it according to the note played (or clicked). The note C3 plays the loop at its original pitch, while notes below C3 transpose the loop down and notes above transpose it up.

■ If you choose 'Play Slices', playing keys within the note range will play the individual slices of the loop. Slices are mapped to keys, in order from left to right. The first slice in the loop is triggered by C2, the second by C#2, the third by D2, and so on. C5 is the top of the on-screen keyboard range, so any slices beyond C5 (the 37th slice) will be out of range. However, you can play any number of slices—well, any number beyond 36 and up to 79 (MIDI note number 127)—by using an external MIDI keyboard controller or MIDI notes from Pro Tools MIDI and instrument tracks. Try using C1 through B1 to trigger stored slicer sequencer patterns (1–12) and set the 'Note Range to Play Slices' (C2–C5+) to interject individual slices.

Controller and Master Sections

The area of the Transfuser window located below the Tracks pane and above the Editor pane contains the controller and master sections.

> **NOTE**
> The controller and master sections can be hidden using the small Show/Hide button located to the right of the Show Preferences button, just above the Show/Hide Editor pane button—see Figure 5.69.

The Controller Section

The controller section (see Figure 5.70) provides six smart knobs, eight trigger pads, and a four-octave keyboard for selecting patterns, triggering, and

Figure 5.69
The Show/Hide button for the controller and master sections

transposing sounds. To the right of the keyboard you will see the MIDI input channel displayed for the selected Transfuser track, with a fader for crossfading between Transfuser Output Busses 1 and 2 positioned to the right of this.

Figure 5.70
The controller
section

Figure 5.71
Assigning a groove
control to Smart
Knob 1

Smart Knobs

To assign a smart knob, Right-click the control you want to assign and choose an assignment from the pop-up menu that appears—see Figure 5.71.

The six smart knobs can each be assigned to control multiple controls in Transfuser. So, for example, you could have the groove controls of the sequencer modules on all the tracks assigned to Smart Knob 1.

> ### NOTE
> Smart knobs apply to the assigned controls regardless of whether or not those controls are on tracks set to receive different MIDI channels. This is especially useful if you are using different MIDI channels to control different tracks, but you want certain controls on those tracks to have a master control.

> ### TIP
> You can always assign MIDI CCs (continuous controllers) on an external MIDI controller (such as the knobs and switches on a keyboard controller) to any of the six Transfuser smart knobs. Smart knobs can also be controlled by Pro Tools automation.

Pattern Switch Keys

The pattern switch keys map to the corresponding pattern keys in the sequencer editors. Clicking a pattern switch key (C1–B1) in the controller section, select the corresponding sequencer pattern.

The pattern switch keys only affect those tracks assigned to the same MIDI input channel as displayed in the channel indicator.

Note Range Keys

The note range keys can be used to play the sequencer modules on tracks. How the different sequencer modules respond to the note range keys depends on the option selected in the note range selector in each sequencer editor.

The note range keys only affect those tracks assigned to the same MIDI input channel as displayed in the channel indicator.

> **TIP**
> Set different tracks' Lo Key and Hi Key settings to different octave ranges for each track. This way you can control multiple tracks independently based on what octave range you play in on the note range keys (as well as when using an external MIDI keyboard).

> **NOTE**
> When playing the Transfuser pattern switch keys or note Range keys from an external MIDI keyboard or from another Pro Tools MIDI track, be sure to select the corresponding MIDI channel for the 'Transfuser **n**' port from the Pro Tools MIDI track output selector (where **n** is the numbered instance of a specific Transfuser insert). To play a synth module directly, select the corresponding MIDI channel for the 'Transfuser **n** Synth' port.

Channel

The channel indicator displays the MIDI input channel for the current (or last) selected track. The on-screen keyboard and trigger pads only affect those tracks assigned to the MIDI input channel displayed in the channel indicator. To change the MIDI channel, select a track assigned to the MIDI input channel you want.

Trigger Pads

Transfuser provides eight trigger pads, each of which can be assigned to trigger sequencer patterns in specific sequencer modules, and also to send MIDI notes (assignable between 48–71) to sequencer and synth modules.

NOTE

If you are using an external MIDI controller with Transfuser, the trigger pads map to MIDI notes as follows: 1 = C, 2 = C#, 3 = D, 4 = D#, 5 = E, 6 = F, 7 = F#, and 8 = G. You can set the default MIDI input channel and octave range for the trigger pads in the Transfuser Preferences.

To assign a trigger pad, Right-click the trigger pad you want to assign. From the pop-up menu that appears, choose the track and the module you want and select the pattern or MIDI note you want to trigger. For example, you might choose C2 (48) to trigger a kick drum.

TIP

You can assign the same trigger pad to trigger another pattern or MIDI note in other track modules as well. A single trigger pad can always be assigned to multiple target modules to trigger different patterns in different modules simultaneously, for example, or to play sampled drum pads in drums synth modules on different tracks.

X-Fade

The X-Fade slider lets you crossfade between all tracks set to Output Bus 1 and all tracks set to Output Bus 2, much like a DJ mixer. This way you can set up one groove on Output Bus 1, and a different groove in Output Bus 2, and then fade one in while the other fades out (and then cue up the next groove to fade in), or mix the two grooves together, or whatever. Assign the X-Fade control to a MIDI CC to control it from a knob or fader on your MIDI controller. It is assigned to MIDI CC 1 (mod wheel) by default.

The Master Section

The master section (see Figure 5.72) provides access to the master groove, to effects ends 1 and 2, to the main effects inserts, and to the recorder. It also provides pitch and volume sliders for the main output, play and stop controls for the master transport, and a button to switch the metronome click on or off.

> **TIP**
>
> You would use the Transfuser metronome click when you want to play or record with the drum sequencer, for example, and the Pro Tools transport is not running, so you can't hear anything to play along with from Pro Tools and no other Transfuser tracks are playing for you to hear the beat. The Transfuser metronome will always run at the Pro Tools session tempo.

Figure 5.72
The master section

Getting the Groove

When the Groove button is highlighted in the master section, the display to the left of this shows the master groove controls. You can apply varying degrees of the 'groove' that you set up here to any of Transfuser's sequencer modules.

Each sequencer step is normally locked to a strict grid. Applying groove lets you modify the 'micro timing' of each step. The master groove editor controls each of the 16 possible steps within a single bar in the sequencer module and lets you adjust the timing of each of these steps. Any of the Transformer sequencers can be set locally to use either this master groove or another local groove preset instead of the strict grid settings for each step.

To adjust the timing of a step to create your own 'groove', click on the step whose timing you want to adjust and the mouse cursor will change to a pencil tool—see Figure 5.73. Drag this up to adjust the timing so that the sequenced note plays just after the grid beat and drag it down to adjust the timing so that the sequenced note plays just before the grid beat.

Figure 5.73
Adjusting the groove

If you prefer, you can select one of the preset grooves or import a Pro Tools groove template using the pop-up Groove menu arrow at the top right of the groove controls area.

The presets include sixteenth note, eighth note and quarter note swing patterns, 'Laid Back' or 'Ahead' grooves. The 'Laid Back' groove pattern makes the notes fall a little bit behind the beat while the 'Ahead' groove pattern makes the notes fall a little bit ahead of the beat. The default Pro Tools templates include Cubase, Logic, and MPC-style grooves and Feel Injector templates—see Figure 5.74.

Figure 5.74
Clicking on the pop-up selector arrow to access the Groove menu in the groove controls area of the master section

You can adjust or invert the amount of groove 'micro timing' available to tracks (and to effects such as the Beatcutter or Gater) using the groove amount slider. This acts as a master control for the overall amount of timing displacement from the strict beat grid.

> **NOTE**
> To individually adjust the amount of 'groove' that will be applied to any Transfuser sequence you can use the groove control knob that you will find in the sequencer section of each track module and in the sequencer editor for each track. Even with the Thru module, which has no groove control knob, the master groove still affects any groove-controllable effects on the track, such as Beatcutter or Gater.

Effects Sends and Inserts

When the FX1 and FX2 buttons are highlighted in Transfuser's master section, these let you access two separate effects sends modules that can be applied to tracks, each of which provides four effects inserts that can be routed serially or in parallel. When you click on any of these insert 'slots', a menu appears that lets you load any of the available effects—as in Figure 5.75.

Figure 5.75
Choosing a send effect in Transfuser's master section

> **NOTE**
>
> Each track has two FX send knobs in its mix section that can be used to route audio from the track to send effects 1 or 2. You can also apply effects directly to each track using its own effects section, but you would typically use the track FX sends for more CPU intensive processing, such as reverb, that you want to apply to more than one track.

When the INS button is highlighted in the master section, this lets you insert effects on Transfuser's main audio output. Again, four effects inserts are available, which can be routed serially or in parallel—just like in the track effects modules and send effects modules. See Figure 5.76. You would typically use these inserts for dynamics control, using a compressor or maximizer effect.

Figure 5.76
Insert effects

When you highlight any inserted effect, its controls appear in the Editor pane below the controller and master sections.

The Recorder

Figure 5.77
The Transfuser
recorder

When the REC button is highlighted in the master section, the recorder is revealed in the pane to the left of the button—see Figure 5.77. You can use the recorder controls to capture one, two, or four bar recordings of whatever is playing back through the main Transfuser audio output.

The recorder is extremely simple to use, with just three controls for record, play, and stop and a pop-up selector to let you choose the number of bars that you want to record. A waveform for the recorded audio is displayed in the Recorder pane when you have finished recording.

NOTE
You can click and drag the waveform from the Recorder window to an existing Transfuser track synth module (replacing it) or into the blank area of the tracks section to create a new track. You can also drag and drop the waveform to the desktop or to a Pro Tools audio track or to the clips list for further editing, arranging, and mixing in your Pro Tools session.

Exporting Transformer Sequencer Patterns

You can export sequencer patterns from Transfuser to Pro Tools so that you can develop these further using Pro Tools tracks to control Transfuser synth modules directly.

To export a sequencer pattern to your Pro Tools session as a MIDI clip, choose a sequencer pattern from the sequencer editor pattern keys, click the key, and drag and drop to a Pro Tools MIDI or instrument track, into the tracks or clip list, or to anywhere in the timeline.

You can drag and drop sequencer patterns to the desktop so that you can use these later in other Pro Tools sessions. It is also possible to drag and drop sequencer patterns from one sequencer module to another. If the sequencers are of the same type, the pattern is simply copied. If not, the pattern is converted to a different type of sequencer, such as from a drum sequencer to a phrase sequencer.

Configuration Screen

To show Transfuser's Configuration pane (if this is hidden), click the Show Preferences button—see Figure 5.78—and the Configuration screen will appear in the Editor pane—see Figure 5.79.

Figure 5.78
Clicking on the Show Preferences button

The trigger pads MIDI input lets you set the MIDI input channel for Transfuser's trigger pads and set an octave range for MIDI note control of the trigger pads.

You can set default input and output gain and output bus channels for the two send FX.

You can also set a default velocity for new events created in the sequencer modules and you can set the volume level for the click in the Transfuser metronome.

The Content Location preference lets you choose the directory location for Transfuser content—that is, track and audio files.

The 'Show Warning Messages' setting lets you choose whether Transfuser will warn you before committing to an edit when you are performing destructive edits.

When 'Sequencer Scale Link' is switched to 'On', all tracks and track previews are forced to follow the musical scale manually set in any track's sequencer. When switched to 'Off', tracks and track previews will play in their originally defined scale.

When 'Optional Aux Outs' is switched to 'On', all of Transfuser's auxiliary outputs are rerouted to the stereo master output. This may be helpful when troubleshooting complex output routing scenarios, where many aux outputs are being sent to the host application.

The Transfuser Preferences pane also displays MIDI signal and audio signal flow for Transfuser.

Figure 5.79
The Configuration
screen

Using Transformer in Pro Tools

Transfuser provides a self-contained arranging and mixing environment for its sequencers and synthesizers that can be used in tandem with the much more full-featured MIDI sequencing and audio mixing, processing, and recording capabilities of Pro Tools itself.

Once you have created one or more Transfuser tracks, you can assign each track a unique MIDI input channel number, from 1–16. Then you can send MIDI from multiple Pro Tools tracks on discrete MIDI channels to Transfuser tracks to trigger Transfuser sequencer patterns or to play Transfuser synth modules directly.

You can also assign the audio output from Transfuser tracks to any of the eight separate audio output busses. These can be routed in turn to Pro Tools auxiliary inputs and audio tracks for further mixing, processing, and recording.

> **TIP**
> When you have set up Transformer with its various tracks within your Pro Tools session, you can save the complete configuration to the Pro Tools Librarian plug-in.

There are several ways you can work with MIDI routing to control Transfuser through Pro Tools. Probably the most immediately gratifying way is to route an external MIDI keyboard controller through a Pro Tools instrument or MIDI track to Transfuser.

211

Using Transformer to add Effects to a Pro Tools Audio or Input Track

One of the simplest ways to use Transformer is to process audio from any of your Pro Tools audio tracks, or 'live' input routed from your audio interface via a Pro Tools auxiliary input and in turn into Transfuser.

To set this up, name a bus pair 'To Transformer', then use this for the signal you want to feed into Transfuser, assigning the audio output path of a prerecorded audio track to this bus pair. Insert Transfuser on a stereo instrument track then choose this 'To Transformer' bus pair as the input to that instrument track—as in Figure 5.80.

In Transfuser, drag the Def Audio In (the Default Audio Input) Track preset from the browser to the Tracks pane—see Figure 5.81.

In the sequencer section you will see the Thru-Seq module, which has no controls or editor. This module simply bypasses the sequencer section and passes any MIDI input straight through to the synthesizer section. In this scenario, you would typically use the MIDI input to pass MIDI data from your hardware controllers—knobs and faders, for example—to adjust any of the parameters in the input, effects, or mix sections in real time.

Figure 5.80
Wah Guitar routed into Transfuser for further processing

Figure 5.81
Default audio input track

The input module, located in the synthesizer section, brings in audio from the input of the Pro Tools track that is hosting this Transfuser plug-in.

The input module has an amplitude envelope that affects the audio passing through when triggered by MIDI—the envelope trigger—with three controls that affect the audio that comes in. The 'Attack' control lets you adjust the attack time of the envelope; the 'Release' control lets you adjust the release time of the envelope; while the 'Thru' control lets you adjust the level of the audio passing through the module.

In the Editor pane (see Figure 5.82), you can graphically edit the amplitude envelope by dragging the attack, hold, decay, sustain, and release parameters in the window or by entering specific values for these.

Figure 5.82
Audio input module editor

If you are routing audio that you have already recorded into Transfuser, start playback in Pro Tools. Alternatively, you can play an external source such as a guitar or vocals through the track's input.

The effects module is where all the action is with this setup. Here you can insert up to four effects, in series or parallel, and control these using your mouse or using external MIDI hardware controllers. Try out the tape delay, for example, stepping through its presets using the left/right arrows in the editor—see Figure 5.83.

Figure 5.83
Tape delay editor

And if four effects is not enough, Transfuser provides two separate effects sends for tracks; each send effects module provides four effects inserts that can be routed serially or as two parallel pairs.

Transfuser also provides effects inserts on the main outputs. Four effects inserts are available, which can be routed serially or as two parallel pairs just like in the effects module and send effects modules.

Using a Single Instrument Track

You can easily control Transfuser with MIDI when it is inserted on an instrument track. The MIDI output of the Pro Tools instrument track is automatically routed to Transfuser on MIDI channel 1. By default, the instrument track is set to receive MIDI from all MIDI input devices, so if you have a MIDI controller already connected and configured, you can start playing Transfuser right away.

> **TIP**
> Depending on the setting for the Default Thru Instrument preference, you may need to record enable the instrument or MIDI track to pass MIDI from the track input to its output. Also, check to make sure that MIDI Thru is enabled in the Options menu.

Assigning MIDI Tracks to Transfuser Synth Modules

You can assign multiple Pro Tools MIDI and instrument tracks to bypass the sequencer module and control Transfuser synth modules directly on different MIDI channels. This lets you trigger individual sequencer modules from external MIDI controllers or from sequences on Pro Tools MIDI tracks.

> **NOTE**
> To assign Pro Tools MIDI tracks to control Transfuser sequencer modules, select the corresponding MIDI channel for the 'Transfuser **n**' port from the Pro Tools MIDI track output selector (where **n** is the number of the MIDI node for a specific Transfuser insert).

AIR Music Technology Boom

AIR Music Technology also created Boom for Pro Tools. This is included as part of every Pro Tools system together with AIR's DB33 tonewheel organ simulator, Mini Grand acoustic piano, Vacuum analog tube emulation synthesizer, Structure Free sample player, and Xpand!2 multi-timbral synthesizer workstation.

Boom is a drum machine that has its own pattern sequencer and offers a broad range of electronic percussion sounds that are ideal for creating electronic dance music, hip-hop, house, and other genres. It comes with 10 drum kits inspired by classic electronic drum machines. The Roland TR808 and TR909 are reproduced in all their glory, for example, and the pattern sequencer programming is extremely similar to the programming of the TR808.

You can create your own patterns or select from a useful collection of preset patterns listed in folders according to their suggested BPM speeds—see Figure 5.84. Like the TR808 and 909, you can use the front panel controls to trigger and switch patterns in real time—in this case using the mouse or using MIDI data from a hardware controller.

Figure 5.84
Boom presets

Taking its inspiration from the TR808, each individual sound has volume, pan, pitch, and decay controls that can be manipulated and automated in real time.

Each pattern is one bar long and is subdivided into 16 16th note steps. You can create up to 16 patterns and save these, along with kit and control settings, into individual presets (pattern memories).

Boom's user interface has four main sections: the matrix display at the left; the instrument section to the right of this; the transport controls at bottom left; and the pattern selector/editor to the right of this—see Figure 5.85.

Figure 5.85
Boom user interface

The matrix display—see Figure 5.86—shows an overview of the current pattern so that you can see at a glance what is happening with the pattern. Ten horizontal rows represent the individual instrument channels and 16 vertical columns represent each of the rhythmic steps that make up a pattern. A grid of LEDs that light up within this matrix reveals the steps within the sequence at which the individual notes play and you can add or remove notes from any sequence by clicking on these LEDs. When you first click an LED that is not lit, this step will be set to play at its highest velocity. There are two lower levels of velocity, and clicking a second or third time lets you access these levels, with the LED glowing a little less brightly at each of the lower volume levels to indicate this.

The Pattern field at the top of the matrix display shows which of the 16 patterns in the current preset is active in the matrix display. To the right of this, the Copy and Clear buttons above the matrix are used to copy or erase patterns when in Pattern Select mode.

Figure 5.86
The matrix display

Underneath the matrix display, there are three global controls that affect all the Boom instruments equally. The volume control, as you would expect, sets the overall output level from the plug-in. The swing control adds a variable amount of rhythmic swing to the currently playing pattern and the dynamics control scales the difference in volume between the pattern sequencer's three possible velocity levels. Messing with the swing and dynamics controls affects the way that the pattern 'grooves'. Below the global controls is a pop-up menu that lets you choose which drum kit you want to use.

To the right of the matrix display, Boom has a set of controls for each of the 10 instruments organized in vertical 'strips' to form the instrument section—see Figure 5.87. Each instrument has pan, level, tuning (pitch),

Figure 5.87
The instrument section

217

and decay (length) controls, along with a solo and a mute button.

Figure 5.88
Boom drum kits pop-up selector

A pop-up selector beneath each instrument channel strip—see Figure 5.88—lets you choose the drum kit from which that particular instrument will be chosen— allowing you to mix and match drums from different kits. Below each strip's pop-up selector, in the Instrument Name area, is a button with the name of the drum kit instrument (kick, snare, etc.) that you can click on to play the drum sound.

At the bottom left of the plug-in's window you will find the Start and Stop buttons and a three-position speed switch with an associated Triplet mode button—see Figure 5.89. The speed switch has three positions: set to X1, Boom's sequencer plays at the same tempo as the master tempo in Pro Tools; set to X2, Boom plays twice as fast; and set to X1/2, Boom plays half as fast. The switch to the right of this lets you put Boom into Triplet mode. In Triplet mode, Boom will only play the first 12 steps in the sequence and the last four steps will be greyed out, indicating that they will not be played. These 12 steps then play in the same amount of time that Boom would normally take to play all 16 steps, allowing you to create triplet patterns. You can use the Start and Stop buttons to play patterns when the Pro Tools session is not playing back—which can be more convenient when you are setting things up.

Figure 5.89
Transport controls

To the right of the transport controls is a two-position Edit Mode switch—see Figure 5.90—that lets you choose whether to edit the pattern or to select from the 16 available patterns in the current preset.

To the right of the Edit Mode switch, below the instrument section, are 16 numbered buttons, technically described as the 'Event switches', that form the Event Bar. These can be used to play the drum sounds and to create patterns. In Pattern Select mode, the Event switches let you choose between the 16 patterns in the current preset.

Figure 5.90
Edit mode switch

In Pattern Edit mode, each Event switch corresponds to a 16th note step in the current pattern. If the pattern is empty, none of the Event switches will be lit. To make a selected instrument play at any pattern step position, all you do is click on the Event switch at that step position—and the Event Switch will light up to indicate this.

Clicking on this step position a second or third time will set either of the two available lower velocity levels for that step. A fourth click will turn the instrument and its light off at that step position.

TIP
Right-clicking an Event switch will toggle its on/off state, preserving the current velocity setting.

NOTE
These patterns can also be selected via MIDI from a keyboard or sequenced using a MIDI or an instrument track in Pro Tools via MIDI notes C3—D#4.

Creating Drum Patterns

Creating drum patterns using Boom is very similar to the way you create drum patterns using a TR808 or 909. First you select a pattern and clear this. Switch the Edit mode to 'Pattern Select' and click on 'Event Switch 1', for example, then click the Clear button above the matrix display.

To write new data into the selected pattern, switch the Edit mode to 'Pattern Edit', hit the Play button to start the sequence running, click on any instrument name button, such as kick or snare, to select the instrument you want to work with, and this instrument's channel strip will become highlighted—confirming that it has been selected.

To enter beats onto any of the 16 (or 12) beat subdivisions in the sequence, just click on the appropriate Event switch in the event bar below the instrument section. If you pick the wrong beat or change your mind, just click again to deselect this and choose another beat.

By default the pattern sequencer is divided into one bar of 16th notes at the sessions tempo. You can easily halve or double this value by adjusting the speed selector in the transport section. For example, if you wanted two bars of 8th notes at the session's tempo you could choose 'X 1/2'; you can even set the matrix to triplets. Unfortunately the speed settings affect all 16 patterns, so if you wanted to mix patterns with triplets and patterns without you would need to use two separate instances of the Boom plug-in.

When you have the first instrument playing the rhythm you like (fours on the kick drum, perhaps), choose the next instrument that you want to work with, such as the snare, and add beats to this in the same way. When you have the snare playing what you want, you can continue to add other instruments, such as toms or claps, or go back to the kick or any other instrument to change this.

When you have created Pattern 1, you can create up to 15 more patterns in the same way, selecting these using the other 15 Event switches in Pattern Select mode.

There is also a Pattern Chain feature that lets you chain these patterns together in any order you like, using the mouse and computer keyboard or using MIDI notes. You can turn the Pattern Chain function on or off from the Setup page that you can access by clicking on the 'spanner' icon at the bottom of the plug-in's window.

Playing Boom Patterns via MIDI

Just as you can switch between patterns by clicking on the Event switches in Pattern Select mode, you can also play and switch between patterns using MIDI data that can be recorded into Pro Tools MIDI tracks. Each note between C3 and D#4 will trigger one of the 16 patterns in the current preset, switching between them on the fly.

So you can write your patterns drum machine-style as you would with a TR808, then play these patterns wherever you like and in whichever order you like in your Pro Tools session by recording the appropriate MIDI note into a Pro Tools MIDI track and routing this Boom to control these patterns.

NOTE
The Setup page—see Figure 5.91—contains a sync mode selector that sets the way Boom synchronizes with Pro Tools when patterns are triggered via MIDI notes. When this is set to 'Off', Boom plays the selected pattern in sync whenever it is triggered by a MIDI note, without synchronizing to the Pro Tools transport. When it is set to Beat mode, Boom starts playing the

selected pattern from the step that corresponds with the incoming MIDI note's place in the current bar. The 1/16 mode is a little more involved: Boom starts playing the selected pattern from one of the first five steps in the pattern, corresponding with the incoming MIDI note's place in the current quarter note. Notes played on the first or third quarter note of the current bar will trigger the current pattern from step 1. Notes played on the second or fourth quarter note will trigger the pattern from step 5.

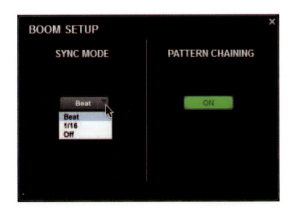

Figure 5.91
Boom Setup Page

Pattern Chaining

When two notes are held down at the same time, Boom by default goes into Pattern Chain mode and plays back two patterns in series, back and forth. You can change this preference in the Settings menu—which you can access by clicking on the small wrench icon at the bottom of the plug-in.

Playing Drum Kit Instruments via MIDI

You can also use MIDI notes to play any of Boom's instruments, each of which is mapped to a MIDI note in the range C1 to D#2—closely matching the General MIDI standard for mappings—and record this MIDI note data to a Pro Tools MIDI or instrument track. This is ultimately a more flexible way of programming the beats because you can bring all the sophistication of Pro Tools MIDI recording and editing features into action this way.

> **TIP**
>
> If you want to get a pattern programmed up simply and quickly, Boom's built-in pattern sequencer is the way to go. When you want to put more complex sequences together, you are better off using Pro Tools MIDI sequencing features. So you might hand program or perform the intros, breaks, and fills, but use the patterns for longer sections.

Recording Separate Drums from Boom into Pro Tools

Boom doesn't feature separate outputs for the individual drum sounds. One way to get around this limitation is to set up audio tracks in Pro Tools for each separate drum sound that you want to record. Then route the output of the instrument track that has Boom inserted to the inputs of each of these individual audio tracks, via an available bus. For example, in Figure 5.92, bus pair 13–14 is used to route the audio output from the instrument track to the audio input of the bass drum audio track.

Figure 5.92
Recording the solo-ed Boom bass drum to a separate track

When you are ready to record the audio from Boom, you can record each drum sound separately by soloing each drum in turn and recording onto the audio track that you have prepared for this. You can record as few or as many separate drum sounds as you like. So, for example, you might have separate kick, snare, and hi-hats with a mix of the rest of the drums appearing in the stereo output from the instrument track.

Summary

If you are working on songs and most popular music, you will probably need drums as one of the basic elements in your musical arrangement. Even if you plan to record real drums, it is often useful to use MIDI drums to get the arrangement started, then replace these later.

For songwriters and composers, Strike is a very useful tool to get things started quickly, and for improvising musicians working in popular musical genres it is much better than playing to click tracks or hastily constructed MIDI drum tracks.

Strike's preset patterns may suffice for many people, and Strike also has a fine selection of popular drum sounds that you can use to write and record your own drum parts. Although Strike has electronic drums sounds, its strength is in the realistic sampled drum kits that it offers.

Transfuser is a very versatile tool to use, especially for electronic dance music production. It not only has drum sounds and patterns but also includes bass, keyboards, and other synthesized sounds and patterns.

You can use MIDI tracks from Pro Tools to trigger various Transfuser sequencer patterns at various times throughout your song using multiple Transfuser tracks set to different MIDI channel inputs set up in one inserted Transfuser plug-in. If you need more than, say, 16 sets of patterns to work with, then you can insert Transformer plug-ins on new tracks as necessary.

At the same time, you can have other MIDI tracks playing Transfuser synthesizer modules directly, using longer, more ambitious MIDI sequence patterns recorded onto your Pro Tools tracks.

Boom—'just like that'—gets you started with its selection of electronic drum kits, inspired by the Roland R808 and its booming bass drum. If you already know how to use a TR808, you know most of what you need to know about Boom. For new users, Boom is not difficult to learn how to use, and you can get satisfying results very quickly.

in this chapter

CHAPTER 6

Working with Virtual Instruments

Introduction

AIR Music Technology offers five powerful virtual instruments for Pro Tools including Structure 2.0 Sampler Instrument Workstation; Hybrid 2.0 High Definition Analog Synth; Velvet 2.0 Vintage Electric Piano Instrument; Vacuum Pro 1.0.1 Polyphonic Analogue Tube Synthesizer; and Loom 1.0.1 Modular Additive Synthesizer.

Spectrasonics offers Stylus with its massive library of drum sounds and loops; Trilian, which features the best sampled and synthesized bass sounds available; and Omnisphere with its comprehensive library of synthesizer sounds.

Arturia's V-Collection of classic synths and keyboard emulations includes the Mini V, Modular V, CS-80V, ARP2600 V, Prophet V and Prophet VS, Jupiter 8-V, Oberheim SEM V, Wurlitzer V, and Spark Vintage—a comprehensive collection of 30 classic drum machines.

Structure 2.0

Overview

Structure 2.0 is a sample playback virtual instrument plug-in that comes with a massive 37 GB premium factory sound library. This library includes AIR's exclusive orchestral elements (strings, brass, woodwinds, and percussion), acoustic and electronic drums (including Nashville Signature drum kits captured from sessions at the famed Blackbird Studios), hand and pitched percussion, pianos and keyboards (including a new high-definition grand piano), choral sounds, guitars, basses, synth pads, synth leads, surround sounds, and atmospheres.

Structure can also use SampleCell I and II, Kontakt 1–3, Tascam Gigasampler, and EXS24 libraries, which many users will already have. If you import Gigasampler patch files, for example, the sample files are copied to your designated local disk drive and a structure patch is created.

Structure allows you to import most types of digital audio files. All common bit depths, sample rates, and surround formats (up to 24-bit/192 kHz/7.1

surround) are supported, along with file formats including WAV, AIFF, SD1, SD2, MP3, and WMA. REX 1 and REX 2 files can be imported and played back using Structure's integrated REX file player.

When you load a REX file, Structure automatically creates a new patch module with two parts: a REX player containing a MIDI part that plays back these slices in the correct tempo and order, and a sampler part that holds the slices of audio.

Structure supports both streaming samples from disk and playback from RAM, and is optimized for the Pro Tools disk engine to allow playback of the greatest possible numbers of simultaneous sampler voices and Pro Tools audio tracks. Structure's database and file browser will help you to quickly find and load samples from your collection that you can tweak to your satisfaction using its integrated sample editor. Structure also features integrated multi-effects processors that have more than 20 effects algorithms—in both stereo and surround formats. A variety of reverbs (including stereo and surround convolution reverbs), delays, compressors, EQs, and other effects are provided. You can insert an unlimited number of effects within each patch, and route any patch, part, or individual zone within any given part of a patch to four global effects sends—each of which features four effects inserts.

The User Interface

Looking at Structure's main window, you will see that it is subdivided into three sections: the patch list at the left, the parameter panel to the right of this with the five page tabs above it, and the smart knobs and keyboard section below.

Just click on any loaded patch to select it for editing in the parameter panel on the right. You can create, edit, and save patches using the Patch menu at the top of the patch list. There is also a Part menu that lets you add various parts to a patch and a View menu that lets you view all or just selected groups.

> **TIP**
> In complex setups with lots of patches it can be useful to organize patches in view groups for a better overview—hiding those you don't need to view and displaying those you need visible. For example, you might put a piano patch and a synth patch into one view group called Keys and a drum patch and a loop patch in a view group called Drums.

At the bottom left-hand corner of each patch displayed in the patch list is a small arrowhead pointing to the right. If you click this, it points downward and reveals a list of parts contained within the patch. You can create, select, move, or edit parts here. In case you were wondering, a part can be a sampler, an audio FX, a MIDI processor, or a sub-patch. The vertical order of the parts reveals the signal flow of the audio within the patch. For example, the audio output from a sampler part may be fed through an audio FX part below it. The audio output from the lowest part in the patch list is sent to the patch output.

A sampler contains a multisample along with its mapping information, metadata, and several sound shaping options, such as filters, envelopes, and modulation. An audio FX part contains an insert effect with its parameter and output settings. A MIDI processor contains a MIDI processor with its settings, for example, a tuning scale. MIDI processors are placed before sampler parts and change the way a sampler part is played. A sub-patch groups multiple sampler parts, insert effects, or MIDI effect parts within a patch. For example, in a patch that holds piano and string sampler parts, you might want to route only the piano through a reverb effect. In this case, you would group the piano part and an effect part using a sub-patch.

In the lower part of the window are six smart knobs and a master volume control, with an 88-note keyboard section below these. You can use these keys to play and control structure by clicking the keys, or using MIDI input from a MIDI keyboard, or using MIDI data from an instrument or a MIDI track in Pro Tools. The parameters these smart knobs control are stored with each patch, so their functions will change each time you select a different patch. Within a patch, each smart knob can be assigned to control one or more parameters simultaneously, which can be useful for quickly adjusting a patch during a performance.

Above the parameter panel you can choose from five page tabs—Main, Effects, Database, Browser, and Setup.

Main Page

The Main page provides controls for all the playback parameters of the currently selected patches and parts—see Figure 6.1. This main page has four tabs to let you access the different controls. The Play tab lets you make settings for pitch transposition, voices, and ranges.

Figure 6.1
Structure user interface showing the main page with the Play tab selected

The Control tab gives you access to three further tabs that you can use to set the smart knob assignments, the MIDI continuous controller assignments, and the key switch assignments—see Figure 6.2.

Figure 6.2
Structure main page with the Control tab selected

The Modulation page is where you can adjust the LFO control settings for the Mod wheel and the Aftertouch—see Figure 6.3.

Figure 6.3
Structure main
page with the Mod
tab selected

The Output tab lets you access controls for the four FX sends, a level trim control for the output, and an output bus pop-up selector—see Figure 6.4.

Figure 6.4
Structure main
page with the
Output tab selected

The Effects page provides four global effect slots with four inserts each—see Figure 6.5. Audio from each patch, part, or zone can be sent individually to these on its output sub-page. Positioned to the right of these, you will find Structure's main output control, which also allows you to insert up to four effects.

Figure 6.5
Structure Effects
page

The Database page lets you look for files previously registered in the database—see Figure 6.6. Metadata for patch files, part files, and samples (audio files), including comments, manufacturer, and ranking for each file, is displayed in the three columns in the upper half of the window while patches matching the selected criteria are displayed in the result list in the lower half of the window.

The browser lets you access all the drives connected to your computer—see Figure 6.7. You can think of this as similar in some ways to the Macintosh Finder or Windows Explorer.

Using the browser, you can navigate through the disks attached to your computer and display any patches, parts, and samples located on your

Figure 6.6
Structure Database
page

Figure 6.7
Structure Browser
page showing
attached disk drives

computer's local file system—see Figure 6.8. The browser then allows you to load these directly into Structure using drag and drop.

Figure 6.8
Structure Browser page showing files and folders

The Setup page has four tabs that all have controls for adjusting Structure's basic configuration. The Global tab has settings for MIDI note display, smart knobs, Pro Tools integration, and interface options (see Figure 6.9).

Figure 6.9
Structure Setup page

The Engine tab lets you configure disk streaming and choose resampling quality. The Instrument tab has pitch transposition, polyphony, and CPU usage controls, and the Content tab lets you choose the location for imported samples and the folders to search for samples.

The Editor Window

Each patch module in the patch list has an edit button that you can click on to access the Editor window—see Figure 6.10. This lets you display and modify individual samples within the sampler parts in a patch. You can edit the waveform, loop the sample, create and modify sample mappings, and adjust individual playback, filter, and amp settings for each sample zone.

Figure 6.10
Structure editor
showing the
Treeview and wave
editor

The Treeview section on the left lists the names of the sampler parts of the patch and the names of all the samples contained in these parts. Just click a part or sample zone in the Treeview to select it for editing in the mapping and edit sections on the right.

The mapping section in the lower half shows the selected part's zones and their keyboard mappings as rectangular zones above the mini-keyboard at the bottom of the window.

The edit section in the upper half can be switched to display the Zone, Wave, Browser, or Database pages. The Wave Editor displays the waveform of the sample in the selected zone in the Treeview or in the mapping section and provides tools for editing and looping. The Zone page provides controls for playback parameters such as pitch, filter, amplifier, and output assignment—see Figure 6.11.

Figure 6.11
Structure Zone
Parameters
sub-page

Structure's Browser and Database pages are also accessible within the Editor window so that patches, parts, and samples can be loaded from here by dragging and dropping into the patch or part list.

Key Switches

Key switches are special MIDI notes or keys that are assigned to switch control values instead of triggering notes. For example, they can switch between different smart knob settings for a patch or mute certain parts within a patch. Almost every control in Structure can be assigned to a key switch and many of the preset patches already have key switches assigned. To see how these work, check out Patch 04, the electronic drum kit that you will find among the Quick Start patches. This has five keys starting from C0 defined as key switches. These are shaded blue while the currently activated key switch is shaded green to distinguish it from the others.

The different effect parts in Patch 04 are not audible initially, but various mix parameters and smart knobs are assigned to the key switches so you will hear the effects when you select the appropriate key by clicking on it on-screen or by playing it on your MIDI keyboard. When you click on a key switch, it stays selected until you select another, and a short description of what it does appears in the display above the smart knobs. Patch 04's key switch E0, for example, adds phaser and delay effects to the TR808-style electronic drum sounds.

Configuring Separate Outputs and Routing to Pro Tools Tracks

If you are using more than one patch module within Structure, you should take the trouble to assign each module in the part list to a separate audio out—see Figure 6.12.

Figure 6.12
Assigning separate
audio outputs

During your Pro Tools session, you will need to create enough new stereo auxiliary input (for monitoring and mixing) or audio tracks (for recording and mixing) for each of the audio outs used by the parts in Structure. Then, using the Pro Tools track input selectors, route the Structure audio outs to the inputs of the corresponding Pro Tools tracks.

> **NOTE**
> The audio outs from Structure are stereo, so if you want to mix each part to a mono track in Pro Tools, assign pairs of parts to the same audio out (for example, assign both the first and second part to Out 4). Then for each pair, pan one part hard left and the other hard right. You can then select the left or right channel of a single Structure out as the input of a mono track in Pro Tools.

The Bottom Line

Like all of the AIR plug-ins, Structure is very easy to use, offers depth of features without these becoming too overwhelming, and has a beautifully designed user interface. The nearest comparison would probably be Native Instruments Kontakt, which offers more programming features and a more comprehensive library of sounds and is also relatively easy to use and well designed. Where Structure beats Kontakt for Pro Tools users is in its closer integration with the Pro Tools environment. And during busy sessions, the fact that Structure is not overburdened with features and has such a clearly laid out user interface helps to make it the sampler of choice for many Pro Tools users.

Hybrid

Hybrid is a virtual synthesizer instrument that uses a hybrid of classic 'subtractive' and digital 'wavetable' synthesis methods to give you a very wide palette of sounds—everything from synth pads, poly synths, and synth leads to keyboards, percussion, basses, and effects.

You can also do some really serious synthesizer programming with Hybrid if you have the inclination, yet it still manages to retain the ease of use that is a major advantage of the AIR user interface designs.

> **NOTE**
> Subtractive synthesis generally starts with one or more oscillators producing electronic waveforms with rich harmonic content. Successive modules (such as filters and amplifiers) then shape this sound by varying the harmonic content and level. Additional modules (such as envelope generators and low-frequency oscillators) are then used to shape and vary the sound over time (a process known as *modulation*). Wavetable synthesis became popular with the first digitally controlled synthesizers. A wavetable is a file that consists of 64 different single-cycle waveforms. A digital oscillator runs through these waveforms to create complex and vivid sounds. Wavetable synthesis is usually combined with subtractive synthesis for additional sound shaping.

Hybrid comes with more than 1,000 preset patches—see Figure 6.13—and has a browser that you can search by category to find what you need.

Figure 6.13
Hybrid presets

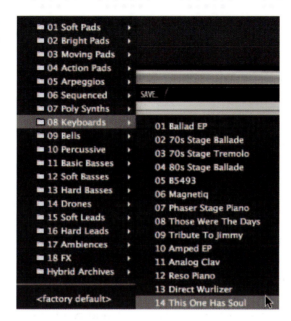

The User Interface

Hybrid's user interface can be broken down into three sections. The setup section at the top of the window lets you load and save Hybrid patches and navigate to any of the control pages—the Common page, the Effects page, or the Part Presets page. The control pages area in the main part of the window shows which of the seven control pages is currently selected. This is where you can program Hybrid's sounds and effects. The master section at the bottom of the window has global controls including master volume and the 'morph' knobs. This section also has two buttons to let you switch Parts A and B on or off, and a master output volume control.

The Main Window

To allow you to create things like velocity layers and keyboard splits, or to spread different sounds across the stereo field, Hybrid has two completely separate synthesizers that work together. These are referred to as 'Part A' and 'Part B'.

The Main window displays the controls for Part A by default—see Figure 6.14. Click on 'B' in the setup section to display the controls for Part B.

The controls for Parts A and B are identical: there are three oscillator modules at the top, with a filter section below this, and an amplifier section at the bottom. The signal flow runs from the oscillators at the top down through the filter into the amplifier section then to the output. To the left of the filter module is a freely assignable modulation envelope, and to the left of the amplifier module is an assignable LFO module. To the right of the filter and amplifier modules are dedicated envelope controls for these.

TIP

You can save individual parts on the Part Presets page. Alternatively, you can save complete patches (including both parts and all other Hybrid parameters) by using the Save button at the top right of the instrument.

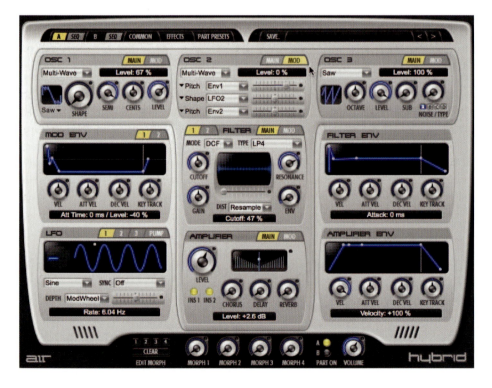

Figure 6.14

Hybrid

The two main oscillators, OSC 1 and OSC 2, both have tuning controls calibrated in semitones and cents and can generate classic analog waveforms including saw sync, saw cross modulation (which modulates the pitch of a sawtooth wave with the output of a triangle wave for complex tones), saw multi wave (a stack of seven sawtooth waves in unison), square sync, square cross modulation, and square pulse width modulation (PWM). Alternatively, you can choose from 100 wavetables, each with up to 64 single-cycle waveforms. The main oscillators also have a shape control that lets you customize the waveform even further once you have chosen the basic waveform or wavetable.

> **NOTE**
> Oscillator 3 primarily supports Oscillators 1 and 2. It transposes in octave steps and provides classic waveforms such as sawtooth, square, and triangle. It also has a sub-oscillator and a noise generator. The sub-oscillator has a square waveform and follows the pitch of Oscillator 3 but one octave lower. The noise generator outputs white noise. A blend of Oscillator 3 with noise also serves as a modulation source for the filter.

Each oscillator also has a modulation section that is revealed when you click on its MOD button—see Figure 6.15. This lets you control the pitch, shape, or level using LFOs, envelopes, or MIDI controllers.

A drop-down menu at the top left lets you select the oscillator type (the basic waveform and algorithm). Below this are three rows of controls. Each row

Figure 6.15
Oscillator
modulation section

lets you choose a destination by clicking the small triangle at the far left; select a modulation source, such as an LFO or envelope, using the drop-down menu in the middle of the row; and use the fader at the far right of the row to determine how strongly you want the modulation source to affect the destination.

Further down the main window are sections for the filters and the amplifier, the filter and amplifier envelopes, the two modulation envelopes, and the various LFOs.

The filter and amplifier modules can be switched to display modulation controls for these—see Figure 6.16.

The two independent filters (each of which can be accessed by clicking the buttons marked 1 and 2 at the top left of the module) let you control the tone color of a part by removing or accenting particular frequencies. You can choose the filter mode or type from the pop-up lists provided. All the filter types have cutoff and resonance controls. The distortion section adds color to the filtered sound by using the integral distortion of the filter. There are seven distortion types, each with a distinct tonal character. The drop-down menu lets you select the distortion type. The horizontal slider above the menu sets the amount of filter saturation.

Figure 6.16

Filter and amplifier modulation controls

> **NOTE**
>
> By default, cutoff is modulated from the dedicated filter envelope, located directly to the right of the filter module. The filter's Env (Envelope) knob adjusts the depth of this modulation. The modulation section lets you control these filter parameters in turn using the LFOs, the step sequencers, or the envelopes—or using MIDI controllers.

LFOs

LFO is the common abbreviation for low-frequency oscillator—one of the basic building blocks used in subtractive analog synthesis. LFOs are typically used to modulate pitch, loudness, or filter cutoff, producing vibrato, tremolo, or electronic sweeps.

The LFO section, located to the left of the amplifier section, provides four LFOs, each of which may be viewed for editing by clicking the selector buttons at the top right of the screen—labeled 1, 2, 3, and Pump.

- LFOs 1 and 2 are monophonic, so each LFO produces just one modulation signal that is distributed to all the voices within a part—ideal for creating vibrato or tremolo.

■ LFO 3 is polyphonic, so it produces one modulation signal per voice of a part resulting in a richer sound. The speed of LFO 3 can itself be modulated.

■ Pump is a special type of envelope that simulates the 'pumping and breathing' effect commonly heard in popular dance music. Normally, this is achieved by keying the gain reduction of a mix-bus compressor using the kick drum: Hybrid's pump LFO can be routed to various parameters to create similar effects.

> **NOTE**
>
> By default, the LFOs are not assigned to a modulation destination. You can use all LFOs as modulation sources on all mod pages. All LFOs produce bipolar modulations, modulating the destination parameter above and below its current value.

Eight modulation types are available from the pop-up menu at the left underneath the LFO waveform display. These range from sine and triangle waves to random, 'drift', and sample-and-hold modulation.

LFO Sync Modes

The LFOs also have six different synchronization modes to determine how rate and phase respond to your playing on the keyboard, or to the tempo of the song, or to the internal step sequencer. This way the modulation can be matched to an envelope sweep, or to the beat of the song, or to a sequencer phrase.

To activate LFO synchronization and bring the Sync mode and the phase control into effect, you can choose a mode from the drop-down list. Depending on the chosen mode, the rate display switches from values in Hertz (Hz) when freely running to values in note length or multiples of step sequencer steps when running synchronized. The various modes work as follows:

■ When the Sync mode is set to 'Off', the LFO is freely running with the frequency set by the rate control.

■ When the Sync mode is set to 'First Note', the LFO restarts when a note is played while no other notes are held. The rate parameter indicates values in units of Hertz or seconds. You can use this method to synchronize the modulation to an envelope sweep.

■ When the Sync mode is set to 'Each Note', the LFO restarts whenever a note is played. The rate parameter indicates values in units of Hertz or seconds and you can also use this method to synchronize the modulation to an envelope sweep.

■ When the Sync mode is set to 'Step Seq', the LFO restarts with Hybrid's step sequencer and the LFO rate is specified in multiples of step sequencer steps. You can use this method to produce sweeps that match the length of a sequencer phrase.

■ When the Sync mode is set to 'Note+Tempo', the LFO rate is specified in fractions of a beat (note length) and restarts whenever a note is played. You can use this mode to match the modulation to the tempo of a song with a specific note length.

■ When the Sync mode is set to 'Beat+Tempo', the LFO rate is specified in fractions of a beat (note length) and synchronizes to the bars and beats of the song while the transport is running. You can use this mode to match the modulation to the tempo and meter of a song.

> **NOTE**
> Because of its polyphonic architecture, LFO3 offers only three synchronization modes: Each Note, Step Seq, and Note+Tempo.

Sequencer

Each part (A or B) has a 16-step sequencer that can be used as a step sequencer, arpeggiator, or MIDI phrase generator. To access these sequencers, just click on the SEQ button next to the part A or part B button. The display switches to reveal the step sequencer—see Figure 6.17—where you can select notes, set velocities, choose control parameters, and adjust performance parameters such as 'Swing'.

In the step sequencer section are four different displays with sequencer lines. These lines carry values for note, velocity, and modulation that you can adjust using the mouse. Below the Step Position LED are 16 gate buttons to play or mute each step of the note and velocity sequences. A pop-up selector lets you choose from nine playback modes, and five performance controls let you adjust the gate length, swing factor, and other performance parameters.

Figure 6.17
Hybrid step
sequencer

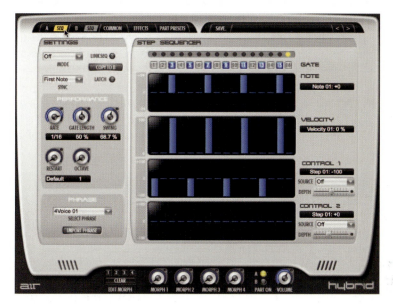

Common Controls

A click on the Common button at the top of the window switches the display to let you access global controls for tuning, pitch bend range, and so forth—see Figure 6.18. Additional settings for Part A and Part B are also available here.

Figure 6.18
Hybrid common
controls

Effects

Similarly, a click on the Effects button switches the display to allow you to access the effects sends—see Figure 6.19. Hybrid provides two insert effect sends for each part and there are more than 40 effect types to choose from. A separate master effects section lets you select reverb, delay, and chorus effects that you can add to any of Hybrid's sounds at the patch level.

Figure 6.19

Hybrid effects

Part Presets

The Part Presets page provides controls for management of the presets for each part and gives you access to their most commonly edited synthesizer parameters—see Figure 6.20. This page is divided into two. At the left you will see the preset browser and synthesizer parameters of Part A. At the right is an identical set of controls for Part B. You can create new patches by choosing presets for Parts A and B and using the controls below the browsers to adjust the sounds.

Figure 6.20
Hybrid Part Presets
page

Morph Controls

At the bottom of the plug-in's window are four 'soft' knobs labeled Morph 1 to 4 that you can use to control 'morph' groups to modify sounds on the fly—see Figure 6.21. The way this works is that you assign any control on the plug-in's interface to one of these four 'morph' groups, each of which can be controlled in turn by its own 'soft' knob. Each morph group can then control all the parameters that you have assigned to it simultaneously. You can adjust these morph groups using the morph controls on the interface or using a MIDI controller. So, for example, you could adjust pan, filter, and oscillator shape from an assignable knob on your MIDI keyboard while you are playing 'live'.

Figure 6.21
Hybrid morph
controls

Assigning Controls

By default, most of the controls on AIR instruments are assigned to MIDI continuous controllers. You can change these assignments whenever you like, or assign physical controls from a MIDI device or control surface using a 'learn' feature, by Control-clicking (on the Mac) or Right-clicking (on Mac or Windows) to access a pop-up menu that lists the options—see Figure 6.22. You can also use these pop-up menus to assign parameters to the morph controls.

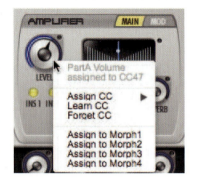

Figure 6.22
Assigning a continuous controller to an oscillator shape control using the 'learn' feature

Automating via Pro Tools

Because you can control Hybrid's parameters using MIDI continuous controllers, this allows you to automate the controls by recording changes to these controller values into any Pro Tools MIDI track.

You can also automate any of Hybrid's controls using the standard Pro Tools automation facilities: click on the Plug-in Automation enable button above the plug-in's window in Pro Tools—see Figure 6.23—then select the controls that you want to work with using the Automation Enable dialog that appears.

Figure 6.23
Auto Enable button

Finally . . .

Despite its complexity, Hybrid's user interface is visually attractive and clearly labeled, making it very easy to work with. The presets are very usable and include lots of pads, basses, leads, and arpeggiated and sequenced effects, but Hybrid really shines when it comes to providing tools for the 'tweakheads' who like to 'roll their own' synth patches from scratch!

Velvet

Velvet is a virtual electric piano instrument that emulates the sounds of four legendary electric pianos: the Fender Rhodes Suitcase, Fender Rhodes MK I and MK II Stage Pianos, and the Wurlitzer A200.

The plug-in's user interface is instantly reminiscent of these classic electric pianos, making it immediately familiar if you've ever played the real instruments. And its controls work very similarly to the originals, allowing you to quickly change the sound using a familiar knob or slider.

The tremolo and autopan effects are exactly like those of the original Rhodes and Wurlitzer electric pianos. Velvet also features a built-in preamp section to let you shape the sound using tube overdrive, compression, and a custom three-band EQ with a parametric mid band. Many electric piano players use effects pedals on stage or in the studio to get the sounds they want from their instruments, so Velvet provides all the classic distortion, wah/filter, chorus, flanger, phaser, and tape delay effects to let you further shape the sounds the way you like.

Velvet has more than 100 electric piano presets to choose from, so it should be easy enough to find something close to the sound you are looking for. And whether you want to tweak a preset or build your own sounds from scratch, the controls are there to let you do this. Try out different settings for the velocity curve to get the playing feel just the way you like it. Turn up the tube drive controls in the preamp section to warm up the tone; tweak the tone controls to give the sound more definition; or adjust the velocity response and the timbre. Engaging the 'Vintage Mode' switch simulates the sonic characteristics of many vintage electric piano recordings by adding a gentle low shelf boost to the overall sound.

> **TIP**
>
> The electric pianos in Velvet have the same key range as the original models (A0–C6 for the A200, E0–E6 for the others) to guarantee authenticity. When the 'Key Extension' switch, located at the top left of the keyboard, is in the 'Up' position, you can play all the notes outside this range.

The User Interface

Velvet's user interface has three separate sections, each of which provides controls to modify the electric piano's sound—see Figure 6.24.

Figure 6.24
Velvet

The setup section at the top provides control over the basic sound and behavior of the selected electric piano model. You can manipulate and tune Velvet in this section—for example, mix in mechanical noises or adjust the velocity sensitivity.

The preamp/EQ and FX section is located in the middle. The left half of this section provides controls to adjust and shape the sound of the electric piano using a one-knob compressor, tube drive, and equalizer. To the right of these, six classic vintage effects let you shape the sounds even further. Click the buttons for distortion, wah, modulation, cabinet, reverb, or delay to reveal its controls in the area below. Each effect has an on-off switch: just click the small light to the right of each effect button. A seventh button, marked FX, can be used to switch all the effects on or off.

The piano front panel in the lowest section provides a piano model selector, master volume control, and tremolo/autopan controls. You can play Velvet by using MIDI input from a MIDI keyboard or an instrument or MIDI track in Pro Tools, or by clicking any of the 73 keys on-screen.

> **NOTE**
> You can assign standard MIDI controllers to virtually any parameter so that you can control Velvet from a MIDI controller in real time. To do this, Right-click (Windows or Mac) or Control-click (Mac) on any control, select 'Learn'

from the pop-up menu that appears (see Figure 6.25), and then move the desired control on your MIDI controller. The parameter will automatically be assigned to that control. Alternatively, you can select the desired MIDI controller from the list of commonly used mappings available in the Assign sub-menu.

Figure 6.25
Velvet control
Right-click menu

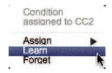

Most virtual instruments use either a sampling technique (every note is real) or a modeling technique (the behavior is more realistic) to emulate sounds and maintain performance efficiency. Because Velvet uses a combination of sampling and modeling techniques to create the sounds, it is much more responsive than sampled instruments. The designers have really gone the extra mile, even recreating the mechanical sounds that you hear if you turn the volume off on your Rhodes or Wurlitzer piano and just play it acoustically—so you hear the sounds of the keys and of the internal mechanisms.

Velvet Preferences

To make adjustments to the default settings, click on the small wrench icon at the bottom left of the keyboard—see Figure 6.26—to open the preferences page.

Figure 6.26
Velvet preferences

By default, the mechanical noises from the internal mechanism are not passed through the FX section—as this most closely models real-world recording conditions. You can set this 'Mechanics Through FX' preference to 'On' if you prefer. By default, 'Sustain Pedal Noise' is set to 'On'. Velvet also routes its signal through distortion insert effects before the wah effect by default. To route the piano through the wah first, set 'Wah Before Fuzz' to 'On'. By default, Velvet's tremolo effects are applied after the FX inserts. To change this, set 'Tremolo Before FX' to 'On'.

All the sample data for Velvet's modeled instruments are stored in a large file called 'Velvet Data.big'. If Velvet is unable to locate this file at startup (for example, if you have moved the file to a new hard disk), you can manually locate the new file by clicking 'Content Location' and responding to the dialog instructions.

Velvet Sounds

Velvet provides lots of great presets—see Figure 6.27. These are subdivided into four main folders for the different instruments emulated, namely 'Mark 1' for the Fender-Rhodes piano Mk. I; 'Mark II' for the Mk. II Rhodes piano; 'SC73' for the Rhodes Suitcase 73 model; 'Model-T' for the Hohner Pianet; and 'A200' for the Wurlitzer EP 200A electric piano.

Figure 6.27
Velvet sounds

Velvet's MK I emulates the Fender Rhodes Mark I electric piano—as can be heard on Chick Corea's 'Spain' or Herbie Hancock's 'Headhunters' album. In this model, the tines have been moved close to the pickup to achieve a full, harmonically rich timbre with a very hard sound at high velocities.

Velvet's MK II models the bright Rhodes piano sound that became popular in the '80s—think Whitney Houston's 'Saving all my love for you'. These pianos were electronically modified to achieve a very bright sound accentuating the metallic attack of the tines, further improved by tines set close to the pickups.

Velvet's SC73 recreates the sound of the Rhodes Suitcase 73, as can be heard on Stevie Wonder's 'Sunshine of my life' or The Doors' 'Riders on the storm'.

Velvet's Model-T recreates the sound of another small 'suitcase-style' piano, the Hohner Pianet. This had a unique tone due to the ground stainless steel reeds, a pickup using variable capacitance, and leather-faced activation pads. The instrument was manufactured from the 1950s to the early 1980s and was used on many hit recordings from the 1960s and 1970s, most notably The Zombies' 'She's not there', The Kingsmen's 'Louie Louie', and The Lovin' Spoonful's 'Summer in the City'.

Velvet's A200 emulates the venerable Wurlitzer EP 200A electric piano, which is the main alternative to the Rhodes electric pianos on high-quality recordings to this day. Supertramp's 'Dreamer' is one of the most famous tracks featuring this, but there are many, many more outstanding examples, including Marvin Gaye's 'I heard it through the grapevine', Three Dog Night's 'Mama told me not to come', Ray Charles's 'What'd I say', and many of Norah Jones's album tracks.

Finally . . .

According to AIR, the model names of the instruments don't 'refer to the exact original pianos that were studied during the development of Velvet, but rather give you a hint to which kind of Rhodes or Wurlitzer sound is widely associated with the model'. For me, compared with previous modeled electric pianos available from Avid's competitors, Velvet sets new standards for realism—with lots of constantly changing movements within the timbres, especially when you hit the keys at different velocities. Other than using the real instruments, Velvet is as good as it gets and is a whole lot faster to set up and use than the originals. Apart from some of the quirks of the original instruments, the main thing that you will miss is the feel of the original keyboards, so it is always a good idea to use a high-quality MIDI keyboard controller with some kind of weighted action when you are playing Velvet.

Vacuum Pro

Vacuum Pro is a polyphonic, vintage-style, analog synthesizer virtual instrument. It does not model any particular synthesizer, but instead takes its design inspirations from the classic analog synthesizers.

Vacuum Pro features two parts—so it can be used to create fat, multilayered sounds. More than 400 presets are included—many of these containing doubles and layers. The Smart page has eight macro knobs, each mapped to numerous parameters on the Part page to allow for intuitive, musical editing of the instrument sound. All parameters can be mapped to a connected MIDI controller and all are automatable.

Vacuum Pro is available as a 64-bit VST, 64-bit AU, and RTAS plug-in for Mac OS X and Windows.

The User Interface

The plug-in's window is divided into three main areas: the master section at the top, the synthesizer section in the main part of the window, and the performance and effects section at the bottom of the window—see Figure 6.28.

The master section provides controls for loading/saving sounds and adjusting the volume between the two parts that make up a sound. The synthesizer section presents all of the elements that you can use to create sounds, including the oscillators, filters, envelopes, LFOs, and other controls. The performance and effects area presents the performance-related controls including the keyboard, pitch and modulation wheels, and arpeggiator, together with various effects including chorus/phaser and delay.

Figure 6.28
Vacuum Pro

The Master Section

Figure 6.29
Vacuum master
section page
controls

At the far left of the master section, the Page Select buttons let you choose which page shows in the window (see Figure 6.29).

A click on the Smart button switches the display to hide the synthesizer controls, replacing these with the eight smart knobs (see Figure 6.30).

The Part A and Part B buttons can be used to switch the synthesizer area between the pages that display each of the two independent, separately programmable parts that make up the overall sound you hear.

To the right of the page controls you will find the parts controls (see Figure 6.30). Looking along from the left of this section, there are two buttons to switch each part on and off, each with a slider that can be used to set the level of each part.

Figure 6.30
Vacuum master
section parts
controls

Figure 6.31
Selecting and
loading a preset
into Part A

The Part A/B load section lets you load individual parts by clicking the name display—see Figure 6.31.

The ' = B' button can be used to copy Part B into Part A—see Figure 6.32. Similarly, the ' = A' button can be used to copy the contents of Part A into Part B.

Figure 6.32
Copying Part B into
Part A

You can load the previous or next preset into Part A or Part B by clicking the previous/next arrow icons next to the display—see Figure 6.33.

Figure 6.33
Loading the previous preset

The Rand button lets you quickly randomize all of the parameters of a part (see Figure 6.34). By default, it randomizes all parameters in all sections of the synthesizer (oscillator, filter, envelope, modulation, or output amplifier).

However, it is possible to lock a section of the synthesizer to prevent changes to that section. To do this, click the buttons (VTO, VTF, ENV, MOD, VTA) and make sure the backlight is off so that randomization is not enabled.

Figure 6.34
Enabling the Part A oscillators for randomization

Part A/B Rand—This button lets you randomize the parameters of a part. If you'd like to prevent a section of the synthesizer from being randomized, ensure that randomization enabling is switched off for that section.

Global Controls

At the top left of the global controls section is a Save button that lets you save any changes you make to a preset—see Figure 6.35.

Figure 6.35
Vacuum master section global controls

To the right of the Save button, a pop-up menu lets you select a preset—see Figure 6.36—with a pair of previous and next preset buttons to the right of this.

At the far right of the global controls section, you will find buttons for Global Unison Doubling On/Off and Global ECO (Economy) mode.

> **NOTE**
>
> Unison doubling creates a 'fatter' sound, but places higher demands on your CPU. Economy mode can reduce CPU load by up to 35 percent. This is achieved through a slight (often unnoticeable) reduction in the sampling rate of the Vacuum Pro sound engine. If you have an older or less powerful computer, you can engage this mode to play complex sounds with higher polyphony.

Underneath these is a pop-up selector for global polyphony that allows you to choose between mono, two, three, four, or six voices.

Figure 6.36
Vacuum Pro presets

Load Locking is a unique feature that allows you to lock certain sections of Vacuum Pro as you load new sounds. For example, if you lock the VTO (oscillator), VTF (filter), and ENV (envelope) sections while loading other sounds, those sections will not change as new sounds are loaded. By default, Load Locking is set to 'Off'. To enable Load Locking, click on the button for the section that you want to lock—see Figure 6.37.

Figure 6.37
Locking the state of
the oscillators

> **TIP**
>
> If you load a sound that has some characteristics that you like but other characteristics that you do not, lock the parameters that you like and try loading other presets. This lets you quickly construct a hybrid sound from many presets using only the parts of each sound that you like.

Synthesizer Area

The synthesizer area—see Figure 6.38—is where the actual sounds are generated using subtractive synthesis with a design similar to that of many classic synthesizers from the 1970–80s.

Figure 6.38

Vacuum synthesizer

The Oscillator Controls

Each part has two oscillators, located at the left of the synthesizer area. Each oscillator has a continuously variable waveshape that can be chosen using the shape knob and the second oscillator can be synced to the first using the Sync Mode switch.

Each oscillator has a Quad mode on/off switch. This is a special mode in which four oscillators play together to create a thick super-saw-style sound. However, unlike some classic synths, you are not limited to using saw waveshapes in this mode—any waveshape can be quadrupled—and you can use the detune knob to vary the thickness of the sound by detuning each of the four oscillations.

Each oscillator also features a very short (one-cycle) delay line with knobs to set delay time and delay amount. This short feedback loop creates interesting tonal characteristics by simulating the electro-acoustical feedback that occurs while playing instruments like the electric guitar. If Re-Pitch is switched off, it takes a brief instant for the feedback spectra to shift to a new note that is played. This results in a more 'organic' transition. If switched on, Re-Pitch immediately resets the feedback spectra at the start of each note, which some describe as cleaner but more digital sounding. This can be set according to your needs.

The Mixer Controls

The audio from the oscillators is fed into a mixer that lets you set the levels of each input. The drive knob simulates tube saturation and the RingMod knob lets you blend in a classic ring-modulated sound.

The Filter Controls

The voice path has two filters—a low-pass/band-pass filter and a high-pass filter. The routing switch lets you use these filters in Serial, Parallel, or Complex mode. The Serial and Parallel modes work as expected. Complex is a proprietary routing developed by the AIR team in which the filters work primarily in parallel but a small amount of output from each filter is fed back into the other filter. This results in an interesting sound due to the unusual way the two filters interact with each other. Finally, each filter also has a sat knob to set the amount of saturation.

The Age Controls

The age controls model the effects of wear and tear classic synthesizers experience. The drift knob sets how much random variance ('drift') you will hear in pitch. When set fully counterclockwise, there will be no pitch variance with each note strike. When set fully clockwise, you may hear a great deal of random variance in pitch with each note strike. The dust control models the effects of dust and corrosion on the contacts of keybed. This results in specific sonic artifacts that appear at the start of each note.

The Output Amplifier

The output of the filters is passed to an output amplifier. This amplifier is controlled by Envelope 3 and its A, D, S, and R knobs. The output amplifier also provides additional wave-shaping controls through the shape knob.

The Setup Controls

The setup controls affect the behavior of the synthesizer. The glide time knob sets the amount of time it takes to portamento (or glide) from one note to the next. The Glide All switch determines whether notes glide at all times or if gliding only happens when one note is already being held down. The Env Retr switch determines whether the envelopes are retriggered with each key strike or if they begin with the first note strike and continue until all notes are released.

The Envelope Controls

Vacuum Pro has a number of envelopes, LFOs, and other modulation sources to help you control the various parts of the synthesizer in new and interesting ways.

Vacuum Pro has four envelopes. Envelopes 1 and 2 share one set of knobs and are used to control the first filter (HP), the second filter (LP), or both. The third envelope exclusively controls the output amplifier. You can assign the fourth envelope to control various parts of the synthesizer by using the Destination drop-down menu.

Vacuum Pro has a unique, 'flat' fourth envelope that is quite useful for pitch effects. This flat envelope does nothing in its default state (as shown earlier). The Att Lvl knob sets the target level (positive if turned clockwise, negative if turned counterclockwise) that will be reached over the time defined by the Att Time knob.

The slope knob sets the post-attack behavior of the envelope while a note is being held. If this knob is greater than zero, the envelope will drift upward; if it is lower than zero, the envelope will drift downward. The Rel Slp knob performs the same function as the slope knob, except once a note has been released.

The LFO Controls

Vacuum Pro has two low-frequency oscillators. The shape of the LFOs can be set using the Shape menu and the LFOs can be synchronized to the master clock of your DAW by using the Sync drop-down menu. Finally, the fade knob allows you to fade in and out (the default '12 o'clock' position produces no fade-in or -out; turning counterclockwise sets fade-out whereas turning clockwise sets fade-in).

The Modulation Controls

The mod section allows you to select a modulation source and destination using drop-down menus. The depth knob sets the amount of modulation that takes place. Three velocity knobs use the velocity of incoming MIDI note data to control the filters and output amplifier sections.

Performance and Effects Area

Located at the bottom of the plug-in's window, the performance and effects area has an arpeggiator section at the left, a keyboard section in the middle, and an effects area to the right—see Figure 6.39.

Figure 6.39
Vacuum
performance and
effects area

Arpeggiator

The arpeggiator controls (see Figure 6.40), located at the far left of the performance and effects area, include an On/Off switch with an associated

Figure 6.40
Vacuum
arpeggiator

LED that lights when the arpeggiator is running and pulses in time with this.

The rate knob lets you set the speed of the arpeggiator in musical divisions of your Pro Tools session's BPM.

There is also a mode knob that lets you choose how the chord will be arpeggiated:

- 'UP' steps notes from bottom to top and repeats.

- 'DOWN' steps notes from top to bottom and repeats.

- 'U&D' steps notes from bottom to top to bottom and repeats.

- 'RND' plays all of the notes in your chord in random order.

Keyboard and Pitch/Mod Wheels

The keyboard, Pch (pitch), and Mod (modulation) wheels—see Figure 6.41—respond to incoming MIDI data from Pro Tools tracks or from an external keyboard routed via Pro Tools. You can also manipulate these controls using your computer's mouse.

Figure 6.41
Vacuum keyboard

Effects

The effects area—see Figure 6.42—has controls for chorus/phaser, delay, and master width and level.

The chorus/phaser controls include a three-position selector switch that lets you choose Chorus, Phaser 1, or Phaser 2. The depth knob sets the intensity of the selected effect; the rate knob sets the speed of the selected effect; and the mix control lets you choose how much of the effect is mixed in with the original audio.

The delay effect has controls for LPF and HPF to let you set the delay effect's low-pass and high-pass filter cutoff frequencies. The time control lets you set the delay time. Engaging the Sync switch synchronizes the time parameter to the master clock of your DAW. If sync is on, this parameter is displayed as beat divisions. If sync is off, this parameter is displayed in seconds.

The Fdbk control determines how much of the outgoing signal is fed back into the delay module; the mix control determines how much of the effect is mixed in with the original audio; and the L-R switch engages a 'ping-pong' stereo delay mode.

The master width control lets you set the 'width' of the stereo sound field, from mono (fully counterclockwise) to stereo (fully clockwise), and the master level control lets you set the final output level of Vacuum Pro.

Figure 6.42
Vacuum effects area

The Smart Controls

The Smart Controls page provides macro-level control, making it easy to take any existing sound and quickly modify it to suit your needs—see Figure 6.43.

The knobs on the Smart Controls page are connected to all of the appropriate controls on the Part A and Part B pages. So turning one smart knob will cause several related knobs on both part pages to change simultaneously to help you achieve a desired sound. This saves you from having to manually locate and change the controls on each part page.

When working with the smart controls, you are always imparting modifications to the parameters on the part pages without actually overwriting those parameters. As you work with the smart controls, you may come across a heavily modified sound that you wish to use as a starting point for even more modifications. To do this, simply click the Apply button.

When you click the Apply button, all of the temporary modifications imparted from the smart controls become permanently imprinted into the part pages. The smart controls are then reset to their baseline ('12 o'clock') settings. You can then save the sound or continue using the smart controls for even more modification.

Figure 6.43
Vacuum Pro Smart
page

The Setup Page

Clicking the wrench icon at the lower right corner of the performance and effects area brings up the Setup page—see Figure 6.44. The Pitch Bend Range control lets you choose how many semitones up or down your note will shift when you use your MIDI controller's pitch wheel. If this is set to zero, Vacuum Pro will ignore Pitch Bend MIDI messages.

Tooltips are helpful hints that appear when you hold the mouse over an on-screen control for one second. You can show or hide the tooltips using the pop-up selector provided.

Most of Vacuum Pro's controls can be reassigned (remapped) to the buttons, knobs, or sliders of your MIDI controller by Right-clicking the on-screen control and selecting 'Learn MIDI CC'. If you use more than one MIDI controller with Vacuum Pro, such as a small, lightweight keyboard for live performance and a large 'master' keyboard in your studio, you can use the Save and Load buttons provided here to quickly save or load 'learned' MIDI assignments for each controller. The Reset button removes all custom assignments and returns Vacuum Pro to its default state.

The 'Parts Inherit Preset Names When Saved' feature applies if you create new sounds using parts of existing sounds. If 'No' is selected, the names of the original parts will remain in the new sound that is created and saved. If 'Yes' is selected, the name of each part will be changed to match the name that is chosen for the sound.

Figure 6.44
Vacuum Pro Setup
page

Finally . . .

Vacuum Pro is a very powerful polyphonic synthesizer with an excellent library of presets ranging from bright leads to classic pads, from arpeggios to basses, from sweeps to SFX. Using the subtractive synthesis method, it has all the controls you need either to edit existing presets or to build new sounds from scratch.

Loom

Loom is the most recently developed modular virtual instrument from AIR. It uses the additive synthesis method, combining up to 512 sine waves to build different spectra, allowing the creation of incredibly complex and musical waveforms. More than 350 preset patches are included—and you can quickly create variations of these using Loom's intelligent randomization features.

The User Interface

Loom has a master section at the top of the window and you can choose whether to display the Edit or Morph pages in the main part of the window—see Figure 6.45.

Figure 6.45
Loom

The master section provides high-level global features such as loading/saving sounds and lets you switch between the Morph and Edit pages.

The Edit page lets you create entirely new sounds from scratch (or make major changes to the existing sounds).

The Morph page not only allows for quick editing and modification of the preset parameters, it can also be used as a live performance tool to visually add motion and movement to your tracks.

The Master Section

Two buttons are provided at the far left of the master section that allow you to select whether the Morph or Edit page is displayed in the main part of the window below. To the right of these are three controls provided for the partials that Loom generates to create each single 'voice' that you hear—see Figure 6.46.

Figure 6.46
Loom master section Morph/Edit buttons and partials controls

The Per Voice drop-down menu lets you set the maximum number of partials each 'voice' can contain and the 'Current' indicator shows how many partials are currently being synthesized.

The 'Synced' parameter is 'On' by default and synchronizes the start of each partial when you play a note. This results in a tight and focused sound. Switching this parameter to 'Off' lets the partials run freely without resynchronizing when a note is triggered. The result is more phasing among the partials and a less focused, more 'blurry' sound that may be perfect for certain applications. There is no correct way to set this parameter—it depends on the sound you wish to achieve.

> **NOTE**
> The Per Voice parameter lets you set the number of partials available for each voice. Increasing this parameter places greater demands on your CPU but allows you to construct certain types of sounds that require a large number of partials to be heard correctly (for example, bright bass sounds with lots of high-frequency partials). If you would like to use a large number of partials while minimizing CPU load, you can switch on the Eco mode.

To the right of the partials controls is a global octave shift pop-up selector that lets you shift the synthesizer up or down by four octaves, with a global polyphony pop-up selector next to this that lets you choose the number of notes the instrument will play (from 1 to 6). See Figure 6.47. Next to these pop-up selectors is a slider control that lets you set the glide time (in seconds) that it takes for a monophonic sound's pitch to portamento (or slide) from one note to another. For example, if this parameter is set to zero the pitch will instantly jump from one note to another as you hold one note and press another note; if you set this parameter to three it will take three seconds to transition from the first note to the second.

NOTE

Glide time only applies to monophonic sounds (i.e., if the poly parameter is set to Mono). If the poly parameter is set to a value of two or greater, Loom becomes a polyphonic instrument and the glide time parameter is ignored.

Figure 6.47
Loom master
section octave, poly,
and glide controls

At the right-hand side of the master section you will find the Save button, a pop-up menu that allows you to load preset patches, previous and next arrow buttons at the right of the patch name, and a Random button that creates new Loom sounds by randomizing all of the modules and parameter settings—see Figure 6.48.

NOTE

Holding the shift key while clicking the Random button randomizes all parameters except for those of any modules that have been selected on the Edit page.

Figure 6.48
Loom master
section

Finally, clicking on the small wrench icon at the far right of the master section opens the Configuration page.

The Morph Page

The Morph page has controls that let you quickly modify any of the existing patches. This page also contains the Morph X/Y Pad, which lets you transform any static sound into a complex, evolving sound in seconds.

The Morph page is subdivided into three separate sections: the macro controls, the Morph X/Y Pad, and a graphic display—see Figure 6.49.

Figure 6.49
Loom Morph page

The Macro Controls

The macro control knobs are automatically connected to appropriate controls on the Edit page when you select a preset patch to allow quick and easy editing of the sounds. Turning one macro knob, such as tone, will change several controls on the Edit page simultaneously—saving you from having to go to the Edit page and manually locate and modify the various individual parameters relating to the tone of the sound.

The knobs in the macro control section are grouped into sound, dynamics, modulation, and master/FX subsections.

Sound/Dynamics/Modulation

The sound controls group includes character, complexity, tone, emphasis, and contour. Dynamics controls include punch, length, and time FX, and depth and speed controls are provided for modulation. See Figure 6.50.

Figure 6.50
Morph Page macro
sound controls

Master/FX

The master/FX controls include pre gain, dist(ortion) mix, mod mix, delay mix, reverb mix, width, and master volume. A row of on/off buttons above the dist mix, mod mix, delay mix, and reverb mix controls allows you to switch the corresponding effects on and off. See Figure 6.51.

Figure 6.51
Morph Page macro
master/FX controls

Located at the top right of the master/FX section, you will find the Apply button. If you have made changes to a preset and would like to make these permanent, just click the Apply button and all of the modifications you have made using the macro controls will be applied to the patch. The macro controls are immediately reset to their baseline settings.

Morph X/Y Pad

The Morph feature lets you create two or more edited versions of a patch then 'morph' between these versions in a variety of interesting ways.

The Morph X/Y Pad can hold up to four variations of a patch in the corners labeled A, B, C, and D. To store a variation, simply Right-click A, B, C, or D and select 'Store State' from the pop-up selector that appears.

When you click and drag the diamond-shaped 'Now Sound Point' closer to one or other of the four corners marked A, B, C, or D in the morph display, dragging the diamond closer to A will make the resulting sound more like the sound of the edited patch version stored in slot A—and less like the versions stored in slots B, C, or D.

> ### TIP
>
> To control the 'Now Sound Point' using a MIDI controller, Right-click anywhere in the Morph X/Y Pad and select either Morph X > Learn MIDI CC or Morph Y > Learn MIDI CC (see Figure 6.52), then move a knob or slider on your MIDI controller. Your chosen knob or slider will move the 'Now Sound Point' along the X- or Y-axis, respectively. You can also control the 'Now Sound Point' using the appropriate MIDI CC parameters in a Pro Tools track's automation view—drawing in changes using a mouse.

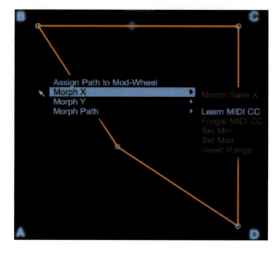

Figure 6.52
Assigning a MIDI controller to Morph X

One of the most powerful features of Loom is the ability to create a morph path that the 'Now Sound Point' follows automatically—allowing you to quickly create complex, evolving sounds.

To create a morph path, double-click anywhere within the Morph X/Y Pad to create your first morph path point, then double-click somewhere else to create your second and subsequent points—which will automatically be connected to the previously created point. You can then move points by clicking and dragging to create exactly the morph path you want—as in Figure 6.53.

> **TIP**
>
> You can use your MIDI controller's modulation wheel to move the 'Now Sound Point' along the morph path. To do this, Right-click anywhere in the Morph X/Y Pad and select 'Assign Path to Mod-Wheel'. This lets you 'play' the morphing of the sound using the modulation wheel.

Figure 6.53
Morph XY display

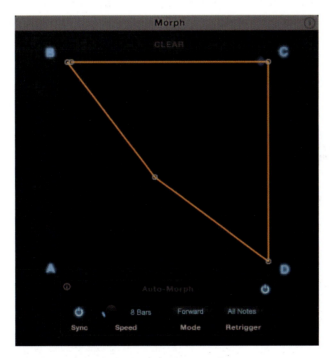

Auto-Morph Controls

Below the Morph X/Y display is a panel containing various auto-morph controls—see Figure 6.54.

The sync on/off parameter synchronizes the movement of the 'Now Sound Point' along the morph path to your Pro Tools session's tempo. When sync is set to 'Off', the 'Now Sound Point' still moves, but it is not synchronized with the Pro Tools tempo.

The speed knob determines the speed at which a note moves along the morph path.

NOTE

If the sync parameter is on, speed is displayed as bars and beats. If the sync parameter is off, the speed display shows information from 0–99 (slowest to fastest).

TIP

You can speed up or slow down an individual section of the morph path by holding down your keyboard's Control key while dragging a morph path line segment up or down.

The mode parameter sets the direction of movement along the morph path— forward, backward, alternating, or random.

When Retrigger is set to 'All Notes', the morph will start from the beginning of the morph path each time a note is played. When Retrigger is set to 'Off', the 'Now Sound Point' moves along the morph path regardless of what is played. It is also possible to assign any of the C-notes on your keyboard (from C-2 to C7) to retrigger the morph path.

Finally, the AutoMorph On/Off button switches the AutoMorph feature on and off.

NOTE

The sync on/off, speed, mode, and retrigger parameters are only active when AutoMorph is switched on.

Figure 6.54
Turning off
Auto-Morph

Figure 6.54
Turning off
Auto-Morph

Morph Graphic Display

The graphic display area displays the spectrum of the sound as this is played—see Figure 6.55.

Figure 6.55
Morph graphic
display

The Edit Page

The Edit page (see Figure 6.56) is where the synthesizer's 'patches'—the connections between the modules and the parameter settings that define the sounds—are created. There are two rows of sound generation and effect processing 'modules' with a smaller row of 'modifiers' below these. The various modules generate and process the sounds while the modifiers (envelopes, LFOs, pitch controls) modulate various parameters of the modules.

The Edit page contains 12 separate modules, each of which adds an element to the overall sound. The first and last modules are fixed, while the 10 middle modules can be chosen from the 30 types available—enabling Loom to create a very wide variety of sounds.

The first module lets you distort and modulate the distribution of frequencies among all of the partials. The user-selectable modules impart their effects on the distorted/modulated output of this first module.

The last module contains four effects—distortion, modulation, delay, and reverb—with master controls for stereo width and output level below these.

> **NOTE**
> An LED next to the level knob lights up when Loom's output limiter is active. This limiter is always on, but only affects signals when they exceed the maximum output level. This limiter is included as a safety mechanism because additive synthesis can unexpectedly become very loud.

The Modules

Each module has a main knob in the top left corner. This knob controls the module's primary function and can be automated using the envelopes and LFOs. Many of the modules also have sub-functions that can be controlled by up to four sliders to the right of the main knob.

Each module has a set of three modulation controls that let you apply the envelope modulation, LFO modulation, or controller (velocity, mod wheel, aftertouch, or key) modulation to control the main knob of the module. Three drop-down menus let you select the modulation sources, and a slider directly below each drop-down menu lets you set the amount of modulation that will be applied.

The Modifiers

At the far left of the modifiers row, the pitch section lets you connect an envelope or LFO to the global pitch of your instrument. The Env and LFO drop-down menus let you select which envelope or LFO you would like to apply to the global pitch. The Env depth and LFO depth knobs directly below the drop-down menus determine the amount of modulation to be applied.

Loom has three AHDSR envelopes, each of which can be modified by dragging breakpoints in its graphical display, and a slope envelope designed to work with pitch and similar parameters. Loom also has three LFOs, each with a fade-in/out time control, a rate control to set the speed of the LFO, a pop-up mode selector that lets you select an LFO waveform, and a pop-up sync selector.

Figure 6.56
Loom edit page

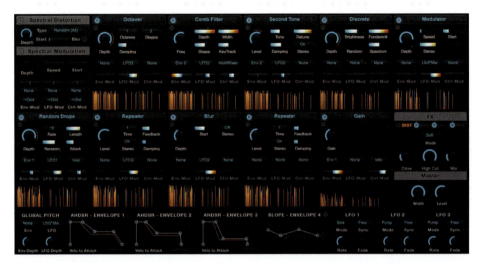

The Settings Page

Clicking the wrench icon at the right of the master section brings up the Configuration page—see Figure 6.57.

Setting the Eco Mode parameter to 'On' engages Economy mode. This slightly reduces the sample rate of Loom's sound engine, reducing CPU load by up to 35 percent. If you have an older or less powerful computer, you are recommended to use this mode to play complex sounds with high polyphony or a large number of partials.

The 'Pitch Bend Range' control lets you choose how many semitones up or down your note will shift when you use your MIDI controller's pitch wheel. If this is set to zero, Loom will ignore Pitch Bend MIDI messages.

Tooltips are helpful hints that appear when you hold the mouse over an on-screen control for one second. You can show or hide the tooltips using the pop-up selector provided.

> **TIP**
> Most of Loom's controls can be reassigned (remapped) to the buttons, knobs, or sliders of your MIDI controller by Right-clicking the on-screen control and selecting 'Learn MIDI CC'.

If you use more than one MIDI controller with Loom, such as a small, lightweight keyboard for live performance and a large 'master' keyboard in your studio, you can use the Save and Load buttons provided here to quickly save or load 'learned' MIDI assignments for each controller. The Reset button removes all custom mapping and returns Loom to its default state.

Figure 6.57
Loom configuration page

Finally . . .

Loom is perfect for creating atmospheric, ambient sound that can be creatively modified using its innovative morph features. The library of presets provides everything from bells to swells to moving pads and wobble sounds—see Figure 6.58.

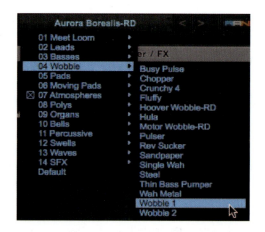

Figure 6.58
Loom presets

Spectrasonics Omnisphere, Stylus RMX, Trilian

Spectrasonics, Eric Persing's creation, produced high-quality synthesizer patches and sample libraries before Eric designed his own suite of virtual instrument plug-ins to make his sounds available to anyone who uses a mainstream DAW such as Pro Tools.

Stylus RMX is a 'cool' groove machine, with its vast libraries of drum sounds and loops, and works well in the Pro Tools audio and MIDI environment.

Where there's drums there needs to be bass, so Spectrasonics developed Trilian, which makes the perfect partner to its 'drummer sibling'. Trilian has a 'to die for' acoustic bass patch together with lots of great electric bass and synthesizer bass sounds.

And then there is Omnisphere! Offered as Spectrasonics' flagship product, a giveaway clue in itself, Omnisphere reveals where Eric's true interests actually lie—it's all about the synthesizer programming, isn't it, Eric? As you can imagine, all the sounds here are suitably impressive with beautiful strings, voices, choirs, and more, and the icing on the cake is its tight 'groove lock' arpeggiator integration with both Trilian and Stylus RMX.

Stylus RMX

According to Spectrasonics, Stylus RMX is a 'groove-based' virtual instrument plug-in—hey, it's a drums and percussion 'machine' implemented in software that plays back sampled sounds!

It comes with a 7.4 GB core library containing thousands of cutting-edge 'grooves', each broken down into individual tracks called *elements* to provide infinite combinations of rhythms that can be played back from your MIDI tracks.

The library also has thousands of kit modules and 250 kit patches that can be played using a MIDI keyboard—you can use these when you want to program your own drums or percussion tracks.

Stylus RMX is multi-timbral with eight parts, each of which can be played from a different MIDI channel via its Multi page—or you can create up to eight layers of sounds played on one MIDI channel.

In the Core Library, you will find subdirectories containing RMX grooves, classic Stylus grooves from the original Stylus, and groove elements that present all of the groove elements (the individual drum sounds) in categories such as kicks or hi-hats. Other categories include break beats, electronic, club, urban, and swing.

Three more directories contain sound menus, example groove menus, and utilities.

The sound menus items present the sounds mapped across the MIDI keyboard so that you can play these from an external MIDI keyboard. Using these sound menus, you can sequence the sounds one instrument at a time. There is a sound menus item that just contains snare drums, for example, and another with guitar bits. The Example Groove Menus directory is intended for use with the Groove Menu MIDI mode—which assigns grooves to different keys for playback from a MIDI keyboard. The utility directory includes clicks and tones.

In the user libraries, you will find any of your own imported audio or REX libraries.

The MIDI Modes

The Groove Menu MIDI mode—see Figure 6.59—lets you play a groove element using a single MIDI note. Every single groove element in a suite has a MIDI note assigned to it: the top element will be assigned to Middle C, for example. The next element down in the list will be assigned to the next key up above Middle C, so to D, and so forth.

One way to use Stylus RMX is to play notes from your MIDI keyboard to trigger elements at different places. Playing these slightly ahead of the beat, Stylus RMX plays the next pattern exactly on the beat.

> **TIP**
> When using the MIDI file drag and drop feature, use Immediate Trigger mode rather than 'Next Beat Trigger'.

In Slice Menu MIDI mode, each MIDI mode plays a slice of the currently selected groove element. Each slice is assigned to a MIDI note in ascending order from C1 on the keyboard. An obvious application is to take the slices and make up your own patterns using the Pro Tools sequencer. You can apply swing feels and so forth.

Figure 6.59
Stylus Groove Menu

The Browser

When you first open Stylus RMX, the SoundCheck preset is loaded ready to play when you click on the Play button lower down on Stylus's main page.

This repeatedly plays a test tone so that you can make sure that the audio is all routed correctly.

To access the sound libraries—the Core Library and user favorites—click on the header (where it says SoundCheck), or on the folder icon to the right of this (see Figure 6.60), to open the browser that lets you navigate through the libraries of sounds.

Figure 6.60
Stylus header with
folder icon

When the browser is open, you can select directories from the pop-up menus at the top left of the window. By default, the Core Library is selected. Scroll through the list of 'suites' in the Core Library and choose one. You can click on any of the elements within the suite, listed at the far right, to hear it play back—see Figure 6.61.

Figure 6.61
Stylus Core Library
suites and elements

> **NOTE**
>
> Stylus always plays back at the tempo of the Pro Tools session—not at the tempo of the chosen groove.

There are three different kinds of elements—grooves, sound menus, and kit modules:

- Grooves have the tempo listed in the first three characters of the name.

- A sound menu is a menu of sounds mapped across the MIDI keyboard. The names of these always start with MENU.

- When Stylus is in Kit mode, you will only see kit module elements in the browser. These kit module elements all have descriptive names at the beginning—for example, 'CLAPS—Funky', 'HIHATS—Tight', 'KICKS—Boomy', and so forth.

> **TIP**
>
> Kit modules only appear in Kit mode. You can change from Multi mode to Kit mode on the Mixer page—see Figure 6.62.

Figure 6.62
Changing to Kit mode

Kit Mode

On the Mixer page the Multi and Kit mode buttons are near the top left of the page.

Working in Kit mode, you can load up to eight kit module elements into these slots. When you first switch to Kit mode, the header displays the name 'Untitled'. Click on this to access the Factory and User Kit menus where you can choose an existing kit—see Figure 6.63.

Figure 6.63
Stylus kit modules

These kit module elements contain individual sounds such as snares or kicks. Elements listed with a <> symbol following the name contain multisample sets of the same sound—see Figure 6.64.

Figure 6.64
Choosing a Stylus
multisample kit set

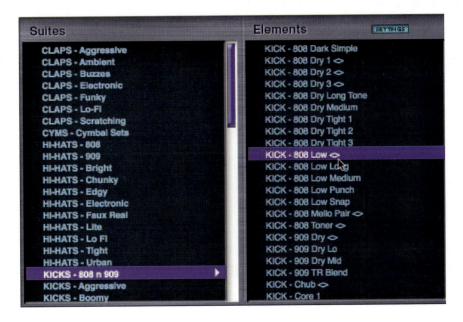

Typically, you will play these kits from a MIDI keyboard, so the different sounds are mapped out using the General MIDI standard with the bass drum on B0 and C1, the snare drum on D1 and E1, and so forth.

Using a pop-up selector near the bottom right of the Mixer page, you can select a MIDI mode where the kit elements in the eight slots can all be played using MIDI Channel 1, or 10, or where slots 1 to 8 are played using MIDI Channels 1 to 8—see Figure 6.65.

Figure 6.65
Choosing a MIDI mode

Stylus Edit Page

On the main Edit page (see Figure 6.66) you can alter the pitch and the filters and do lots of other cool stuff.

Figure 6.66
Stylus Edit page

Stylus Easy Page

The Easy page (see Figure 6.67) offers a subset of these controls laid out for ease of use.

Figure 6.67
Stylus Easy page

Stylus Chaos Designer Page

The footer area lets you switch to the Chaos page—see Figure 6.68. This allows grooves to improvise by throwing some chaos into them.

Figure 6.68
Stylus Chaos page

Stylus FX Page

On the FX page (see Figure 6.69) you can open up to three effects from a long list of FX types that includes compressors, EQ, and delays.

Figure 6.69
Stylus FX page

Stylus Time Page

On the Time page (see Figure 6.71) you can apply feel templates, choosing from those listed in Figure 6.70.

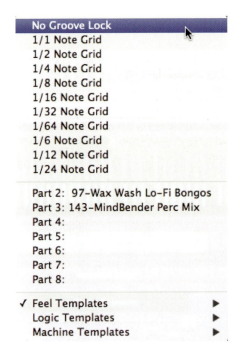

Figure 6.70

Stylus feel templates

Figure 6.71

Stylus Time Designer

Stylus Mixer Page

You can use the Mixer page, working in its default Multi mode, to overlay up to eight grooves to build up extremely intricate rhythms—see Figure 6.72. Here you can load different grooves into any of the eight available slots. Click on any of the empty slots to open the browser, where you can choose your grooves. Then you can use the faders, pan controls, and aux sends for each of the eight parts to mix these together.

Figure 6.72
Stylus Mixer page

TIP

Each of the eight parts in the mixer has its own set of edit parameters, so you can change the pitch of the snare, bass drum, or whichever, for one part without affecting the others. Click on Mixer channel strip 1 to select this, then click on the Edit button to go to the Edit page and edit any of the parameters. Go back to the Mixer page and select another mixer channel to edit its parameters in the same way.

Stylus Multis

Using the Multi display menu (see Figure 6.73) you can choose from the preset 'multis'. The multis let you load up to eight parts that can be played back as composite grooves.

Figure 6.73
Stylus Multis

Routing Stylus RMX Audio into Pro Tools

To get audio from Stylus RMX into Pro Tools, you can always route the audio output from the instrument track to an audio track in Pro Tools and record this directly to disk.

A more flexible way is to transfer the MIDI data that plays back a Stylus groove into a Pro Tools MIDI or instrument track by dragging the currently selected MIDI file from Stylus and dropping it onto the track. Then you have it under your control in Pro Tools.

A third option is to set up separate outputs for each of the eight parts from a multi using the pop-up selectors at the left of each part's channel on the

Mixer page. These all default to OUT A, which allows you to balance the parts using the Stylus RMX mixer. When you assign these outputs separately to Pro Tools tracks, you can mix these using the more powerful mixing features in Pro Tools.

> **TIP**
>
> You can also create a separate audio output for an edit group. First identify, say, a snare by playing a MIDI note from one of the grooves; then go to the Edit page and use the Assign pop-up menu to create an edit group from the note you just played and assign this to, say, output F. This allows you to send a single snare drum from a stereo audio loop to its own Pro Tools fader where you can add compression, EQ, and reverb or whatever.

Omnisphere

Spectrasonics' flagship virtual instrument, Omnisphere, combines a wide variety of hybrid real-time synthesis techniques with a vast library containing more than 40 GB of samples and inspiring patches.

Omnisphere offers a host of hybrid synthesis and control features, including variable wave-shaping DSP synthesis, granular synthesis, timbre shifting, FM, polyphonic ring modulation, high-resolution streaming sample playback, harmonia, dual multimode filter structure, chaos envelopes, an advanced unison mode, and the innovative flex-mod modulation routing system.

Although Omnisphere focuses on providing cutting-edge timbres, the library also includes plenty of conventional sounds. These include beautiful orchestral string ensembles; a Celeste, church bells, glass armonicas, and guitars; and a fine selection of different choirs and human voices.

The Main Page

Omnisphere's beautifully designed user interface (see the main page in Figure 6.74) lets you delve as deeply as you like to craft new sounds—which advanced sound designers will fully appreciate—while allowing you to play back the presets with a minimum of fuss.

Figure 6.74
Spectrasonics
Omnisphere Main
page

The Edit Page

The Edit page (see Figure 6.75) has controls for modulation, LFOs, filters, and envelopes, and you can switch the central area to display controls for the sample or synth used as the sound source. You can also switch between this Layer A page and an identical page for the second layer—Layer B—using the buttons marked 'A' and 'B', respectively.

Figure 6.75
Spectrasonics
Omnisphere Edit
page

The FX Page

The FX page (see Figure 6.76) has four 'racks' into which you can place any of the 33 effects available. These effects include limiters, compressors, EQ, distortion, phasing, flanging, delays, and reverberation.

Figure 6.76
Spectrasonics
Omnisphere FX

The Arpeggiator

One of Omnisphere's best features is the Arpeggiator's groove lock integration with Stylus RMX: groove lock enables the Arpeggiator patterns to instantly lock to the groove or feel of any RMX drum loop—see Figure 6.77.

Figure 6.77
Spectrasonics
Omnisphere
Arpeggiator

Live and Stack Modes

Omnisphere also offers a Live mode that facilitates instant patch switching and layering, and a Stack mode for powerful performance mapping.

IPAD CONTROLLER

The Omni 'Touch Remote' (TR) app for the iPad provides hands-on interactive performance for Spectrasonics instruments. With Omni TR, you can create Omnisphere setups and sound modifications on the fly, activate sounds, tweak filters, remix, bend, and spin amazing performance variations!

With its high-contrast interface, Omni TR is perfect onstage for live performance or even in the studio where the iPad sits at the controller keyboard located away from the studio computer. Best of all, the Omni TR app is extremely easy to use, offering full two-way communication with Omnisphere with just a simple wireless connection to the computer running the plug-in.

Trilian

Trilian is undoubtedly one of the most versatile virtual bass instruments available today, with its 34 GB core library of expressive acoustic and electric basses and cutting-edge synth bass tones. More than 60 different four-, five-, six-, and eight-string electric basses are included—in fingered, picked, fretless, slapped, tapped, and muted technique variations—providing a huge selection of sounds to suit any musical genre.

NOTE

Trilian makes an ideal partner for Omnisphere, as its complete sound library can be opened within Omnisphere, allowing Trilian's sounds to be used within Omnisphere to set up keyboard/bass splits—or for more detailed sound design.

Trilian uses acoustic and electric bass samples that have been captured with the highest levels of detail. For example, Trilian's stunning Acoustic Bass uses four audio channels and more than 21,000 samples to create this single instrument! This attention to detail has really paid off—Trilian's now legendary

Acoustic Bass is definitely the closest thing to a real acoustic bass that I have ever heard!

Here are some of the secrets of Trilian's success: on playback, extensive 'round-robin' variations of the sounds produce more natural-sounding bass lines with repeated notes. Also, Trilian automatically selects legato and release articulations as the user plays, providing a very dynamic and subtle playing experience. Multisampled dynamic slides let notes realistically slide from one to another and the balance between direct and phase-locked microphone/amplifier versions of the sounds can be adjusted to your personal taste. Using these controls, it is easy to create your own customized versions of any of the core library sounds and save these as user presets.

> ### TIP
> Highlights of the electric basses include a Music Man five-string studio bass, Chapman Stick, Lakland Rock P-Bass, Clean Fender Jazz Bass, a Hardcore Rock Bass, and a retro 1960s Epiphone Viola Bass.

Main Page

The custom controls interface on the main page of each patch (see Figure 6.78) presents the most useful and interesting sound modification controls for that specific patch. So, for example, the Acoustic Bass has controls for the mix between the microphone and pickup and any noises that you want to blend in. It also has controls for EQ and compression.

> ### NOTE
> Trilian's controls are all automatable and MIDI learnable and you can create your own configurations by Right-clicking on any control and choosing appropriately from the pop-up menu that appears.

Any one of these custom controls can be used to modify several parameters to create unique effects by repeatedly choosing the 'Assign this control to next touched audio control' command from the control's Right-click menu and touching the relevant control. With the Acoustic Bass, for example, you can balance the microphone and pickup levels together with the amount of 'noises' in the mix, or introduce some 'humanization' to make the intonation a little less than perfect—therefore more 'human' sounding.

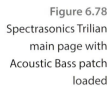

Figure 6.78
Spectrasonics Trilian
main page with
Acoustic Bass patch
loaded

Edit Page

Trilian's Edit page (see Figure 6.79) provides lots of controls for detailed sound design, including the 'FlexMod' modulation system, dual filters with lots and lots of filter types, versatile multistage looping envelopes, six full-featured LFOs, dual morphing modulation—and much more.

Hundreds of sound sources were sampled from more than 30 highly regarded hardware synthesizers—everything from the original Moog Taurus pedals to the latest boutique analog modular synthesizers. For synthesizer sounds, Trilian uses its four-pole 'Juicy' and 'Power' Filter algorithms with oscillating resonances—ideal for creating electronic bass sounds.

Figure 6.79
Spectrasonics Trilian
edit page

Effects Page

Trilian's FX racks (see Figure 6.80) include all of the 33 FX processors used in Omnisphere and Stylus RMX, including limiters, compressors, EQ, distortion, phasing, flanging, delays, and reverberation.

Figure 6.80
Spectrasonics Trilian
Effects page

The Arpeggiator

Trilian's Arpeggiator (see Figure 6.81) features the 'groove lock' technology used in Omnisphere and in Stylus RMX's Time Designer—so the feel of bass patterns created with Trilian's Arpeggiator can be perfectly synchronized between all three Spectrasonics instruments. Just drag the MIDI file from Stylus RMX to the Arpeggiator in Trilian and the two grooves will be locked together—and to Omnisphere.

Figure 6.81
Spectrasonics Trilian
Arpeggiator page

The Multi Page

Trilian also offers eight-part multi-timbral operation. You can play back each part on a separate MIDI channel if you like, or you can layer up multiple sounds and play these back on the same channel—see Figure 6.82.

Figure 6.82
Spectrasonics Trilian
Multi page

Live Mode and Stack Mode

Live mode lets you switch to different articulations using designated keys outside the range you are using to play. This lets you throw in a gliss or harmonic on the fly during a performance. Using Stack mode, you can create your own custom articulation switching using velocity, or do crossfades, or whatever.

Arturia V-Collection

This suite of virtual instrument plug-ins—see Figure 6.83—includes 10 software instruments: Mini V, Modular V, CS-80V, ARP2600 V, Prophet V and Prophet VS, Jupiter 8-V, Oberheim SEM V, Wurlitzer V, and Spark Vintage—a comprehensive collection of 30 classic drum machines (including the TR-808, TR-909, LinnDrum, and other favorites).

Moog Modular V

The first modular Moog appeared in 1965. Large modular synthesizers like these were expensive, cumbersome, and delicate, and not ideal for live performance. Nevertheless, several artists (including Emerson, Lake & Palmer; Klaus Schulze; Tangerine Dream; and Hideki Matsutake with the Yellow Magic Orchestra) successfully toured with Moog modular systems—despite their unsuitability for going on the road.

Figure 6.83
Arturia V-Collection
www.arturia.com

NOTABLE MOOG MODULAR USERS

Recording musicians Paul Beaver and Micky Dolenz used the Moog on The Monkees' fourth album, 'Pisces, Aquarius, Capricorn & Jones', released in November 1967. But it was Wendy Carlos' 1968 'Switched-On Bach', featuring Carlos' custom-built modular synthesizer as the only instrument on the recording, that truly announced the Moog synthesizer to the world. Shortly after, Keith Emerson, Jan Hammer, Tangerine Dream, The Beatles, and The Rolling Stones also bought modular Moogs.

Arturia, in partnership with Bob Moog, has made a faithful reproduction of the Moog Modular Synthesizer 55 that has authentically recreated the sound of the original, adding contemporary features including polyphony, MIDI control—and the all-important ability to save and reload the patches!

See Arturia's Modular V keyboard and control section in Figure 6.84.

A click on the small icons at the right of the toolbar opens up additional sections that contain the programmable modules—see Figure 6.85. Here you will find the oscillators, filters, envelopes, LFOs, effects, and sequencer controls that are used to create its sounds.

To save you from having to reinvent the wheel, Arturia provides a tremendous selection of presets created by several leading synthesizer programmers—so you won't need to 'roll your own' to get started. And I can guarantee that you will not be disappointed with the presets!

Figure 6.84
Modular V keyboard
and control section

Figure 6.85
Modular V modules

mini V

The minimoog monophonic analog synthesizer was invented by Bill Hemsath and Robert Moog. It was released in 1970 by R. A. Moog Inc. (Moog Music after 1972), and production was stopped in 1981 after sales of around 12,000 units had made it the most popular synthesizer throughout the '70s. The minimoog was designed for use in rock and pop music, incorporating the most important parts of the modular synthesizer in a compact package, without the need for patch cords.

With its 24dB/octave filter, its three oscillators, and tuning instabilities that tend to keep the oscillators moving against one another, the minimoog can produce extremely rich and powerful bass sounds, for which purpose it remains in high demand with producers and performers of electronic pop and dance music to this very day.

NOTABLE MINIMOOG USERS

Jazz composer and bandleader Sun Ra used a prototype minimoog that Moog loaned to him in 1969.

Keith Emerson was the first musician to tour with a minimoog, in 1970, during 'Emerson, Lake & Palmer's Pictures at an Exhibition' shows. Emerson was the first to use many of the characteristic pitch-bending techniques that minimoog players use, many of whom learned how to pitch bend by following his example.

'Yes', keyboardist Rick Wakeman said of the minimoog: 'For the first time you could go on [stage] and give the guitarist a run for his money.'

Funkadelic and Parliament both featured a minimoog played by Bernie Worrell, who used the minimoog to play the bass line on the hit song 'Flash Light', for example.

Kraftwerk's cofounder Ralf Hutter used a minimoog on the classic album 'Autobahn' and used it extensively on many subsequent albums including 'Man Machine' and 'Computer World'—inspiring an entire generation of electronic musicians.

Pink Floyd's Richard Wright used a minimoog on 'Shine on you Crazy Diamond'.

Manfred Mann's Earth Band made the minimoog an integral part of its sound in the mid-1970s.

Jan Hammer demonstrates pitch-bending technique using the wheel on Jeff Beck's album 'Wired'. The Mahavishnu Orchestra also featured keyboard playing by Jan Hammer, who prominently featured the minimoog on the album 'Birds Of Fire'.

Michael Jackson used a combination of two minimoogs to create the bass sound on the 'Thriller' album's title track, released in 1982.

Arturia, in partnership with Bob Moog, has recreated the minimoog in software as the mini V—see Figure 6.86. Of course, this now has MIDI, can save and load patches, and has other welcome additions including an advanced automation mode, an arpeggiator, chorus, and stereo delay effects, and vocal filter formant-based effects. As with the Modular V, the range of preset patches supplied is very broad and the quality of these is stunningly good.

Figure 6.86
Arturia miniV showing additional controls at the top of the window

ARP 2600 V

The ARP 2600 was a semi-modular analog subtractive audio synthesizer, designed by Alan R. Pearlman with Dennis Colin, and manufactured by his company, ARP Instruments, Inc. as the follow-on version of the ARP 2500. Unlike other modular systems of the time, which required users to individually purchase and wire the modules, the 2600 was semi-modular with a fixed selection of basic synthesizer components internally prewired. The 2600 was thus ideal for musicians new to synthesis because of its ability to be operated either with or without patch cords, and was, upon its initial release, heavily marketed to high schools, universities, and other educational facilities.

Three basic versions of the ARP 2600 were built during ARP's lifetime from 1971 until 1980. The first, dubbed the 'Blue Marvin', was housed in a light blue/grey metal case with a keyboard that mated to the synthesizer, and was assembled in a small facility on Kenneth Street in Newton Highlands, Massachusetts, during ARP's infancy as a company. It was often mistakenly referred to as 'Blue Meanie', but 'Marvin' is the correct name, after ARP's CEO Marvin Cohen. Later ARP 2600s used vinyl-covered wood construction with metal corners for both the synthesizer and keyboard, making it a more durable and portable instrument. Early versions contained an imitation of Robert Moog's famous four-pole 'ladder' VCF, later the subject of an infamous threatened (though ultimately nonexistent) lawsuit. Finally, to fit in with the black/orange theme of ARP's other synthesizers, the ARP 2600s were manufactured with orange labels over a black aluminum panel. The mid-production grey 2600 models featured many changes. Changes in circuitry and panel lettering provided at least three different grey panel models.

In 1972, ARP launched the Odyssey, which would be in direct competition with Moog Music and its minimoog released one year earlier. In 1976, ARP released a small 16-step sequencer in the form of two independent eight-step sequences. This became famous and is still very sought after (it is emulated in the ARP 2600 V.)

ARP was the market leader in synthesizers during the '70s with around 40 percent of the market share. Alan R. Pearlman was just as innovative as a salesman as a synthesizer designer. He provided synthesizers to well-known musicians, such as Pete Townshend, Stevie Wonder, and Herbie Hancock, each in exchange for his endorsement as a professional user. An ARP 2600 was even used to create the voice of R2-D2 in the *Star Wars* movies!

NOTABLE ARP 2600 USERS

Jean Michel Jarre, Ian Underwood with Frank Zappa, Steve Miller, and many other '70s musicians used the ARP 2600 on hit albums during the 1970s. Joe Zawinul usually played two with Weather Report, one for each hand, and Edgar Winter prominently featured the ARP 2600 on his classic single, 'Frankenstein'.

For the ARP 2600 V, Arturia has recreated the styling of the ARP 2600P version 4 produced in 1974—see Figure 6.87.

Figure 6.87
ARP 2600 V
keyboard and
sequencer controls

The upper section of the ARP 2600 V can be revealed by clicking on the Up button near the right of the toolbar. Here you will find the oscillators, filters, envelope controls, sample and hold, noise generator, and other subtractive synthesis building blocks together with chorus and delay effects—see Figure 6.88.

As with the Modular V, Arturia has provided a truly comprehensive set of preset patches programmed by expert users that can be used in many different musical contexts—and, of course, to provide sound effects for film scores and wherever creative new sounds are required.

Figure 6.88
ARP 2600 V

CS-80 V

The Yamaha CS-80 polyphonic analog synthesizer was first released in 1976 and production ceased in 1980 after about 3,000 had been manufactured. This was the heaviest synthesizer that I ever had the misfortune to carry in and out of a studio (always with another person to help) back in those days—it weighed around 200 pounds (91 kg), or more when with a case! My back still hurts today. . . .

The CS-80 caused a big stir among musicians when it first came out because it was one of the first relatively affordable truly polyphonic high-quality analog subtractive synthesizers. The CS-80 was eight-voice polyphonic (with two independent synthesizer layers per voice) and featured a primitive patch memory system that used a bank of micro potentiometers (rather than the digital programmable presets the Prophet-5 would feature soon after) to provide six user-programmable presets. The CS-80 had a total of 22 presets, including the six user presets, each of which could be selected using a front-panel button.

The 61-key weighted keyboard had excellent performance features. The splittable keyboard was both velocity sensitive, like a piano's, and pressure sensitive, with 'polyphonic' after touch that could be applied to each individual key—one of very few synthesizers that offered this feature (the Yamaha DX-1 the only other I am aware of). The CS-80 also had a ribbon controller allowing for polyphonic pitch bends and glissandos. This can be heard on the *Blade Runner* soundtrack by Vangelis, in which virtually all the sounds were created using a CS-80.

NOTABLE CS-80 USERS

The CS-80 was used by the BBC Radiophonic Workshop's Peter Howell in creating the 1980s version (used until 1985) of the *Doctor Who* theme music. Other famous users include The Who on their 1978 album 'Who Are You', Stevie Wonder, The Yellow Magic Orchestra, Peter Gabriel, the Electric Light Orchestra on the 1979 album 'Discovery', and Brian Eno, notably on his album 'Before and After Science'. Michael Jackson used the CS-80 on his album 'Thriller'; Michael McDonald used a CS-80 on 'What a Fool Believes'; and Toto used the CS-80 extensively, notably on 'Africa' and 'Rosanna' on Toto IV (the glissando effect is heard on the keyboard solo for 'Rosanna' and is also seen in the music video).

Arturia's CS-80 V—see Figure 6.89—provides all the features of the original CS-80 and also features a unique Multi mode and a modulation matrix to create entirely new sounds. The CS-80 V allows the creation of eight parallel voices, so, in theory, eight different sounds can be played at the same time. With Multi

Figure 6.89
CS-80 V

mode, you can assign each of these eight voices to four keyboard zones and to four different MIDI channels. The eight voices can also be superimposed across the entire keyboard, in Unison mode, to create very rich and expressive composite sounds. The CS-80 V comes with about 800 presets programmed by experts, with lots of great lead sounds, bass sounds, pads, and effects.

Prophet V

The Prophet-5 is a polyphonic analog synthesizer manufactured by Sequential Circuits in San Jose, California, between 1978 and 1984. Introduced at the Winter NAMM show in January 1978, the Prophet-5 broke new ground as one of the first analog synthesizers to implement patch memory: a feature that stored the user settings of every parameter on the synthesizer into internal memories. It was also one of the first polyphonic synthesizers, with a maximum polyphony of five voices (which means that up to five notes can play at the same time). Like the minimoog, the pitch wheel was not spring loaded, but had a dented mechanism that clicked as it was centered so that the musician would be aware of this.

Three revisions were produced, the first two (commonly referred to as Rev 1 and Rev 2) using oscillators manufactured by Solid State Music (SSM), and the last one (Revision 3) using Curtis CEM chips from Curtis Electromusic Specialties. The total number of Prophet-5 synthesizers manufactured, including all revisions, totaled almost 6,000 units.

The Prophet-5 was particularly valued for its poly-mod feature. This routed the output of the filter envelope generator and the second oscillator in each voice through two mixer knobs, which could then be connected to the pulse width and pitch controls on the first oscillator, to the filter cutoff frequency control, or to all three at the same time. Because the second VCO was not limited to being an LFO, this allowed the Prophet-5 to generate two-operator FM synthesis and ring modulator-style effects, as well as complex sweeping sounds.

The Prophet VS, a 61-note, five-octave, eight-voice polyphonic, programmable synthesizer with a velocity-sensitive keyboard introduced in 1987, uses 'Vector Synthesis' (VS). Capable of emulating a Prophet-5, an Oberheim, a PPG, or a Jupiter 8, the VS can also create sounds that are unmistakably DX7-like, and still others that are completely unique to it. The 'vector' in Vector Synthesis comes into play when the VS's joystick is manipulated. Its several tasks include providing real-time control over oscillator mix, with the joystick at the center of two axes, A-C and B-D. Wiggle the joystick to change the perceived level of the oscillators. During patch editing, the joystick can also be used to customize waveforms. You select a waveform for each oscillator slot, give it a frequency value (which multiplies a waveform's fundamental frequency), and tweak the joystick. Listen to the way the waveforms interact with each other—frequency values have an effect on this interaction—and when you hear something you like, store the result as a custom waveform. It's even possible to use custom waveforms as the basis for more custom waveforms, further increasing complexity and depth. Additional flexibility is offered by joystick control of each stage of the VS's mix envelope.

NOTABLE PROPHET-5 USERS

The Prophet-5 has been used by Thomas Dolby, the Doobie Brothers, on Michael McDonald's big hit 'Yah Mo B There', and by Wally Badarou, Vince Clarke, Soft Cell, Ryuichi Sakamoto, Pet Shop Boys' Paul Hardcastle, Pat Metheny, Phil Collins, Peter Gabriel, Kim Carnes, Depeche Mode, Dr. Dre, Donald Fagen, Eurythmics, Hall & Oates, Genesis, Soft Cell, The Prodigy, and too many others to mention. The Prophet VS has not been as widely used, but does add very useful new sound dimensions.

Arturia's Prophet V virtual instrument replicates the sounds of both the Prophet-5 and the Prophet VS. You can switch between these using the buttons provided in the toolbar at the top of the virtual instrument's window. The first button, marked '5', presents you with controls for all the features of the original Prophet-5—see Figure 6.90.

Figure 6.90
Prophet V keyboard
and controls

The second button, marked 'VS', presents you with controls for all the original features of the Prophet VS—see Figure 6.91.

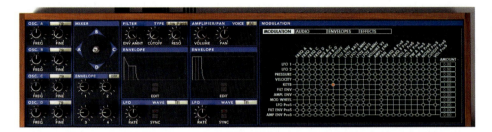

Figure 6.91
Prophet VS controls

The third button, marked 'HYBD' for Hybrid mode, presents both sets of synthesizer controls in a single window, with the VS controls above the Prophet-5 controls and the keyboard at the bottom.

> **NOTE**
>
> The hybrid interface offers a combination of the two synthesizers stacked one on top of the other. It permits you to create a large variety of sounds never before heard thanks to the mix of subtractive synthesis of the Prophet-5 and the wavetable synthesis of the Prophet VS.

Jupiter-8V

The Jupiter-8, or JP-8, is an eight-voice polyphonic analog subtractive synthesizer introduced by Roland Corporation in early 1981. Each voice features two VCOs with cross-modulation and sync, pulse-width modulation, a nonresonant high-pass filter, a resonant low-pass filter with two-pole (12 dB/octave) and four-pole (24 dB/octave) settings, an LFO with variable waveforms and routings, and two envelope generators (one invertible). Features include adjustable polyphonic portamento and a hold function for infinite sustain of notes and arpeggios.

The Jupiter-6 was released two years after the JP-8 as an attempt at a more affordable version of Roland's flagship. It features a similar voice architecture and appearance. It stored fewer patches and had six voices. The Roland MKS-80 'Super Jupiter' is a MIDI-controlled, rack-mountable sound module with a similar voice architecture to the Jupiter-8. However, its first released incarnation in 1984 (revision 3 and 4) used hardware identical to that of its predecessor, the Jupiter-6 (which had a combination of Curtis VCO and VCA chips combined with Roland's own proprietary filters). In 1985, Roland released another revision of the MKS-80, known as 'Rev 5', which used different VCO, VCA, and filter circuits. As a result, the MKS-80 Rev 5 can sound quite different from its predecessors. The Rev 5 filter was also used in the JX-8P, JX-10, and MKS-70 synthesizers.

NOTABLE JUPITER-8 USERS

The Jupiter-8 can be heard on Harold Faltermeyer's 'Axel F' (featured on the soundtrack for *Beverly Hills Cop*), on Enya's album 'The Celts', on *Frankie Goes to Hollywood*'s hit single 'Relax', on Dire Straits' 'Brothers in Arms' album, on Thomas Dolby's 'The Golden Age of Wireless', on Duran Duran's 'Hungry Like the Wolf', and on hits by Heaven 17, Human League, Howard Jones, and Giorgio Moroder. Queen used the Jupiter-8 on 'Queen on Fire—Live at the Bowl' (1982/2004), 'The Works' (1984), and 'A Kind of Magic' (1986); The Cars featured the Jupiter-8 on 'Heartbeat City' and 'Door to Door'; Simple Minds used Jupiter-8 sounds on 'New Gold Dream' and 'Don't You (Forget About Me)' and on most of their 'Once Upon a Time' album. Most recently, Lady Gaga used it on her hit album 'The Fame Monster'.

Arturia's Jupiter-8V—see Figure 6.92—recreates all the features of the original Jupiter-8 and adds a 32-step 'step sequencer'; the 'Galaxy' module (a special module that allows very complex triple-LFO modulations); and an effects interface that allows access to chorus/flanger, parametric EQ, phaser, and ring modulator effects. It comes with 400 factory presets.

Figure 6.92
Jupiter 8 V

Oberheim SEM V

In May 1974, Tom Oberheim showed the SEM (Synthesizer Expander Module), which he designed with the help of Dave Rossum of E-mu Systems, at the Audio Engineering Society Convention in L.A. Although SEM modules and other external controllers could be interconnected, the SEM is not a patchable synth module like other modular synths of the time: its signal routing was hardwired. The SEM itself was a keyboard-less module with two analog oscillators, a two-pole multimode filter, ADR envelopes, and an LFO. Each of the two oscillators offered triangle or rectangular waveforms. The SEM's 12dB/octave filter gave it a different sound to the ARP and Moog models. Unlike a 24dB/octave minimoog, ARP 2600, or Odyssey filter, it didn't self-oscillate, which made it impossible to create certain synthesizer sounds. However, the SEM filter offered four modes (low-pass, high-pass, band-pass, and notch), which allowed the SEM to deliver a diverse range of sounds not available from its competitors.

Oberheim entered the synth market in 1975—coupling SEMs with a keyboard and an analog sequencer into compact performance synths, the Oberheim 2-voice (TVS-1) and Oberheim 4-voice (FVS-1). Interestingly, the Oberheim 4-voice synthesizer was the one of the first electronic instruments to boast multi-timbrality. Oberheim's Polyphonic Synthesizer Programmer, introduced in 1976, could be used with the SEM and the 2-, 4-, and the Oberheim 8-voice model, which was released in 1977. Uniquely, this programmer could store the settings of a SEM module—which no other synthesizer manufacturer could offer at this time. Subsequently, Oberheim Electronics released the OB-1, OB-X, OB-Xa, OB-8, Xpander and Matrix6, and 12 models—all highly regarded and influential synthesizers of their time—until it closed down in 1985.

NOTABLE OBERHEIM SEM USERS

Oberheim SEMs have been used by Jan Hammer, Josef Zawinul, Herbie Hancock, and filmmaker/composer John Carpenter.

Faithfully reproducing the Oberheim SEM sounds, the Oberheim SEM V (see Figure 6.93) comes with a great collection of preset sound programs from several of the world's top synthesizer sound designers. The 12dB/octave multimode filter helped define the sonic identity of the Oberheim SEM because it shaped a diverse range of sounds almost impossible to create on other synthesizers. The SEM V carefully models the four modes of this filter (low-pass, high-pass, notch, and band-pass). The exact response of the ADS envelopes is also accurately reproduced, together with the VCF mixer, modulation knobs and switches—allowing you to nail any of the classic SEM tones. Reproducing this unique SEM architecture, with its oscillator sync, characteristic filter shape, and specific envelope response, the SEM V is perfectly able to generate the fat basses and other sounds that have made the Oberheim synthesizers so popular over the years.

The SEM V adds a host of additional features, including white noise to mix into your etheral pads and FX sounds; a sub-oscillator with sawtooth and sine waves, for added growl and bottom end; a second LFO with sine, sawtooth, and square waves, for more complex modulations; overdrive, delay, and chorus effects; an arpeggiator, for instant bass lines and inspiring melodic patterns; and portamento to unlock more expression in your playing!

The SEM V also adds three completely new modules to extend its sound palette even further, including keyboard follow, a voice programmer, and a modulation matrix.

- The modulation matrix offers eight simultaneous modulations that you can set between eight different sources and more than 25 destinations—helping the SEM V to deliver more powerful lead sounds and effects.

- Keyboard follow lets you draw modulation curves on the fly and you can map up to six parameters of your choice, providing even more expressive sound possibilities.

- Inspired by the multi-timbrality of the Oberheim 2-, 4-, and 8-voice models, the voice programmer allows you to edit no less than seven additional virtual SEM voices, making the SEM V equivalent to eight SEM modules! With this module, you can think of the Oberheim SEM V as an Oberheim 8-voice synthesizer, where each voice can play a different sound, opening up an entire world of multi-timbrality.

With its polyphonic capability (up to 32 voices in total), arpeggiator, voice programmer, modulation matrix, and effects, the SEM V can create an incredible array of new sounds, while keeping the ease of use and authentic original sounds of the Oberheim SEM architecture.

Figure 6.93
Oberheim SEM V

Wurlitzer V

The Wurlitzer electric piano was one of a series of electromechanical pianos manufactured and marketed by the Rudolph Wurlitzer Company of Corinth, Mississippi, and Tonawanda, New York. The earliest models were made in 1954 and the last model was made in 1984.

The Wurlitzer piano is a 64-note instrument with a keyboard range from A an octave above the lowest note of a standard 88-note piano to the C an octave below the top note of an 88-note piano.

Tone production in all models comprises a single steel reed for each key, activated by a miniature version of a conventional grand piano action and forming part of an electrostatic pickup system. Wurlitzer electric pianos use flat steel reeds struck by felt hammers. The reeds fit within a comb-like metal plate, and the reeds and plate together form an electrostatic or capacitive pickup system, using a DC voltage of 170v. This system produces a very distinctive tone—sweet and vibraphone-like when played gently, and developing a hollow resonance as the keys are played harder.

The earliest versions were the '100' series: these had a case made from painted fiberboard or wood and were fitted with a single loudspeaker mounted in the rear of the case. Models produced until the early 1960s used vacuum tube circuitry; the 140B was the first solid-state model, introduced in 1962. The model 145 was tube and came out around the same time as the 140 solid-state pianos. Both were replaced in 1968 by the much lighter, plastic-bodied 200, which had two small loudspeakers facing the player. This model was updated as the 200A in 1972 and continued in production until 1982. The 200 was available in black, dark 'Forest Green', red, or beige. The 200A was only available in black and avocado green. The Wurlitzer EP 200A electric piano is the main alternative to the Rhodes electric pianos and is used on high-quality recordings to this day.

NOTABLE RECORDS FEATURING WURLITZER ELECTRIC PIANOS

Ray Charles—'What'd I Say'

The Chantays—'Pipeline'

Marvin Gaye—'I Heard It through the Grapevine'

The Small Faces—'Tin Soldier', 'Lazy Sunday'

The Faces—'Stay With Me'

Queen—'You're My Best Friend'

Joni Mitchell—'Woodstock' on Ladies of the Canyon

Supertramp—'Dreamer'

Steely Dan—'Do It Again'

Three Dog Night—'Joy to the World', 'Mama Told Me Not to Come'

King Harvest—'Dancing in the Moonlight'

The Carpenters—'Top of the World', 'Superstar', 'Ticket to Ride'

Pink Floyd—'Time', 'Money', 'Have a Cigar', 'Shine on You Crazy Diamond (Parts 6–9)'

(Continued)

The Beatles—'I Am the Walrus'

The Rolling Stones—'Miss You'

Norah Jones—'What Am I to You'

The Wurlitzer V (see Figure 6.94) is a software recreation of the classic Wurlitzer 200A electric piano that uses physical modeling. The Wurlitzer V comes loaded with more than 200 presets and can also simulate Fender Rhodes and organ sounds, with examples provided among the presets. However, most of the presets offered are varieties on the Wurlitzer electric piano sound—ranging from 'Dry' and 'Distorted' to 'Chill' and 'Bizarre'.

Figure 6.94
Wurlitzer Electric
Piano V

As do the rest of the Arturia virtual instruments, the Wurlitzer V offers a number of additional features, including a vibrato rate control, a dynamics control, pickup distance and pickup axis controls, octave stretch, hammer hardness and hammer noise, 'note off' noise, sustain pedal noise, harmonic variation, and impedance controls—see Figure 6.95. There is also an adjustable velocity response curve and a 10-band graphic equalizer.

Figure 6.95
Wurlitzer Electric
Piano V Open

The Wurlitzer V includes emulations of 11 classic stompbox-type effects including screamer overdrive, analog phaser, wah wah pedal, autowah, analog flanger, compressor, chorus, vocal filter, analog delay, pitch shifter/chorus, and digital reverb. It also has four guitar amplifiers—Fender Deluxe Reverb Blackface, Fender Twin Reverb Blackface, Fender Bassman, and Marshall Plexi—and a Leslie speaker. Modeled microphones include the Shure SM57, Sennheiser MD 421, and Neumann U 87.

The Wurlitzer V lets you choose between three output modes: Studio, Stage, and Rotary.

In Studio mode, the Wurlitzer V and the effects are connected to a direct box that enables you to listen to the pure sound directly on the output—see Figure 6.96. A reverb follows the direct box, as would be typical in a studio.

Figure 6.96
Wurlitzer effect

In Stage mode, the concept was to recreate the sound of a Wurlitzer in a typical 'garage band' situation, so Arturia added a guitar amp simulator with multiple mic and speaker options and a spring reverb—see Figure 6.97.

The third output mode features a Leslie-type rotary speaker—see Figure 6.98. Rotary speakers are normally used with organs but this works great with the Wurlitzer V!

One of the many impressive things about the Wurlitzer V is the user interface design—a model of simplicity that is a joy to use. The control provided over the sounds goes much further than on the original instrument, and the addition of the modeled stomp boxes, guitar amps, and speakers, microphones, and reverb units plus the Leslie speaker makes this package hard to beat!

Figure 6.97
Wurlitzer amp

Analog Lab

Arturia has successfully emulated the Moog Modular, minimoog, ARP 2600, Yamaha CS-80, Roland Jupiter-8, Oberheim SEM, Wurlitzer EP 200A, Sequential Circuits Prophet-5, and Prophet VS. Arturia's virtual instrument versions of these are the Moog Modular V, minimoog V, ARP 2600 V, CS-80V, Jupiter-8 V, Oberheim SEM V, Wurlitzer V, Prophet V, and Prophet VS.

Figure 6.98
Wurlitzer Leslie

Analog Lab brings 5,000 presets from these classic instruments together into one virtual instrument with powerful browser facilities and data filtering that makes it easy to find presets of interest.

A basic level of sound editing capability is always available within Analog Lab but if you do own additional Arturia virtual instruments, Analog Lab also allows access to the full editing capabilities of each additional instrument.

The User Interface

The user interface (see Figure 6.99) is divided into three sections, with tools and buttons in a narrow 'toolbar' area along the top, and with the main interface underneath these. Below the main window is a controls and envelopes section with pitch and mod wheels, a master volume control, and 10 rotary and nine sliding controls that may be assigned to different parameters to control different presets. At the bottom of the window, a keyboard can optionally be displayed.

Figure 6.99
Analog Lab

The Main Window

The main window, below the toolbar, can be switched between three modes of operation: Sound, Multi, and Live.

The Sound page lets you browse through the sounds, then make basic edits to the sounds you have found using the controls provided by Analog Lab. You can make more detailed edits if you also own the Arturia virtual instruments that originally created the Analog Lab presets.

The Multi page allows you to quickly create splits and layers using simple drag and drop functionality.

The Live page allows you to organize your sounds and multis so that you can recall them quickly via program change messages.

Sound Mode

When you click the Sound tab at the top left of the window, this opens the Sound mode preset manager and displays a list of all the available presets. From this window you can select single presets and play them via the virtual MIDI keyboard display or via an external MIDI source.

All the patches from the different synthesizers are listed in the left half of the main window by default. You can click on any of the graphical representations of the various synthesizers pictured in the studio view in the right half of the window to restrict the list of presets to those of your chosen synthesizer—see Figure 6.100.

Figure 6.100
Analog Lab studio view

For more sophisticated filtering, click the button marked Filter View at the top right of the main window. See Figure 6.101. Here you can not only filter by instrument, but also by type (bass, pad, lead, etc.) and by characteristics (bright, dark, funky, bizarre, etc.).

When you find a preset that you like, just click on its name to load it. If it is highlighted, it has been loaded.

TIP

One of the great features of Analog Lab is that it is possible for any of its presets to be opened and edited using the original Arturia soft synth on which it was created, as long as you have purchased and installed a copy of that synthesizer on your computer. That being the case, clicking on the picture of that synthesizer inside the preset details window will open the current preset inside that synth.

Multi Mode

A multi is a combination of any two sounds that can be triggered at the same time from a single keyboard. They can be layered together or split, with one on the left side of the keyboard and the other on the right.

Each sound in the multi (known as a *part*) has an independent setting for level, stereo panning, two FX sends, and transposition (both chromatic and/or by octaves). You can also determine which part will be affected by MIDI messages such as pitch bend, mod wheel, aftertouch, sustain, and expression.

The multi itself allows you to select two different effects and edit them to suit your needs. There is also a master section with control over the levels of the FX returns and the overall level of the multi. And just like in Sound mode, there is a large selection of ready-made multis that you can use as is or as a basis for your own multis.

To get into Multi mode, click on the Multi tab at the left of the toolbar—see Figure 6.102. From this window you can select the multi presets and play them using the virtual keyboard or an external MIDI source. As with Sound mode, the selected multi will be highlighted in blue. Also as with sound mode, you can sort the list of multis according to the columns on the top of the Multi window: Preset Name, Favorite, Rating, Designer, Genre, and Factory/User.

The preset details window on the right side of the screen tells you a lot at a glance: Part 1 and Part 2 are clearly marked, and you can see the name of the sound that was used for each part inside the smaller windows. Part 1 is outlined in red, while Part 2 is outlined in yellow. This color-coding scheme is used elsewhere in the details window, such as in the 'MIDI and Split' section: you can see immediately that this is a split multi, with Part 1 covering the upper half of the keyboard (the thick red line) and Part 2 assigned to the lower half of the keyboard (the thick yellow line). The key ranges, MIDI channels, and transposition information for each part are also color coded in the boxes to the right of the small keyboard graphic.

Figure 6.102
Analog Lab multi setup

Live Mode

There's a third performance option for Analog Lab positioned next to the Sound and Multi mode buttons: Live mode. This mode allows you to construct a set list, for example, by assigning sounds and/or multis to particular MIDI program change numbers for instant recall from a MIDI controller keyboard or from MIDI tracks in Pro Tools.

Click on the Live tab at the left of the toolbar to open the Live mode display. There are three main areas here: the preset selection section in the left half of the window, the sound/song details section at the top right, and the program map in the lower part of the right-hand section—see Figure 6.103.

Figure 6.103
Analog Lab live
setup

In the preset selection section you can select a multi or sound preset from the respective preset list and drop it into the program map.

In the sound/song details section you can view details about the current sound or multi and enter the name of an associated song.

Mostly, you will be setting up the program map in this display. You can drag and drop multis and sounds to the program map in any order you like so that they can be called up using MIDI program change numbers. There are 128 MIDI program change numbers, so the program map has 128 slots for you to fill with your sounds and multis.

First decide whether you want to place a multi or sound preset into your set list by selecting one or the other category at the top of the preset list. Then click on any of the preset names and drag it over to the program map section on the right side of the Live mode window. To select a different item, either drag another sound from the sound presets list or switch to the multi category and drag and drop a multi preset.

For example, you might place Synchro_Bass at the top of the program map. Now when you want to select that sound from an external MIDI source or a DAW, just make sure that Analog Lab is in Live mode and send it MIDI Program Change #1. Synchro_Bass will be selected. Then you might place a multi preset into program map slot 2. MIDI Program Change #2 will select that item, and so on.

Saving Presets

If you own the instrument you have been editing, then once a preset has been modified to fit your needs it can be saved as a user preset. After this you'll be able to recall it as easily as any other preset. You can even add filtering options so it shows up along with other presets of similar characteristics.

When you click the Save As button in the toolbar, a dialog window will appear asking you for the name of your preset. Type its new name and hit 'Save As'. You will now see your new preset in the preset list of the Sound Search window.

The Controls and Envelopes

The controls and envelopes section presents the most important and useful parameters of the selected preset, ready for you to tweak. The rotary controls are assigned to parameters that control the oscillators, filters, and so forth, while the sliders control the envelopes.

Snapshots Tab

The snapshot feature gives you even faster access to your favorite presets. If you find yourself frequently using a particular sound or multi, save it as a snapshot. You can store up to 10 snapshots, depending on the selected controller type, and they will be reloaded when you launch Analog Lab for instant recall.

You'll find the Snapshot tab at the left of the virtual keyboard's top panel. When you click on the blue tab area, this will open the snapshots section. This section opens to the right, obscuring the performance wheels and some of the rotary controls. Just click the blue tab to put it away again.

To store the current preset as a snapshot, simply drag it onto one of the snapshot buttons. The same holds true for the multis: grab one or more favorites and drag each to its own snapshot button. Now you can recall it at any time by simply clicking on the same button, even after quitting and restarting the software.

The snapshots can be easily accessed from a MIDI controller. So the user can store 10 favorite presets, together with any modifications, and recall these at the touch of a button during a live session without referencing the computer screen. This feature is especially useful when an artist plays live.

It can also be useful for comparing modifications to sounds to choose which fits best into a musical production. You could, for example, store 10 different states of the same preset: first state or original, second with the cutoff applied, third state using the delay, and so forth.

Chord Pads

As with the Snapshots tab, the vertical blue label on the right side of the keyboard hides the pad group for the virtual keyboard. The number of available pads varies for each keyboard option. Click this label and the snapshots panel will open, revealing the snapshot buttons. Click it again and the pads panel will close.

The pads allow you to specify and trigger chords when you click the pad. To do this, click inside each field and select the values you want: for example, if you want the pad to trigger a C Maj 7 chord in the middle of the keyboard range, select C3 from the Chord Root menu and Maj7 from the Chord Type menu.

Virtual Keyboard

Analog Lab has a virtual keyboard at the bottom of the window underneath the controls section. The keyboard section can be hidden when not in use by clicking on the KEYB button in the toolbar.

The default keyboard controller section has the largest number of controls: 10 knobs, nine sliders, 10 snapshot buttons, and 16 pads for chords. Keyboard controller interfaces are also provided for Arturia's various hardware controllers.

Finally . . .

If you cannot justify buying the whole suite of Arturia virtual instruments, you can just buy the Analog Laboratory to get access to thousands of the best presets: this single virtual instrument gives you access to the great bass sounds of the Moog Modular, the brass and strings of the Prophet, the pads and FX of the ARP 2600, and the rest!

Spark Vintage

Instrument manufacturers first developed drum machines back in the 1970s when devices such as the Roland Rhythm 77, also known as the 'Transistor Rhythm' TR77 (see Figure 6.104), Roland's first commercial product that was first available in 1972, were sold as accessories for people who played home organs. I bought one of these in the late '70s to use while practicing and writing songs.

Figure 6.104
Roland Transistor
Rhythm TR77

These Roland devices, or similar models, quickly found their way onto popular recordings of that time, most famously Timmy Thomas's 1973 hit, 'Why can't we live together', and George McCrae's 1974 hit, 'Rock Your Baby'. They were also used by pop and rock groups such as Ultravox with 'Hiroshima Mon Amour', and even Phil Collins—who used the Disco-2 pattern on the similar-sounding Roland CR78 on the introduction to his 1979 recorded hit single in 1981, 'In the Air Tonight'. Roland continued to develop drum machines throughout the 1980s, including the TR808, TR909, and TR707 and 727 models. Other manufacturers followed suit and released classics such as the Linn Drum and the Emu SP12.

The sounds of all these vintage models, and many more, are available again— courtesy of Arturia—which introduced the Spark Creative Drum Machine in 2011. Spark Vintage not only includes emulations of the classic analog drum machines, but also includes acoustic drums, physical models, and powerful electronic kits—all combined in one easy-to-use interface.

With in-depth control over your sounds, eight velocity-sensitive touch pads, an advanced loop mode, and an XY touchpad with eight real-time effects,

Spark allows you to create unique beats using its three separate drum engines: analog synthesis, sampling, and physical modeling. Automation is available on all parameters and every kit is easily tunable and customizable.

Aimed at the producer and the live beat maker, Spark offers a unique workflow mixing the 16-step style of sequencer programming with the live approach of pads, filtering, slicing, and looping functions.

The User Interface

Spark's user interface is divided into three main panels: top, center, and bottom. The center panel appears by default and you can bring the top or bottom panels into view by clicking on the relevant button in the toolbar at the top of the window.

The Center Panel—The Spark Controller Panel

The center panel, Spark's controller panel, as its name implies, is where you will find Spark's main controls—see Figure 6.105. There's a row of sequencer step buttons at the top, with a digital display below this. To the right of the digital

Figure 6.105
Spark Vintage

display is a circular arrangement of controls for songs and patterns with a jog wheel in the center that can be used to make selections. In the lower half is a row of global controls and buttons, with eight sets of individual controls and pads for the drum sounds below these.

The Top Panel

When you click on the Top button in the toolbar the pattern window opens by default. The top panel lets you display the pattern, song, or preferences panels. At the bottom of this panel you will find buttons for the pattern panel and the song panel button at the left with a Preferences button at the right.

The Pattern Panel

The pattern panel (see Figure 6.106) can be used to create or edit sequences. When you open this panel, you can only see eight of the 16 available instrument tracks. To see the other eight, you will need to use the scroll bar.

The sequencer is very easy to use—press Play in the Spark controller to run the sequencer, then just click in any of the beat subdivisions for any of the instruments to enable that note to play in the sequence. Just click again to stop the note playing. So you might work on the bass drum first, followed by the snare drum, adding a hi-hat later—and you can just work out your patterns by trial and error until they sound good to your ear!

Figure 6.106
Spark Vintage drum machines (top section) showing the pattern panel

Now how many beat subdivisions are there in a bar? Well, in 4/4 time signature, one bar can be subdivided into four quarter notes and each quarter note can be subdivided in turn into four 16th note steps. So there are a total of sixteen 16th note steps in a bar—which is exactly the number of steps available in Spark's sequencer....

The Song Panel

A song is an editable ordered sequence of several patterns that defines the structure of your song. The song panel (see Figure 6.107) lets you chain up to a maximum of 64 patterns together to create a song.

Figure 6.107
Spark Vintage drum
machines (top
section) showing
the song panel

The Bottom Panel

The bottom panel features the studio, the mixer, and the library, with buttons to let you switch between these located at the bottom left of the window.

The Studio Panel

The studio panel (see Figure 6.108) displays all your instruments with their parameters, lets you load instruments into slots, and lets you apply filters and effects.

323

Figure 6.108
Spark Vintage drum machines (bottom section) showing the studio panel

The Mixer Panel

The mixer panel (see Figure 6.109) has 16 mixer channel strips, with two auxiliary returns and a master track at the right of the window. You can insert up to two effects on each channel. Aux 1 sends to the reverb by default, and Aux 2 sends to the delay.

Figure 6.109
Spark Vintage drum machines (bottom section) showing the mixer panel

The Library Panel

The library panel is divided into two main sections: the current project section in the upper half of the window and the library section in the lower half—see Figure 6.110.

Figure 6.110
Spark Vintage drum machines (bottom section) showing the library panel

The library section is further subdivided into library, project, bank, pattern, and instruments sections.

To create a new project, click on the New Project button at the left of the library section.

To load a factory or user project into the current project, double-click on an instrument kit in the project section of the Library window. This will load the entire kit to your current project and switch the current project in the upper half of the Library window to its edit display—see Figure 6.111.

Figure 6.111
Spark Vintage drum machines (bottom section) showing the current project panel

To load one or more instruments from a kit, go to the instruments section of the Library window and double-click on an instrument to load it into your project. When you double-click on an instrument, it is loaded into whichever slot is currently selected in the loaded instruments section of the Current Project Edit window. To select a slot, click on the slot number in the loaded instruments section.

> **TIP**
>
> An easier way to load an instrument is to drag and drop the instrument from the Instruments window to any slot in the Loaded Instruments window.

You can also load banks or patterns from the library section to the current project section.

- To load a bank or a pattern from the library to the same bank or same pattern number in your current project just double-click on it.

- To load Bank A from the Library to Bank D of your current project, drag and drop Bank A on to Bank D.

- To load pattern A1 to pattern C3 of your current project, drag and drop A1 to C3. Pattern C3 will light up to confirm your selection.

Spark DubStep

Spark DubStep (see Figure 6.112) is squarely aimed at DubStep producers. Featuring a comprehensive library of 30 kits and 480 instruments created in partnership with Sample Magic, Spark DubStep gives DubStep producers everything they need—from atomic wobble basses to filthy FX and dirty drums—in one easy-to-use interface. The kits range from 'pure wobble anthems to revisited EDM and Hip Hop tracks on dub steroids' (as Arturia puts it) and there are more than 960 preprogrammed MIDI patterns that you can customize to your needs.

Using Spark's integrated 16-track mixer, you can automate levels, pans, and sends and fine-tune your mix using a useful complement of effects including multiband compressor, reverb, sub-generator, destroyer, bit crusher, multiband EQ, chorus, delay, distortion, phaser, plate reverb, flanger, space pan, limiter, EQ 10 bands, analog delay, pitch shift chorus, vocal filter, analog chorus, and Leslie.

NOTE

Perfectly integrated with Pro Tools, each Spark DubStep virtual instrument can route up to 16 independent audio outputs into the Pro Tools mixer if more sophisticated mixing capabilities are required.

TIP

To export your patterns in .wav and MIDI directly into your Pro Tools session, simply drag and drop any pattern from your current kit and it will be automatically rendered in the format of your choice.

Spark DubStep has lots of great features for 'live' performance as well: you can play the drum sounds using any MIDI keyboard or drum pads or using Spark's FX Pad, Looper, and Slicer features; and real-time automation is available for all parameters.

Figure 6.112
Spark DubStep

With its easy-to-program pattern and song sequencers, Spark DubStep provides its users with a complete rhythm track construction kit for building the rawest-sounding DubStep tracks.

Spark EDM

Put together especially for electronic dance music producers, Spark EDM (see Figure 6.113) contains 30 kits with 480 instruments and 960 patterns of hard-hitting drums and filter-modulated mayhem.

With everything from big room floor-filling stormers to underground tech house gems, Spark EDM has all the drums, stabs, one shots, and effects that you need to compete with successful producers creating dance floor hits.

Spark EDM uses samples and custom-programmed virtual analog synths, drums, and effects to provide its authentic, high-quality sounds. Style categories include house, dub, rock, electro, creative, and FX.

With its easy-to-program pattern and song sequencers, Spark EDM provides users with a complete rhythm track construction kit for building 'hit-sounding' EDM tracks.

Figure 6.113
Spark EDM

Summary

AIR Music Technology's suite of virtual instruments was designed specifically for Pro Tools systems. All of the instruments have first-rate technical specifications, plenty of well-designed preset sounds, and user-friendly and efficient interfaces. Velvet has the best Fender Rhodes simulation I have heard, and I am looking forward to working more extensively with the two most recent additions—Vacuum Pro and Loom—now that I have sampled their 'delights'!

The Spectrasonics suite is just perfect when you are searching for bass, drums, and a main keyboard pad for your next song production, and it's just as useful if you are working on music to picture, with masses of percussion sounds, exotic bass sounds, synthesizer effects, drones, ambiences, and sequences.

Now as far as Arturia's V-Collection is concerned, all I can say is that 'No professional studio should be without one of these'! The number of hit records that the original versions of these classic instruments have helped create is truly legendary. The Wurlitzer piano sounds excellent and the classic drum machines recreated for Spark Vintage include all the most popular models.

in this chapter

CHAPTER 7

ReWire Applications

ReWire

Propellerhead Software developed ReWire to make it possible to transfer MIDI and audio data between applications running on the same computer without using any external connections—ReWire makes virtual connections internally.

So how does ReWire let you route audio into Pro Tools? Well, compatible ReWire client applications are added to the list available and Pro Tools and its plug-ins automatically detect them. These ReWire plug-ins then allow you to route audio from the ReWire 'client' application directly into your Pro Tools mixer channels.

> **NOTE**
>
> When you insert a ReWire plug-in into the Pro Tools mixer, Pro Tools automatically launches the corresponding ReWire 'client' application, such as Reason, if the client application supports this feature. If not, as with Live, you must launch the ReWire application manually.

And how does the MIDI work? It couldn't be easier for the user: when the ReWire 'client' application has been launched, the MIDI inputs for that application automatically become available as destinations in the MIDI track output selectors in Pro Tools.

The timing of the linked applications (Pro Tools and the ReWire client) is synchronized with sample accuracy and you can use the transport controls on either application to control the other. Pro Tools transmits both tempo and meter data to the ReWire client application, allowing the ReWire application's sequencers to follow any tempo and meter changes in the Pro Tools session. For example, you may have recorded sequences using Reason that you want to play back in sync with your Pro Tools session.

> **NOTE**
>
> With the Pro Tools Conductor button selected, Pro Tools always acts as the tempo master, using the tempo map defined in its tempo ruler. With the Pro Tools Conductor button deselected, the ReWire client acts as the tempo master. In both cases, playback can be started or stopped in either application.

Of course, once the audio outputs from Reason or Live are routed into Pro Tools you can process these incoming audio signals with plug-ins, automate volume, pan, and plug-in controls, and record to Pro Tools tracks or use the 'Bounce to Disk' command to 'fix' the incoming audio as files on disk.

ReWire 2 Features

■ Real-time streaming of up to 256 individual audio channels (up to 64 with ReWire 1) and up to 4,080 individual MIDI channels (255 MIDI buses with 16 channel per bus) from one application to another.

■ Common transport functions—if both applications have built-in sequencers of some sort, you can play, stop, rewind, and so forth in any of the applications and they will both locate to the same position.

■ High-precision synchronization—complete, glitch-free sync between the two applications, with no settings to make and no parameters to worry about.

■ Additional querying—one application can ask the other about audio channel names and so forth.

Looping Playback when Using ReWire

You can set a playback loop either in Pro Tools or in the ReWire client: if you want to loop playback in Pro Tools, just click and hold the mouse button and drag across the ruler in the Edit window to select the time range that you want to loop in the Pro Tools timeline before starting playback. Or you can set loop or playback markers in the ReWire client sequencer before you start playback.

> **NOTE**
>
> If you create a playback loop by making a selection in the Pro Tools timeline, once playback is started, any changes made to loop or playback markers within the ReWire client application will deselect the Pro Tools timeline selection and remove the loop.
>
> Also, some ReWire client applications, such as Reason, may misinterpret Pro Tools meter changes, resulting in mismatched locate points and other unexpected behavior. So you should be careful about using meter changes in Pro Tools when you are using Reason as a ReWire client.

Reason and Live

Lots of creative musicians use Reason or Live, or both of these, especially when they are songwriting at home or on the road with their iMacs or laptops. Why? Well, you get so much bang for your buck with both of these! They are both relatively affordable and have just about everything you would want for making music on your computer without needing a studio full of hardware!

Reason is a total powerhouse for music production with its virtual rack of interesting MIDI-based synthesizers, samplers, drum machines, and effects modules and its own sequencers. It lets you import audio from your music library and Web formats, and its factory sound bank comes fully loaded with tons of loops and drum kits. You can even turn your own recordings into REX loops and play these back in Reason. You can integrate your own MIDI instruments into Reason using its MIDI out capabilities, so you can use a MIDI keyboard and any other MIDI hardware that you already have. You can record audio tracks as well, so you can add vocals, guitars, or whatever you like, and it has great audio slicing and audio quantize features. Reason also has its own fully fledged mixing desk—modeled after the legendary SSL 9000k console. The mixer comes complete with flexible routing, full dynamics, EQ, advanced effects handling, full automation, groups and parallel channels, master-bus compression, and a spectrum analyzer window with visual EQ.

Ableton Live is essentially an incredibly flexible sequencer with a vast library of sampled and synthesized sounds that can be used to create, modify, and play back loops, phrases, and songs using its 'elastic audio' technology. Live also allows you to import samples and loops from other libraries. Change the tempo, and the loops change tempo automatically. Choose a new loop running at a particular tempo, and this will automatically run at the tempo of your sequence—Live's advanced algorithms for stretching audio are ideal for working with individual tracks or stereo mixes. And Live's MIDI features will let you do just about anything that's possible—you can even extract harmony, melody, or drums from an audio sample and convert it to MIDI. Like Reason, it also lets you work with external MIDI equipment, and you can record and play back linear audio tracks for vocals, guitar, and any other instruments to run alongside your loops and samples. The standard Live package comes with 11 GB of samples and loops, three virtual instruments, and 37 audio and MIDI effects, while the Suite version has a whopping 54 GB of sounds, nine virtual instruments, and 40 audio and MIDI effects and also includes 14 Max MIDI and audio effects! Live also supports VST and AU plug-ins.

Reason and Live both run as stand-alone applications that connect to Pro Tools using Propellerhead's ReWire technology. So you can use Reason or Live for loop- and sample-based composition and have the audio from both piped directly into your Pro Tools sessions. You can even use both at the same time within the same Pro Tools session—although, predictably, this works best on faster CPUs with plenty of RAM.

It is also possible to connect Live and Reason using ReWire—and to most other popular DAWs including Logic, Cubase, Nuendo, Digital Performer, and Sonar, which all support ReWire.

Reason

Overview

Reason is a collection of virtual instruments and audio processors that can be assembled in a window that looks like a rack of MIDI and audio hardware—see Figure 7.1. You can choose what to put into this 'rack' from a menu that contains a range of MIDI synthesizers, samplers, pattern sequencers, drum machines, effects devices, and a mixer.

Figure 7.1
Reason rack showing a small selection of the available devices

Reason has its own main sequencer—see Figure 7.2. Although quite easy to learn, this can be a bit 'fiddly' to use—and its features are no match for those in Pro Tools (or Digital Performer, or Cubase, or Logic, or Sonar).

Figure 7.2
Reason sequencer

All Reason's rack devices operate similarly to their hardware equivalents. For example, The Echo delay (see Figure 7.3) closely resembles a Roland Space Echo. Sound modules include the Thor polyphonic synthesizer, Subtractor analog and Malström wavetable synthesizers, NN19 and NN-XT sampler players, ID8 Instrument Device (pianos, guitar, organ, drums, bass, strings, and other basic

Figure 7.3
The Echo

335

sounds), Kong Drum Designer (16 pads with 16 sounds using analog synthesis, physical modeling, sampling, and REX loops), Dr Octo Rex Loop Player, and Redrum—a TR808-style drum machine. The RPG-8 Arpeggiator lets you create arpeggiated melodic lines from your chords, using its play mode, note insert, and rhythmic pattern features. Reason also provides a wide range of effects including reverbs and delays—and even a vocoder.

The combinator rack module can contain a selection of Reason devices along with associated effects and modulation routings. You can save these combinator patches as 'combis' so that you can instantly recall any setup, whether a complex chain of effects or a split or layered multi-instrument. For example, the MClass mastering suite (see Figure 7.4) is a ready-made combi containing a suite of mastering tools—a four-band EQ, a stereo imager, a compressor with side-chain and soft-knee options, and a maximizer to make your tracks louder.

Figure 7.4
Reason MClass
Mastering Suite
processors

You cannot use VST, AU, or other plug-in formats with Reason, but Propellerhead does allow third-party developers to offer Reason rack extensions, which you can add to Reason. Because these are designed to slot right into Reason's rack, you can load them into combinators, route cables around the back, automate all the parameters, and work with them just as though they were original Reason rack components. So, for example, you can add high-quality signal processors from McDSP and iZotope, or synthesizers from Rob Papen, Korg, or Synapse.

Working with Reason

In Reason, you build your rack as you work on your song, creating instruments and effects and routing them any way you want to. You can add, delete, replace, or tweak anything whenever you like. Just press the Tab key on your computer keyboard to flip the rack over to access the back of the rack with all the cables and connectors. Here you can manually route anything to anything, just like in a real studio. Or you can simply leave it to Reason to automatically connect all your instruments and effects!

All sample players in Reason are samplers. Just hook up a sound source to the rack's sampling input and you are ready to start sampling. Use a mic, a turntable, or an instrument or grab the entire Reason mix. This 'live' sampling capability, together with pitch detection of root key and automatic zone mapping, makes it very easy for you to sample an instrument and map the samples across the keyboard—so you can create your own multisampled instruments for NN-XT or NN-19 really fast.

Adding effects in Reason couldn't be much easier: each instrument in the Reason rack has its own mix channel device. The mix channel has a slot for effects and that's where you drag the effects you want to apply. If you want to change the routing manually, hit the Tab key to access the back of the rack where you can connect anything to anything. Adding effects to the mixing consoles effect sends is just as easy—just Right-click the FX (effects) return section and select the effect you want to add, for example, the Scream 4 Sound Destruction Unit, which is a 'sound mangler' that lets you add all kinds of distortion, or you can use the pulverizer demolition effect, which combines distortion with compression and modulation effects. For serious sound 'sculpting', the Alligator Triple Filtered Gate multiband processor offers pattern-controlled gating, filtering, and envelope shaping for three parallel bands. To tighten up vocals you can use the Neptune Pitch Adjustor, and to keep the guitar and bass players happy, Reason comes with Line 6 guitar and bass POD units installed!

Reason's mixing console (see Figure 7.5) features a warm-sounding EQ section with high- and low-pass filters and high and low shelving filters as well as parametric midrange filters.

> **TIP**
> You can hide or reveal the different main mixer sections using the small buttons located at the lower right of the Main Mixer window.

Figure 7.5
Reason mixer showing the dynamics and fader sections with the master section controls selected (outlined in blue)

Every channel has its own fully featured dynamics section with compressor, gate, and expander and the console has eight dedicated effect sends, each of which can be set to work pre- or post-fader. Group channels give you access to the same processing tools as on individual channels, including sends and inserts, and make it easy to balance the levels of all your instruments.

Reason's Spectrum EQ window (see Figure 7.6) is a very cool feature. This combines a powerful spectrum analyzer with a graphic EQ overlay that lets

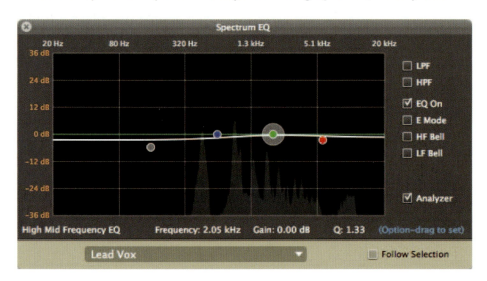

Figure 7.6
Reason Spectrum EQ

you adjust the mixer's EQ, giving you control with instant visual feedback when you select a channel or group—or your final mix. You can open or close this window using Reason's Window menu—or by pressing the F2 key on your computer's keyboard.

When your mix is almost ready, the built-in SSL master bus compressor lets you 'glue' it all together while adding more punch to your drums, more warmth to your bass, and more sparkle to your overall mix.

> **TIP**
>
> Although you can have the Reason rack, sequencer, and mixer all contained within one window, this can get very unwieldy to work with, especially when you are using lots of instruments and effects. To help you to get more organized, especially if you have a large screen, or even two screens attached to your computer, Reason has window menu options that allow you to detach the main mixer, and/or the rack, from the main Reason window—which always contains the Reason sequencer.

So Why ReWire Reason to Pro Tools?

Now that you know about all the wonderful features that Reason has to offer, you may well be wondering why anyone who has this would ever want to use anything else! Well, if you are at home working alone, or in a small studio with your band, you may never need anything else.

But if you want to take your music a lot further, working with other songwriters, producers, arrangers, and recording engineers—and especially if you go to a large professional studio—you are likely to find that many of them will be using Pro Tools.

Reason has become extremely successful as a suite of affordable, popular virtual instruments that can be used alone to develop instrumental arrangements that are then incorporated into more ambitious projects in Pro Tools, where vocals and live instruments can be added and additional virtual instruments and plug-ins can be used to complete the production. And even if you have recorded vocals and live instruments at home or in a demo studio using Reason, you may well wish to rerecord these using better studio facilities elsewhere.

Also, many very talented and creative musicians, arrangers, and producers have started out using Reason and know how to get the best out of this and how to work with it very quickly and efficiently. Rewiring Reason into Pro Tools allows Reason users to work with Pro Tools specialists to get the best of both worlds!

Using Reason with Pro Tools

Step 1. Launch Pro Tools first and open a new or existing session.

Step 2. Add a MIDI track that you will use to play back one of Reason's rack of synthesizers and an auxiliary input or audio track to receive the audio output from this synthesizer—or add an instrument track that handles the MIDI and the audio monitoring on the one vtrack.

> **NOTE**
>
> It is best to use instrument tracks to communicate with individual Reason instruments. If you intend to use several Reason instruments (each controlled by a separate Pro Tools MIDI track) but with all the audio from Reason mixed inside Reason, then you should route Reason's Mix L and Mix R audio outputs to a stereo Pro Tools auxiliary track. If you want to have separate audio outputs from each Reason instrument AND control these separately from individual Pro Tools tracks, then use an instrument track in Pro Tools for each Reason instrument.

Step 3. In the Mix window, insert the Reason ReWire plug-in using the audio or instrument track's inserts section. Reason will launch its default song automatically in the background as soon as you insert the ReWire plug-in.

Step 4. The ReWire plug-in's window will also open—ready for you to choose the outputs that you want to use from Reason—see Figure 7.7. Select the outputs that you want to use by clicking on the pop-up list located at the lower right of the plug-in's window.

Figure 7.7
Reason ReWire plug-in showing Reason's Mix L, Mix R, Channel 3, and Channel 4 (of 64) outputs

Follow this procedure until you have all the MIDI or instrument tracks that you need set up in Pro Tools to control the Reason instruments that you want to use from Pro Tools, with the audio routed from Reason into one or more Pro Tools instrument or auxiliary tracks.

Setting up the Reason Synthesizers

Normally, when you launch Reason, the standard default song document opens. This contains a mixer, various effects modules, and several Reason synthesizer modules—along with the demo song sequences.

If you are using Reason with Pro Tools, it can make sense to avoid using Reason's mixers and effects, because you can use the much more full-featured mixer and higher-quality effects in Pro Tools itself.

In this case you can prepare a custom default song containing just the Reason devices you plan to use, with their outputs connected directly to the Reason hardware interface's audio outputs. This hardware interface then feeds Reason's audio outputs directly into Pro Tools via ReWire.

Step 1. To set this up, you should set the general preferences in Reason to open an empty rack first—see Figure 7.8.

Figure 7.8
Reason general
preferences

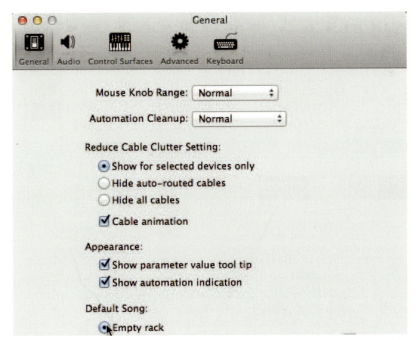

Step 2. Close the default song, if it is open, and select 'New' from Reason's File menu to open a new empty Reason rack.

Step 3. Add a selection of devices to the rack using Reason's Create menu. Each time you create a new device in your Reason rack, a sequencer track with its output set to this device will be automatically added to Reason's main sequencer—as in Figure 7.9.

Figure 7.9
Reason rack with Kong drum designer and Thor polysonic synthesizer modules added and with the sequencer window showing empty tracks for these

TIP

If you are not intending to use Reason's main sequencer, you can click and drag on the dividing line between the sequencer and the modules above it until it is completely hidden from view. You can also collapse Reason's transport controls by clicking the small arrowhead to the far right of the controls.

Step 4. Press the Tab key on your computer keyboard to reveal the back of the Reason rack—see Figure 7.10. Here you can choose which modules are connected to which outputs.

Figure 7.10
Rear of Reason rack showing module output connections

Step 5. Connect each module in turn to individual Reason outputs. You can then route these to individual Pro Tools mixer inputs using ReWire.

> **TIP**
> You may prefer to mix the audio outputs of the various Reason modules inside Reason using one of its internal mixers—the 6:2 line mixer or the

(Continued)

343

14:2 mixer. If so, add this module to your Reason rack, before connecting any modules to outputs, and connect the Reason modules' outputs to this mixer instead—see Figure 7.11.

Then connect the Reason master section outputs to, say, Reason audio outputs 1 and 2, and route these to a stereo Pro Tools track input using ReWire. This way, you can audition all your Reason modules while setting them up and choosing their sounds without having to switch tracks in Pro Tools: just choose a different Reason module as the MIDI destination each time you want to listen to this.

Figure 7.11
Module outputs routed to the Reason 14:2 mixer, which is routed via the main mixer's master section to Reason audio outputs 1 and 2

Step 6. If you will not be using the Reason sequencer, then it makes a lot of sense to hide the sequencer section and the transport controls in Reason.

Step 7. When you have set up your Reason rack the way you want it, you should save this into your Reason folder. You might call this 'MyReasonSynths'—or whatever works for you.

Step 8. Now change Reason's general preferences (see Figure 7.12) to open a custom song as default. In the Preferences window, select the 'MyReasonSynths' song document by clicking on the small folder icon at the right of the Preferences window and navigating through your disk drives and folders until you find the 'MyReasonSynths' song document, which will be in whichever folder you saved it to.

Figure 7.12
Reason's
preferences set
to open a custom
default song

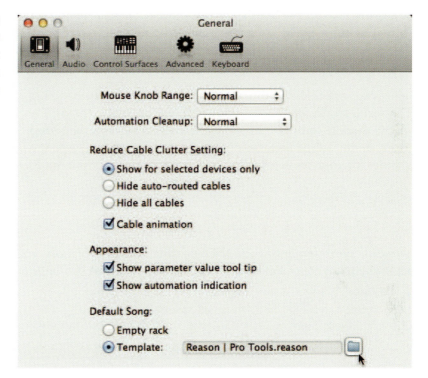

NOTE

Once your custom default song is set up in Reason, this is what will be opened each time you launch Reason or open a new document from Reason's File menu.

Reason uses the default song file as a template on which to base any new Reason document, so the document opens as 'Untitled'. Reason expects that you will record new sequences into its own sequencers and that you

(Continued)

will subsequently save this new document (based on the default song template) as a new file—named 'My New Song' or whatever you want to call your new song.

If you do not record any new sequences into Reason or make any changes to the configuration of the rack modules, then there will be no need to save a new Reason document before ending your Pro Tools session. Next time you open this Pro Tools session, the same 'Untitled' Reason document based on the default song will open, so, unless you have changed the Reason default song in the meantime, the same set of Reason modules will be available.

Of course, if you have selected different patches for your synthesizers, added or removed any modules to or from your Reason rack, or made any changes to the module settings that you want to keep, then you will need to save a copy of this edited Reason song document—preferably into the Pro Tools folder for your session so that it can be backed up (hint, hint!) with the session.

To get the Reason synths to appear as MIDI destinations in Pro Tools, you need to enable keyboard control in Reason, ticking this option in the Options menu—see Figure 7.13.

Figure 7.13
Enable Keyboard
control

Routing MIDI from Pro Tools to Reason

When you have inserted a Reason plug-in into a Pro Tools MIDI or instrument track and Reason has launched, all the available Reason sound modules in the current Reason document will be listed as Pro Tools MIDI track output destinations.

This allows you to use Reason like a rack of virtual instruments, where you sequence MIDI in a Pro Tools MIDI or instrument track, and trigger Reason's synthesizers, drum machines, and samplers via ReWire.

Step 1. Choose a rack module from among the Reason devices listed as MIDI output destinations—see Figure 7.14.

Figure 7.14
Pro Tools MIDI
output destinations
with Reason
modules available

Figure 7.15
Reason NN-XT
sampler set up for
recording into a Pro
Tools instrument
track

Step 2. To play a Reason sound module from a MIDI keyboard connected to your MIDI interface and record the MIDI from this, make sure that 'MIDI Thru' is selected in the Pro Tools Options menu and that the Record button is enabled in the Pro Tools MIDI or instrument track—see Figure 7.15.

Step 3. Now you can go ahead and record some MIDI into your session: press 'Record' then 'Play' in the transport window or hold the Command (Mac) or Control (Windows) key and hit the spacebar to start recording. Press the spacebar to stop.

Recording Audio from Reason into Pro Tools

When you are satisfied with the audio coming into Pro Tools from Reason, you should either bounce it to disk or record directly to Pro Tools audio tracks so that you don't have to continue running Reason together with Pro Tools.

TIP
It is not possible to adjust the mixer or other controls during a bounce to disk, so I usually prefer to record to Pro Tools tracks instead.

Step 1. Use the Track menu to create a new stereo audio track.

Step 2. Choose an unused stereo bus pair as its input.

Step 3. Change the audio output of the instrument or auxiliary input track that is being used to monitor Reason to the same stereo bus pair so that the output of this instrument or auxiliary input track is routed to the input of the audio track.

Step 4. Record enable the audio track.

Step 5. Make sure that the counter is at the location you wish to record from (e.g., press 'Return' if you want to start from the top) and start recording by pressing 'Play' then 'Record' in the Transport window.

Step 6. Pro Tools will record the audio to disk – see Figure 7.16. When this has finished, hit the spacebar to stop recording.

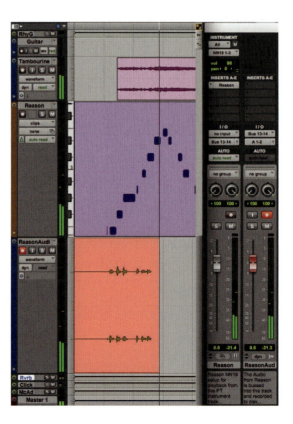

Figure 7.16
Recording the audio output from Reason into Pro Tools

Ableton Live

Overview

So what is Live all about? For me, Live is all about 'Instant EDM (Electronic Dance Music) Gratification'—with its library of electronic sounds that will immediately bring this musical genre to mind!

Live is definitely one of those 'must-have' software applications if you are putting beats and bass lines together with MIDI—building arrangements up with beats and bass line, leads and pads, bells and whistles (or whatever) and dropping in samples all over the place until you hear something you like. Live lets you quickly rearrange the order in which sections of your music play and makes it simple to play parts from one section alongside another section. And when you drop in sampled loops, these automatically play at the correct tempo. All amazingly helpful!

Live's session view lets you do the 'jamming'—improvising arrangements on the fly using short clips that can be overlaid and sequenced—while the arrangement view is used for linear multitrack recording and editing, and you can quickly switch between these using the Tab key on your computer's keyboard.

You can show or hide an info view at the bottom left of the main window. This displays helpful explanations about whatever you point the mouse at in Live's window—making it very easy to learn how to use Live. You can also show or hide a help view at the far right of the main window that lets you access tutorial material that explains in more detail how to use Live.

Live's browser, positioned at the top left of the main window, lets you search through the supplied preset sounds, samples, and effects. For example, the instrument rack presets combine an instrument with a selection of Live's audio effects. When you drag one of these presets into a MIDI track, the track name changes to the name of the preset and the controls for the preset appear in the editor area at the bottom of Live's window.

> **TIP**
> You can also drop a preset into the empty space next to the tracks and Live will automatically create a new MIDI track for the instrument.

Using Live Basics

Programming beats to create 1-bar, 2-bar, or 4-bar (or longer) MIDI clips in Live's session view is one of the most basic ways to get started with Live. Recording linear MIDI tracks in Live's arrangement view is just as easy.

Follow the steps listed here to get a basic idea of how to operate Live.

Step 1. Open the Impulse drum kit folder from Live's browser and drag a preset kit into a MIDI track in Live's session view.

Step 2. The track's Arm button is activated automatically, so it can receive MIDI, and you can go ahead and play your MIDI keyboard or hit the middle row of keys on your computer keyboard to play the drum sounds. The lower-case keys in the middle row of the keyboard, (a, s, d, f, etc.) are mapped by default to Impulse's drum slots.

Step 3. Double-click any empty session slot in the track that contains the Impulse instrument to create an empty MIDI clip to record into. You can create any number of empty MIDI clips in the session view and record into these.

Step 4. Make sure that you are hearing a metronome from Live (or from Pro Tools), and click the Session Record button (the one with the empty circle positioned to the right of the automation buttons at the top of Live's window) to activate recording—then start dancing on those keys! Stop whenever you like and play back to listen to what you recorded. To overdub, just press the Session Record button again.

Step 5. Press the 'b' key on your computer keyboard to enter Draw mode. Now you can draw notes into the MIDI note editor at the bottom of Live's window.

Step 6. Press 'b' again to exit Draw mode then click and drag any note to reposition this. Or click a note to select this and hit the Delete key to get rid of it. You can drag around several notes to select these or press Command-a (Mac) or Control-a (Windows) to select all the notes, then press Command-u (Mac) or Control-u (Windows) to quantize the notes to the current grid settings.

Step 7. One of the most impressive things about Live is the speed with which you can do things. Say you have programmed a one-bar pattern and you want to change this to a four-bar pattern. Just click and drag on the loop length parameter (see Figure 7.18) and it instantly changes to the new length—even while the pattern is playing back. So you can keep on building up patterns while listening to your music build up before your very ears!

Step 8. Press the Tab key on your computer keyboard to switch to Live's arrangement view.

Figure 7.17
Using Draw mode in
Live's session view

Figure 7.17
Using Draw mode in
Live's session view

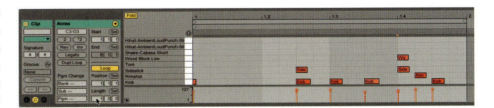

Figure 7.18
Live's MIDI note
editor showing the
clip's loop length
parameter being
edited

Step 9. Drag a piano preset from the Piano and Keys folder in the browser's Electric Instruments folder and drop this onto a MIDI track in the arrangement. The track will be armed, ready to play, and the device editor will appear in the lower part of the window.

Step 10. Now you are all set to record a linear MIDI track lasting the whole length of a song if you like. Just go to the bar from which you wish to start recording, hit the record button that you will find with the other transport controls at the top of Live's window, and play your MIDI keyboard or the keys on your computer's keyboard. You will hear the drum loop that you recorded earlier playing back at the same time.

Step 11. When you have finished recording, double-click the colored bar at the top of the MIDI clip that you have just recorded to open the MIDI note editor in the clip view at the bottom of the window. Here you can edit your performance.

Step 12. Hold the Shift key then press the Tab key on your computer's keyboard to switch the clip view back to displaying the device editor so that you can edit the sound of the instrument or apply effects—see Figure 7.19.

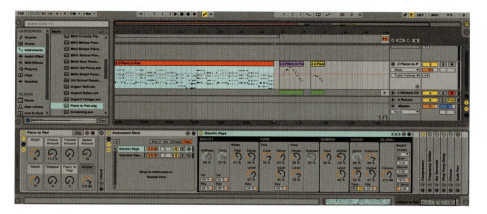

Figure 7.19
Live's arrangement view showing a recorded MIDI track with the device editor displayed in the lower section of the window

More Live Features

Recording bass lines and lead lines is equally simple using the built-in sampler instrument—called Simpler. This has enough bass sounds, pads, choirs, assorted keyboards, and strings to let you put relatively ambitious musical arrangements together on top of your beats. As with Impulse, you can play this from the computer keyboard. This has an octave of notes laid out on the middle row of keys with the sharps and flats on the upper row of keys. You can transpose the range by hitting the 'z' or 'x' keys and change the velocity of the notes using the 'c' and 'v' keys. So you can do lots of stuff with Live just using a laptop—without a MIDI keyboard!

A third built-in instrument, Operator, is a powerful synthesizer that, like Impulse and Simpler, is totally integrated such that every parameter can be automated or performed in real time. The drum rack instrument features an expanded version of Impulse's pad layout, with many more editing and sound design possibilities.

Live also has lots of built-in effects such as phaser, flanger, arpeggiator, and my favorite—beat repeat. This lets you create short loops on the fly, controlling their lengths manually or via random functions for endless variations.

And there's much more! Like Reason's combinator patches, Live's Device Groups feature allows you to save Simpler, Impulse, Operator, or any other instruments, together with chains of MIDI and audio effects attached, as presets. You can also use external MIDI instruments, and support for MP3 and other compressed formats means that you can use just about any type of audio file in Live.

These features have all proven extremely popular with DJs and remixers who appreciate the way that Live lets you make spontaneous creative decisions while building up ideas.

Using Ableton Live with Pro Tools

One way to use Live with Pro Tools is to build sequences and arrangements using Live alone, then open these for playback into the Pro Tools mixer. With this setup, you can conveniently record the output from Live onto Pro Tools tracks, then quit Live to reduce the load on your CPU. With the Live material in Pro Tools, you can then record additional material alongside this.

Another way to work is to build sequences and arrangements in Live while it is being hosted by Pro Tools. Within this scenario you can always record material directly into Pro Tools at any stage during the session. This is a more CPU-intensive way to work, so it works best with faster computers.

Step 1. To use Live with Pro Tools, launch Pro Tools first and open a session.

Step 2. Create an instrument or auxiliary input track to monitor Live's audio output.

> **NOTE**
> You could use a Pro Tools audio track to monitor Live, but this uses a playback 'voice' that you may not wish to lose. Also, if you want to record the audio from Live into your Pro Tools session, then you will need to use a separate audio track as well as the monitor track, because Pro Tools won't allow you to record into a track that is being used to monitor a plug-in.

Step 3. Insert an Ableton Live ReWire plug-in into the audio track's inserts section—see Figure 7.20.

Figure 7.20
Ableton Live ReWire plug-in showing Live's Mix L, Mix R, Bus 03, and Bus 04 (of 64) outputs

Step 4. Choose the Live outputs that you wish to use from the pop-up selector at the lower right of the ReWire plug-in window. The Mix L–Mix R pair is the obvious choice as default.

Step 5. Unlike Reason, Live does not launch automatically in the background as soon as you insert the ReWire plug-in, so the next step is to launch this manually. When Live launches (while Pro Tools is running with an Ableton Live ReWire plug-in inserted), it automatically connects to Pro Tools via ReWire.

Step 6. Now you can press 'Play' in Pro Tools or in Live, and both applications will start to play back in sync.

Figure 7.21
Ableton Live playing
back through a Pro
Tools instrument
track

At this point you could play back an existing Live sequence that you recorded earlier into Live or that you obtained from someone else, and play this in sync with your Pro Tools session. Or you could open a new Live sequence and record new material directly into Live. Figure 7.21 shows a Pro Tools instrument track with Live playing back through this.

With this setup, using just one Pro Tools track to monitor the audio from Live, you would use Live's own mixing features to mix the sounds in Live and then balance the audio coming via ReWire from Live with the audio from the rest of your Pro Tools tracks using the Pro Tools instrument track or auxiliary input fader.

Another option is to set up multiple Pro Tools tracks to accept multiple outputs from Live, each with different sounds, then mix all of these, together with any other Pro Tools tracks, using the Pro Tools mixer.

A third way is to record MIDI into Pro Tools tracks that are set up to play back instruments in Live—using Live as a set of virtual instruments. You can monitor the stereo mix outputs from Live using just one stereo auxiliary input in Pro Tools and use individual MIDI tracks in Pro Tools to play individual instruments in Live, using Live's internal mixing features to adjust the levels, add effects, and so forth.

Figure 7.22
Playing a Live
instrument from a
Pro Tools MIDI track
while monitoring
the audio from Live
through a Pro Tools
auxiliary input

To set this up, open a new Live set and drag a selection of instruments from Live's browser into MIDI clips in Live's session view or into MIDI tracks in Live's arrangement view. These instruments will automatically be listed as Pro Tools MIDI track output destinations. Figure 7.22 shows a MIDI track output selector pop-up with two Live instrument destinations available. The track is armed for recording and an external MIDI keyboard is being used to play a piano instrument in Live, which is being monitored via a Pro Tools auxiliary input that has been set up to monitor the main mix outputs from Live.

> **TIP**
>
> In this scenario, it usually makes better sense to use a separate instrument track in Pro Tools both to record the MIDI and to monitor the audio for each Live instrument that you are using. This allows you to adjust levels and apply effects individually to each instrument using the Pro Tools mixer.

When you have everything that you want playing back from Ableton Live into Pro Tools tracks, you should convert these into audio on separate audio tracks in Pro Tools. It is possible to bounce audio playing through auxiliary inputs to disk by selecting and soloing the audio and choosing the 'Bounce to Disk' command from the File menu. Alternatively, you can record directly to Pro Tools audio tracks in real time.

> **NOTE**
>
> Recording the audio output from Ableton Live into Pro Tools works the same way as recording audio from any virtual instrument into Pro Tools: the auxiliary input or instrument track that you are using to monitor the audio output from Live will not allow you to record this audio to disk. The solution is to route the audio from the auxiliary input or instrument track into an audio track using the internal buses.

Rewiring Reason to Live

Step 1. Open Live first, then Reason, which will open in ReWire Slave mode—you can check this by looking at the hardware interface in Reason—see Figure 7.23.

Figure 7.23
Reason hardware interface displaying ReWire Slave mode status

Step 2. Open a new Live set and make sure that there are, say, two MIDI and two audio tracks for you to work with.

Step 3. Click on the Show/Hide button at the right of the window in Live to reveal the I/O section.

Step 4. Go to the first MIDI track in Live and use the uppermost 'MIDI To' pop-up selector to choose Reason as the MIDI destination. Use the pop-up selector below this to choose the Reason instrument for this MIDI track—for example, Thor. Repeat the same steps with the second MIDI track in Live, choosing a different Reason instrument, such as the ID8, this time.

Step 5. In Reason make sure that you have wired the direct outputs from your Reason instruments to the hardware interface's Audio I/O, avoiding the first pair, Mix L and Mix R, which are normally wired to Reason's main mixer outputs. So, for example, you could wire Thor to outputs 3 and 4, and ID8 to outputs 5 and 6—as in Figure 7.24.

> **NOTE**
> By default, the audio outputs from each Reason instrument are wired into Reason's main mixer and the audio outputs from the main mixer's master section are wired to the hardware interface's audio outputs 1 and 2, which are labeled Mix L and Mix R when they appear in the pop-up selectors in Live (or Pro Tools).

Figure 7.24
Reason wiring with direct outputs from Thor and ID8 to Reason's hardware interface audio I/O. Reason's master section mixer outputs are wired to audio outputs 1 and 2 by default.

Step 6. Now go back to Live to set up the audio inputs. Using the uppermost 'Audio From' pop-up selector, choose Reason. Using the pop-up selector below this, choose the actual outputs from Reason that you want. In this example, you would choose Reason output pairs 3 and 4 for Thor and output pairs 5 and 6 for ID8.

Step 7. Also in Live, you need to switch the monitor buttons on each audio track in Live to 'In' so that you will hear the sounds coming from Reason.

Step 8. When you have got all this set up, you can enable either of the MIDI record buttons in Live to play the corresponding instrument in Reason. You can also record or draw MIDI notes into a Live MIDI track, hit play, and you will hear the notes being played by the corresponding Reason instrument—just as if each Reason device was a stand-alone hardware MIDI sound module.

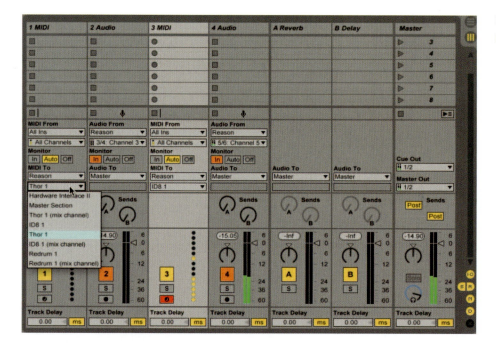

Figure 7.25
Live

Exporting Live ReWire Channels to Audio Files

When you have everything playing back from Reason (or any other slaved ReWire application) into Live, it makes sense to export the individual tracks, or a mix of these, to disk. Then you can quit Reason (to conserve CPU power) and simply replay these audio files using Live audio tracks while you continue to work on your arrangement using Live.

Step 1. Make sure Reason is playing back properly via ReWire.

Step 2. Select 'Export Audio/Video' from the File menu.

Step 3. A dialog opens where you can set a number of options—see Figure 7.26. You can manually select which channels to export from the Rendered Track drop-down menu. If you render the master output the rendered file will contain everything you hear from Live.

Figure 7.26
Live export audio
options

Step 4. Click OK. A file dialog will appear allowing you to select a destination folder for the file to be rendered.

Step 5. Quit Reason and open the rendered file or files in Live audio tracks.

Summary

Reason and Live are both extremely popular, especially among young musicians and those working in electronic dance music, rap, pop, and Top 40 music genres.

ReWire is an important technology that lets you transfer MIDI and audio between Reason, Live, and Pro Tools. The list of software that supports ReWire also includes Sibelius, Digital Performer, Logic, Garageband, Cubase, Nuendo, Melodyne, Sonar, Fruity Loops Studio, Acid Pro, and many others.

Both Reason and Live have lots of 'virtual' instruments, sounds, and samples that can be played directly from Pro Tools via MIDI, and audio from these can be routed back into Pro Tools audio tracks and mixed together with other Pro Tools tracks. So it can make sense to buy copies of one or both of these just to gain access to the sounds that so many others are using on popular recordings—even if you prefer to do your MIDI programming and audio mixing in Pro Tools.

PART 3
Studio Techniques

Chapter 8—Vocal Production, Effects, and Manipulation—covers microphones, microphone preamplifiers, and vocal processors—the analog hardware needed to 'capture' the vocals 'into the box'. Once 'inside the box', you can work on the vocals even more using plug-ins—for de-essing, compressing, EQ-ing, and so forth. Then you can correct the pitch or timing using Auto-Tune or Melodyne if necessary.

Chapter 9—Audio Signal Processing and Effects Creation—delves more deeply into the world of signal processing, looking at the individual processes and offering detailed overviews of many leading plug-ins in each category. This chapter includes a wealth of detail about these products supplied by the various manufacturers and presented more accessibly here.

Chapter 10—Mastering and Delivery—explains all the steps involved in mastering, with detailed sections included about iZotope software, and concise information about CD premastering, the various output formats, premastering for vinyl, delivery via MP3 and AAC, mastering for iTunes, and other useful topics.

in this chapter

CHAPTER 8

Vocal Production, Effects, and Manipulation

Recording Vocals

The first rule for recording vocals is to make the singer feel comfortable with the recording environment as this can be very unsettling for many singers (and musicians) if they are not used to this.

If your role is to be the producer, arranger, or recording engineer, then it is obviously very important that you make the singer feel at ease with you in every way possible while you are working in the studio.

Here is a useful checklist to observe: make sure that there is a music stand available to hold lyric sheets and music manuscripts; have plenty of drinking water available to soothe dry throats; make sure that there is plenty of moody lighting available in the recording space rather than anything too bright or distracting; and don't forget to make sure that the recording space is properly ventilated so that the singer gets plenty of good-quality air into his or her lungs! From a technical point of view, use a good pop shield to prevent annoying plosive sounds spoiling the recording, and consider how close to position the microphone is and how reflected sounds may affect the recording and how to minimize or maximize these.

Microphone Types

All microphones work in a similar way. They have a diaphragm (or a ribbon) that moves in response to changes in air pressure due to sound waves impinging on the diaphragm—in the same way as the diaphragm or 'eardrum' in the human ear responds to sound. The movement of this diaphragm or ribbon then produces a changing voltage, with a positive value (or amplitude) when the diaphragm moves one way and a negative value (or amplitude) when the diaphragm moves the opposite way.

There are three main types of microphone to choose from—dynamic, ribbon, and condenser.

With a so-called dynamic microphone (the word *dynamic* means changing or moving), a small coil of wire, positioned within the magnetic field of a permanent magnet, is attached to the diaphragm. When a sound wave enters the microphone, the diaphragm vibrates back and forth in an analogous way to the movements of the sound wave, producing an electrical 'signal' in the coil that changes in an analogous way to the movements of the vibrating diaphragm. This, by the way, is the origin of the term *analog audio*—the diaphragm's movements are an analog of the air pressure changes due to the sound wave, and the electrical signals induced in the coil when it moves within the magnetic field are an analog of the mechanical movements of the diaphragm.

In the case of a ribbon microphone, the ribbon is made of metal and is suspended in a magnetic field, so, when it vibrates, varying electrical signals are produced. Most ribbon microphones have a so-called figure-of-eight pickup pattern, responding to sounds coming from both the front and the back of the mic (but not from the sides).

In a condenser microphone, also known as a capacitor microphone, the diaphragm acts as one plate of a capacitor, and the vibrations produce changes in the distance between the plates. These changes are translated into varying electrical signals by one of two methods—DC biasing, which is the most common method, or RF biasing, which is used primarily in Sennheiser microphones and allows these to operate successfully in humid or damp conditions that would prevent DC-biased designs from working properly.

Microphone Specifications

The microphone's frequency response tells you the range of frequencies that the microphone can capture, but it does not tell you how well it captures these frequencies. For this you need to take a look at the frequency response graph, normally available from the manufacturer, which shows how the output of the microphone varies at different frequencies.

Self-noise (also known as the equivalent noise level) is a measure of the background noise generated by the microphone itself. This will be very low in relation to the sounds you are recording, but can be a problem if you are recording extremely quiet sounds and the microphone's self-noise is relatively high. Obviously, the lower the number the better. This measure is often stated in dB(A), which is the equivalent loudness of the noise on a decibel scale, frequency weighted for how the ear hears, for example: '15 dB (A) SPL'. The acronym SPL stands for sound pressure level and the A means that the 'A-weighting' adjustment is applied to the measurements to make these correspond more closely to the way the ear/brain perceives sound. A quiet microphone will typically measure around 20 dB (A) SPL.

The maximum SPL that a mic can handle is often quoted for particular values of distortion that this level will produce, such as 142 dB SPL at 0.5 percent total harmonic distortion. Higher values are better here. Sometimes the clipping level is quoted instead. This can be a better measure to use because low levels of distortion quoted at the maximum SPL are unlikely to be audible, whereas clipping is caused either by the diaphragm reaching its absolute displacement limit or by the microphone's electronic circuitry overloading and will produce a very harsh sound.

The dynamic range of a microphone is the difference in SPL between the noise floor and the maximum SPL, and you can easily work this out for yourself by subtracting the self-noise figure from the maximum SPL figure. In practice, the dynamic range is the usable range of sound levels that the microphone can successfully cope with without the self-noise intruding on low-level signals and without distortion interfering with high-level signals.

Microphones have an electrical characteristic called impedance, measured in ohms (Ω), that depends on the design. Anything under 600 Ω is considered to be of low impedance. Between 600 Ω and 10 kΩ is considered to be medium impedance, and anything above 10 kΩ is regarded as high impedance. For best results, the output impedance of the microphone needs to be matched to the input impedance of the microphone preamplifier that you are using to feed the microphone's output into your mixer or recorder.

The sensitivity is a measure of how well the microphone converts acoustic pressure to output voltage and is quoted either as a certain number of millivolts per Pascal (abbreviated Pa, the standard measure of sound pressure) at a frequency of 1 kHz, or as a number of dB referred to a 1V/Pa standard reference. Millivolts/Pa is a very straightforward concept to get your head around: apply one Pascal of sound pressure and out comes that stated number of millivolts from the microphone. The older method of quoting the output voltage in dB referenced to one volt/Pa results in negative values because microphones always output much less than one volt—typically values in the millivolts range—a millivolt being a thousandth of a volt. Typical values here might be −60 dB or −70 dB and in this case −60 means more output than −70 because these are negative numbers. Negative numbers can be a little less intuitive to deal with until you know how these work—then they're dead easy, of course. So, to sum up here, a microphone's sensitivity is really a measure of how much output it will produce. The important thing to realize here is that if a particular mic has a very low output, as with older ribbon models, you are going to need a microphone preamplifier with lots of gain to get this up to the correct level to feed into your mixer, recorder, or whatever.

Polar Patterns

Another typical microphone specification is the pickup pattern, more usually called a polar pattern. This is a circular graph showing how the microphone picks up sound coming from the different directions in a circle around the microphone.

An omnidirectional microphone, as its name suggests, picks up sound equally from all directions. This is particularly useful when you want to capture the ambience of the space around the microphone.

A cardioid microphone, again as its name suggests (*cardioid* means heart-shaped), picks up sounds in a heart-shaped pattern mainly from the front while rejecting sounds coming from the sides and the rear. The advantage here is that this microphone mostly captures sound from the sound source that it is pointing at.

A hyper-cardioid is similar but with a narrower area of sensitivity at the front and a small area of sensitivity at the rear. A variation on this is the super-cardioid, which has an even narrower pickup area at the front and even less pickup from the rear. These pickup patterns can help isolate the instruments that they are intended to pick up even more effectively, and also help to minimize feedback in 'live' situations.

The other main type of pickup pattern is the figure-of-eight or bidirectional pattern in which sound is picked up equally from the front and the rear, while sound from the sides is strongly rejected. Microphones with this type of pickup pattern can be used to record two nearby sound sources on either side of the microphone while rejecting sounds coming from the other two directions.

Here's a useful checklist to remind you of what to look out for:

- Check the frequency response curve: Is it relatively flat, in which case it may sound more natural with instruments, or does it boost the higher frequencies, in which case it may sound better when recording a singer?

- Check the polar pattern: Is this variable and does it have the pickup patterns you need?

- Check the sensitivity: How much gain will you need on your preamp?

- Check the impedance: Will it interface optimally with your preamp?

- Check the dynamic range and self-noise figures: Will it handle the softest and loudest sounds without problems?

Connectors and Cables

Three-pin XLR connectors (or occasionally 1/4" TRS jack connectors) are almost always used on professional microphones while 1/4" mono jack plugs are used on semi-pro microphones. You may also come across 3.5 mm (1/8") mini jack plugs on cheaper microphones. Stereo microphones can have pairs of mono connectors or may use five-pin XLR, or 1/4" TRS (tip, ring, sleeve), or 1/8" TRS stereo mini jack plugs.

Professional XLR and TRS mono connectors use cables with two wires and an earth (ground) connection in a configuration known as a *balanced connection*. When this type of connection is plugged into a microphone preamplifier, any interference from mains power or radio frequencies that gets into the two signal carrying wires is rejected by the input circuitry, so you can use long cables without fear of interference. Cheaper mono connectors use cables with one wire and an earth connection that cannot reject interference in this way, so are only suitable with shorter cable lengths.

> **NOTE**
>
> 'Digital' microphones are also available. Professional models that conform to the AES 42 standard, defined by the Audio Engineering Society, use a male XLR connector to directly output digital audio (the A to D converter is incorporated into the microphone itself). Home recording models, such as the Audio Technica 2020 USB, and professional models, such as the Neumann KM 184 D and TLM 103 D (when used with the DMI-2 digital microphone interface), can be plugged directly into most computers via USB connectors.

Stop the Shocks!

Most of the large diaphragm studio condenser mics and even some ribbon or dynamic mics are supplied with a special anti-shock mounting or you can buy these as optional accessories. They prevent low-frequency sounds getting into the microphone that may come from other instruments being played or from trucks passing outside, or from doors banging if you are recording at home. Vibrations can also travel via the floor into a microphone stand. Some stands are specially designed to isolate the microphone from any such vibrations—or you can buy foam blocks to place under the feet of tripod-style microphone stands, for example.

> **TIP**
>
> Be careful not to let the microphone cable or the singer's headphone cable swing against the microphone stand while you are recording, as sounds or vibrations from this can easily find their way onto your recordings!

Stop the Pops

When a singer sings fairly close to the microphone, it is all too easy for puffs of air to hit the microphone that can produce unwanted noises. Many years ago, the more inventive engineers started to use a wire coat hanger twisted to suspend a piece of nylon stocking attached to a circular wire frame formed by the hanger and positioned a few inches in front of the microphone to stop these pops. Typically called a *pop filter*, these can be bought ready-made these days, with a clip to attach to the mic stand. Stedman makes some superior models that use a metal mesh instead of a piece of nylon that is said to be a big improvement—and won't wear out as fast!

Proximity Effect

Typically, cardioid microphones progressively boost bass frequencies by 10 or 15 dB at around 100 Hz as the microphone gets as close as 6 mm (1/4 inch) away from the sound source. This phenomenon, known as the *proximity effect*, produces a deeper, more powerful sound that many pop and rock singers, radio announcers, and DJs really like. Unfortunately, it requires the vocalist to maintain a consistent distance from the microphone to avoid changing the low-frequency response, so for more sophisticated vocal styles, microphones such as omnidirectional models that have less of this proximity effect are more desirable.

Which Bit Do You Sing Into?

The jargon to watch out for here is *side entry* versus *front entry*, also known as *side fire* versus *end fire*—and before your imagination runs riot, this has nothing to do with sex! Pick up a Shure SM58 and you are not going to be in

Figure 8.1
AKG C12 VR
microphone

any doubt as to which end you sing into—it's the big round mesh on the end opposite the end that the cable attaches to (in case you are still wondering). But with some microphone types, it is not so clear-cut. For example, the AKG C12 VR (see Figure 8.1) has mesh at the end opposite the cable, and mesh all around the circumference of the mic behind the front end—and this is split into two sections.

You never use the front end (opposite the cable) in this case, even though it has a mesh that looks as though you could sing into it. This because the C12 VR is a condenser microphone with a capsule that faces out from one of the sides with the rear of the capsule facing the other side. The Electrovoice RE20, on the other hand, looks as though it might be a side-entry model, but is actually a front-entry design—see Figure 8.2. This all can, of course, be a little confusing for the novice . . .

Figure 8.2
Electrovoice RE20
microphone

The Right Mic for the Job

You need to look for microphones that will work well with the different instruments that you want to record.

Vocals are obviously very important, but a Shure SM58 for about $100 is incredibly affordable and has been used to record many hit records—so you have no excuse for not owning one of these! Ribbon mics such as the Beyer M160 and M260 can also be used for vocals, but studio condenser microphones made by Neumann, Gefell, Brauner, Josephson, and many others are generally thought to give best results.

Bass drums need a very robust mic that can take a lot of level without distorting, such as the AKG D112 or the Electrovoice RE20. Shure SM57s are the classic models to use for the snare. Beyer M201s, AKG C451Bs, C414s, or Neumann KM184s work well as either tom mics or overheads. For a different, darker, more vintage sound, you can use a pair of STC 4038 ribbon microphones as drum overheads, and these sound great for recording brass instruments as well.

Shure microphones always sound great on electric guitars, but for a different sound, try a ribbon mic or a condenser mic such as the Royer R-122 or Mojave MA200 models. All the Audio Technica studio condenser models including the AT4033 and the AT4047 also sound very good with electric guitars. I can also strongly recommend the AEA R84 ribbon mic (see Figure 8.3) for a different sound that really suits electric guitars.

For acoustic guitars, ribbon or small diaphragm condenser mics usually give the best results, bringing out all the nuances. For banjo or mandolin, a small diaphragm condenser model such as the Neumann KM 184, Mojave Audio MA100, or Josephson e22 is best.

Figure 8.3
AEA R84
microphone

Other mics—such as the Beyer M130 figure-of-eight ribbon (see Figure 8.4) or even the Electrovoice RE20—sound great with brass while others—such as the more expensive condenser models from Neumann, Microtech Gefell, Schoeps, and Sennheiser—sound smoother with strings.

Microphones that I Use

In my studio, I use a pair of very high-quality Beyer microphones, the MC834 and the MC740 (now discontinued and replaced by the MC840) to capture the sound of my Leslie rotating speaker cabinet—one for the rotating bass speaker and one for the rotating upper frequency horns. If I had more room, I would use a second MC834 on the upper frequency horns, but my room is not big enough!

Figure 8.4
Beyer M130
microphone

On upright acoustic piano, with the front removed, I sometimes use a Beyer M130/M160 MS stereo combination, which gives a very natural sound. Or I use a spaced pair of AKG C414 ULS mics if I want a 'harder' sound for rock and blues stuff. Currently, I am using a pair of Beyer MC930 small diaphragm condenser microphones, which have a very clear sound.

For my drum kit, a 1960s Ludwig kit with a 22" bass drum, two toms, and a snare with Zildjian hi-hat, crash, and ride cymbals, I use a Shure 545 Unidyne III (forerunner of the SM57) on the snare (see Figure 8.5), an Electrovoice RE20 on the bass drum, a pair of Beyer M201s on the toms, a pair of AKG C414s or the Beyer MC930s overhead, and I sometimes add the Beyer M130/M160 stereo combination, either pointing at the front of the kit or positioned at the back of the kit, above the drummer's head, to catch the sound that the drummer hears.

Figure 8.5
Shure 545D
microphone

For amplified electric bass I might use the Electrovoice RE20, an AKG C414, one of the Shures, or sometimes the Beyer M130/M160 combination to capture the sound in stereo. My favorite microphones for recording amplified electric guitar are the Audio Technica AT 4047/SV large diaphragm condenser microphone, which has a unique audio signature that I find very attractive with electric guitars, or the AKG C12 VR, which always sounds very classy.

Figure 8.6
Mojave MA-200

For individual instruments such as saxophone, acoustic guitar, or flute, I often use the AKG C12 VR or the AKG C414, both of which work well with almost anything. I also have a David Royer-designed Mojave Audio MA-200 vacuum tube large diaphragm condenser microphone with a fixed cardioid pattern that sounds great on acoustic and electric guitars and bass—see Figure 8.6.

For other instruments such as vibraphone, pedal harp, banjos, mandolins, hand percussion instruments, or even snare drums, I have had excellent results with a pair of Mojave Audio MA100 small diaphragm vacuum tube condenser microphones—see Figure 8.7. These have interchangeable cardioid and omnidirectional capsules so they are very versatile and can be used as a stereo pair in X-Y, spaced pair, Mid-Side, ORTF (French Broadcasting Organization), or NOS (Dutch Broadcasting Foundation) configurations, or as individual mono microphones.

Figure 8.7
Mojave MA-200
microphones
in NOS stereo
configuration

Choosing Vocal Microphones

Choice of microphone is a very subjective decision. Many producers use the classic Neumann U47, U67, or U87 large diaphragm condenser microphones that have been studio standards for decades. The U87 Ai model is still currently available from Neumann, but the older U67 and U47 models are no longer manufactured. Neumann does offer two current models based on the U47—the M 147 cardioid and the M 149, which has nine directional patterns.

Figure 8.8
AKG C414 XL II
microphone

For vocals, I normally use the AKG C12 VR large diaphragm condenser microphone, which is a reissue of AKG's classic C12 vintage tube microphone, or the AKG C414 Xlii, a relatively recent version of AKG's classic C414 studio microphone that also uses the C12 microphone capsule and has a frequency response tailored especially for vocals—see Figure 8.8.

My third choice is the Beyer MC840 large diaphragm condenser model—see Figure 8.9. This competes very strongly with the AKG C414 models and sounds a little less 'colored'.

Figure 8.9
Beyer MC840
microphone

Occasionally, I use a dynamic microphone such as the Shure 545 Unidyne III (see Figure 8.10), forerunner of the Shure SM57, or a Shure 565 Unisphere I, forerunner of the Shure SM58, or an Electrovoice RE20, or a ribbon microphone such as the Beyer M260 or the higher-quality Beyer M160. All of these are classic microphones that have been used to record many hit records.

I have also had excellent results using the Calrec Soundfield microphone and the Neumann M149 large diaphragm condenser microphone, which I used for several months while testing preamplifiers.

Shure 565SD

Figure 8.10
Shure 565SD
microphone

Recording with Headphones

Arguably, the most important thing to do while preparing to record is to make sure that everything is okay in the headphones, with just the right balance between the level of the vocals and the level of the music for the singer to feel comfortable. You also need to make sure that the singer is hearing the right amount and type of reverb in the headphones. Some singers like to hear the vocals 'dry', with no reverb added, but more often, singers won't feel comfortable unless you add at least some reverb.

Figure 8.11
Beyer DT100
headphones

Figure 8.12
Beyer DT48
headphones

I use both the standard Beyer DT100 model (see Figure 8.11) and the DT150, which has a more extended bass response that some people prefer. I prefer the sound of the DT100 to the DT150 because I can focus on the details that I need to hear when the bass is not too prominent. I also have an old pair of Beyer DT480 headphones that sound even better. These are not made any more, but look almost identical to the DT100/150 style and use the same transducers as the classic Beyer DT48 headphones that I often use while I am engineering recordings.

Beyer DT48 headphones (see Figure 8.12) are still made, despite costing several times as much as the DT100 models, and have the clearest mid and high frequencies that I have heard—great for hearing fine details of the sound. The low frequencies are not prominent at all, which can be a definite advantage at times. The DT48 headphones are very well constructed, but are not quite as comfortable to wear as the DT100 models.

The other set of headphones that I can recommend for studio use is the Audio Technica ATH-M50. These headphones are very reasonably priced, offer a full-range sound with plenty of bottom end, and allow you to swivel one or other of the earpieces around if the singer wants to work with one ear free. They are similar in terms of comfort to the DT100 and are a little lighter in weight.

NOTE

For studio use, closed-back headphones (as opposed to open-backed headphones) are normally chosen. The closed backs prevent any sounds from the headphones entering the acoustic space outside the headphones and vice versa. Open-backed headphones allow sounds to pass easily between the acoustic spaces on either side—so the headphone mix will leak from the rear of the headphones and will inevitably be picked up by any open microphones in the vicinity—which is not what you want to happen!

Recording without Headphones

Some singers just never feel comfortable wearing headphones. They want to hear the sound of the music all around them in a natural way, as if they were performing 'live'. It is perfectly possible to record a singer without using headphones and many successful records have been made this way. In a recording studio, the singer can record 'live' with the band or orchestra and, depending on where the singer is positioned in relation to the other musicians, there may well be enough separation between the vocals and the music if you use a cardioid microphone or use acoustic screens between the musicians and the singer to reduce the amount of spill from the other musicians into the vocal microphone. Or you can put the singer in a separate room or booth with a loudspeaker in the room so he or she can hear the music at a more reasonable level. In an overdubbing situation, you can have playback loudspeakers in the studio for the singer to hear, again making careful choices of microphone pickup pattern, position of the singer in relation to the speakers, and playback volume levels.

An increasingly popular option is for the singer to sing in the control room area, where it is much easier for the producer, arranger, or recording engineer to communicate with the singer. Here, you can play the music back so that it is loud enough for the singer to work with, but not so loud that it swamps the vocal on the recording.

TIP

Some engineers use a phase cancellation technique to remove the sound of any music that has spilt or bled into the vocal microphone—either by leaking from headphones when using loud playback levels or when recording without headphones. Immediately after recording the vocal, play the music again and record this through the same microphone setup, without altering any aspect of this (ideally with the singer staying in position also), so that you capture this microphone spill or bleed-through on a separate track. In Pro Tools you can insert a trim plug-in on this track and use this to invert the phase of the audio. Play this 'bleed' track along with the vocal track and, assuming that the levels of the bleed-through are about the same, the phase-reversed track should cancel out most, if not all, of the bleed-through from the vocal track when these are mixed together onto a third track.

A variation of this technique is to feed two monitor speakers in a studio with a mono version of the music, position the singer at one apex of an equilateral triangle equidistant from the two speakers, which are positioned at the other apices and at the same height as the singer's head. Then record the vocal with the phase of the two speakers reversed so that any spill from the music, which would otherwise be picked up through the microphone, gets cancelled.

Note that with all of these phase reversal techniques, phase cancellation is never going to be 100 percent, so there will almost always be some audible music left in the vocal track. Obviously, it is very important that there is nothing in the mix that you feed to the singer that will not be used in the final mix, as parts of this may still be heard in the final mix, even at quite low levels.

Microphone Preamplifiers and Recording Channels

In this increasingly digital age of recording technology one of the areas that remains firmly in the analog domain is the path between the source microphones and the recorder. At the very least this consists of a microphone preamplifier to lift the low-level microphone signals up to line level so they can be fed to the A/D converter. Once digitized, EQ, compression, and other signal processing can be applied in the digital domain. Nevertheless, many engineers and producers still prefer to apply such signal processing using analog equipment—appreciating the superior technical performance or wishing to take advantage of the unique colorations of the sound provided by many designs.

Most mixing consoles have mic pre's built in along with EQ and often with compressors, de-essers, or limiters as well. These are the basic tools that you need to get your music recorded. Many smaller studios are dispensing with mixing consoles and doing everything 'inside the box'—in other words, using the mixing, signal processing, and recording features in Pro Tools, Logic, or whatever. But they still need A/D and D/A converters/interfaces to record audio in and to play it back. Often, these are just line-level interfaces—or have one, two, four, or maybe eight low-cost mic pre's built in. So if you need more mic pre's or want higher-quality sound, you are going to have to shell out for stand-alone mic pre's. The basic models just amplify the microphone signals, but others provide a full recording 'channel' with EQ, compression, and maybe de-essing and limiting as well—or any combination of these.

SETTING LEVELS

To get the levels set correctly, you should ask the singer to sing some lyrics from the song at the loudest volume level he or she intends to use during the recording while you adjust the input gain on the microphone preamplifier to make sure that this level does not overload the preamplifier input—or the microphone itself. Many singers will sing louder than this when they are actually recording, so you may have to stop recording and adjust for this. Or you can use compressors or limiters to prevent overloads. Some microphones, and some preamplifiers, allow you to switch a resistance 'pad' into the electrical circuits to drop the level by 10 or 20 dB to help prevent overloads from occurring.

For pop music, you often want to 'color' the sound in the search for that elusive 'hit'. For classical, jazz, and other types of acoustic music, you are more likely to seek transparency. If you do both, you will want a versatile unit that can provide everything in one box—or you buy different boxes for different applications if you have the budget.

Important specifications include the amount of gain provided—ribbon mics and some others have very low output, so you need at least 60 dB of gain for these. If you are using long cable runs and there is any chance of electrical, magnetic, or radio interference, you will need good shielding against these. Generally speaking, transformers color the sound, solid-state circuits can be made more transparent, Class A circuits minimize distortion, and discrete transistor circuits typically offer greater headroom and improved dynamic range.

Does expensive necessarily mean better? Like most things, you tend to get what you pay for. If you want to use long cable runs, then you want high-quality transformers on the inputs and outputs. If you want valve-sound quality, these are not cheap. If you want total transparency, this can cost you lots.

If sound quality is important for you, you should make it your priority to get your voice or instrument recorded the best way you can afford! And, believe me, it is definitely worth paying for the difference you will hear between the standard mic pre's built into typical DAW interfaces and budget-priced mixers and the sound of quality mic pre's from quality manufacturers.

A PRODUCER PERSPECTIVE

To get some real-world user feedback about the subject, who better to comment than top UK producer, John Leckie, who has worked with such diverse artists as John Lennon, George Harrison, Pink Floyd, The Stone Roses, Simple Minds, Suede, The Verve, Radiohead, Los Lobos, Rodrigo y Gabriela, Portico Quartet, Dr. John, and Baaba Maal? 'I always reckon mic pre's are like the lens on a camera,' says Leckie. 'Everyone knows it's easy to expose film, but if you want to take a great photograph of a really important subject you don't use a Kodak Instamatic—you use a Hasselblad or Leica or Nikon. Nothing else is as good for quality. It doesn't matter what film or paper you use, the quality of the lens that first takes in (magnifies) the light always survives and comes through. Now consider this with audio and you will see the importance of the microphone and pre-amp.' Leckie has used many of the best mic pre's in the world during his 44-year career working in the world's top studios—and has learnt from practical experience what works well and what does not. 'You know you used a good mic pre when you start digging in with EQ to create a sound and find that it has great scope, range and possibilities. No matter how good the EQ device is, if the mic pre is shit, the recording will still sound like shit! And with a good mic pre you don't need EQ.' So as far as Leckie is concerned the mic pre is an absolutely crucial part of the recording chain—and most experienced engineers would agree wholeheartedly with this.

Microphone Preamplifiers, Compressors, Equalizers, and Recording Channels

George Massenburg GML 8302

The two-channel George Massenburg GML 8302 mic pre (see Figure 8.13) fits neatly into a 1U rack space. It has just two rotary controls for input gain with associated overload indicator LEDs and a power on/off LED indicator on its front panel. Fifteen to 70 decibels of gain are provided in accurate 5 dB steps and the +24 dBv CLIP indicators warn of any impending overloads. Switchable phantom powering is built in. The discrete bipolar transistor circuit topologies are completely transformerless with no FETs, ICs, or electrolytic capacitors in the signal path. Even the outputs are direct coupled with time-proven active servos that remove DC without adding any artificial color.

The 8302 provides 80 dB of common-mode rejection at 100 Hz and 10 kHz, and accepts +12.4 dBu maximum input level before clipping. With a +20 dBu output level, distortion is 0.0015 percent at 30 dB gain, rising to 0.007 percent with the maximum 70 dB gain. Maximum output level is +27.4 dBu, and effective input noise is −126.5 dBu with a 150 Ohm source. The theoretical limit of −129.5 dBu is reached with a 0 Ohm source. Frequency response is within 0.3 dB between 1.7 Hz and 260 kHz, and phase shift reaches 22 degrees at these extremes.

Figure 8.13
GML 8302

The GML 8302 and the 8304 Series II have been reconfigured with front-panel-mounted phantom power and phase reverse switches, and the +48v phantom supply has been completely redesigned resulting in lower noise. Ten to 65 decibels of gain is front panel selectable on the front panel in precise 5 dB steps. Both GML 8302 and 8304 will occupy 1U rack space. Both versions ship with an external power supply (GML Model 8355) and a five-pin XLR power cable. As always, the 8302 model can be upgraded up to four channels later on.

MY IMPRESSIONS

George Massenburg's GML 8302 contains a pair of the most transparent microphone preamplifiers I have been able to find—after carrying out extensive listening tests with all of its competitors! The GML 8302's extremely wide dynamic range and frequency response are undoubtedly doing the trick here!

George Massenburg GML 2020

Designed by the creator of parametric EQ, the George Massenburg Lab's GML 2020 is a single-channel mic pre combined with a four-band parametric EQ and a full-featured dynamic range controller (that's a compressor to you and me). This is a very sophisticated box of tricks that takes a bit of time to get the best out of. But the mic pre is as good as it gets, the EQ is faultless, and the compressor circuitry can achieve marvelous results (once the controls are mastered)—which is what you would expect if you are going to pay $6,650 for one of these beauties!

The GML 2020 (see Figure 8.14) is a single-channel combination of features from the legendary 8300 Mic Pre (a four-band version of the 8200 Parametric EQ) and the 8900 Dynamic Range Controller. Further enhancements not found in the stand-alone units include: mic, line, and direct instrument input selection, front panel two-pole high-pass filter, phase reverse, and phantom power switches, along with accurate LED input metering. From −10 dB of attenuation up to a whopping 70 dB of amplification, input gain is selectable in precise 5 dB steps accommodating all sources from low output ribbon mics through semi-pro keyboards to inserts in super hot mastering chains with one knob. Both the EQ and dynamics sections may be independently bypassed and a particularly useful feature of the 2020 allows you to place the EQ section either before or after the dynamics section or insert it into the side chain.

Figure 8.14
GML 2020

MY IMPRESSIONS

This is the Rolls-Royce of mic pre's! I found the quality simply stunning when I used it to record my guitar. I even preferred the sound of the processed guitar coming back off disk to the live sound—it was so good, with the GML 2020 adding a clarity, a warmth, an extreme quality—plus a feeling of hearing a very expensive recording!

Millennia ORIGIN STT-1

Priced at just over $3,000 and built like a tank, the single-channel Millennia STT-1 (see Figure 8.15) actually incorporates two separate microphone preamplifier circuits in one box—the acclaimed HV-3 solid-state mic pre and the M-2 vacuum tube mic pre.

The STT-1 also features twin NSEQ-2 mastering-grade four-band parametric equalizers—vacuum tube or solid-state—with 15 dB of boost or cut per band; twin optical de-esser paths—vacuum tube and discrete solid state—and no less than three optical compressor/limiter paths—vacuum tube, discrete solid state, and passive!

Input coupling is switchable between transformer or transformerless and the 1/4" vacuum tube DI input can be routed via tube or solid-state gain paths.

The STT-1's frequency response is +0/−3 dB from below 5 Hz to beyond 300 kHz, the maximum output level is + 32 dBu, and the equivalent input noise is −131 dBu.

MY IMPRESSIONS

Using its solid-state input, the Millennia STT-1 offers similar clarity and definition to the GML 2020 but provides significantly more flexibility by providing both tube and transformer input options. The massive amount of headroom available together with the extremely wide frequency response allows the solid-state signal path to provide extremely natural, uncolored sound quality. Using the transformer input and the tube mic pre, you can add 'color' to the sound as and when you need this. The EQ, compressor/limiter, and de-esser combine to make this one of the most versatile 'channel strips' available.

Figure 8.15
Millennia ORIGIN
STT-1

DESIGNER'S COMMENTS

Asked to comment on users who favor non-colored versus colored performance, Millennia's president John La Grou told me, 'They are all correct—it depends on the production values of the producer. Transformers always add some degree of colouration or distortion. Millennia's basic philosophy is to offer circuits that are open, transparent and accurate. Our primary markets are classical and critical acoustic music—audio engineers who demand purity of sonics. This is why we avoid transformers in our HV3—which uses a discrete hybrid solid-state design. Transformers are especially troublesome at very low frequencies and when dealing with high dynamic program. Some people do want the colouration that comes from these designs, but a good tube circuit and a good solid state circuit should both sound very pure. What people are really hearing in many tube designs is mostly colouration from the transformers—not the tubes. The Origin is our channel strip and this has a switchable transformer for big colour if you want it. It offers the entire range of sonic quality from extreme accuracy to deep colouration. We use Twin Topology where you can switch between all Class A tube or all solid state circuitry in the same box—selectively running the signal through one or other of these paths. We optionally offer ECC83 vacuum tubes made by Telefunken in the 1960s which give a lot more musical colour than other manufacturers' ECC83's. In the HV3 preamp, we also keep the signal entirely balanced from end to end. So pin 2 on the XLR and pin 3 each have their own mirror-matched amplifier paths. We have placed about 7,000 HV3 channels to date to critical users for orchestral work, film scoring, and acoustic recording—Millennia is becoming somewhat of a standard when it comes to accurate music recording. In the UK, we're proud to have attracted top users like Tony Faulkner, Olympic EMI Studios, Mark "Spike" Stent, Kevin Killen and John Pellowe, to name just a few.'

Millennia NSEQ-2

The dual-channel Millennia NSEQ-2 (see Figure 8.16) offers the same Twin Topology EQ as the STT1 Origin Mic Preamp channel strip, including both a Class A vacuum tube parametric equalizer and a Class A discrete solid-state parametric equalizer—using a transformerless design. The technical specifications are exemplary, with a frequency response of +0/−3 dB from below 2 Hz to beyond 300 kHz, a maximum output level of + 28 dBu, and a noise level measured with all the EQ stages in and the controls flat of −106 dBu (solid state) and −94 dBu (vacuum tube).

Figure 8.16
Millennia NSEQ-2

Above each of the four EQ bands is a pushbutton to let you switch this in or out of circuit. The outer two EQ bands, Bands 1 and 4, each have three controls: a rotary frequency select switch that selects fixed high and low band frequencies; a pushbutton peak/shelf switch that selects the high and low band curve shape (shelving EQ at 6 dB per octave with the switch depressed, peaking EQ with a fixed 'Q' of 1.0 when the switch is not depressed); and a rotary +/− 20 dB boost/cut control.

> **NOTE**
> EQ bands have no detectable sonic signature when they are in circuit as long as the boost/cut control is set at zero.

The inner two EQ bands, Band 2 and Band 3, each have four controls: a rotary frequency control, a 'x10' frequency range pushbutton switch, a rotary 'Q' control ranging from 0.4 to 4.0, and a rotary level control. This provides up to +20 dB of boost/cut with 21 detented positions (that can also be configured for +/−10 dB using the master gain range switch).

The frequency control sweeps all center frequencies from 20 Hz to 25 kHz. The low-mid band sweeps 20 Hz to 220 Hz, or 220 Hz to 2.5 kHz, and the high-mid band sweeps 250 Hz to 2.5 kHz, or 2.5 kHz to 25 kHz—depending on the status of the frequency range switch.

> **NOTE**
> 'Q' is defined as the ratio of the center frequency to the bandwidth. For example, a filter setting with 3 dB down points near 100 Hz and 1,000 Hz exhibits a 'Q' of approximately 0.4.

In the central section are three pushbutton switches for each of the two channels:

The top pair—the master channel buttons—lets you switch the left and right EQ channels out of the circuit. When this switch is not engaged, or when the NSEQ-2's power is off, the EQ channel is completely bypassed via a relay and the inputs become hardwired directly to the outputs.

The Twin Topology pushbutton switches in the central position let you select either vacuum tube or solid state topology for each EQ channel. Both designs use Class A amplifiers—one based on twin triode vacuum tubes, while the other uses all-discrete J-FET servo amplifiers.

The '10 dB Gain Range' switches at the bottom let you boost or cut the gain ranges for the left or right channels by 10 dB.

Focusrite Red 8

The Red 8 (see Figure 8.17) has two perfectly matched mic preamps—ideal for stereo recording. Using the same circuit topology as the original Focusrite ISA microphone preamplifier designs, the Red 8 offers two channels of ultra-high-quality Focusrite microphone amplification.

Used with high-quality ribbon, valve (vacuum tube), or condenser mics, the Red 8 obtains outstanding results with any sound source, but especially voice, piano, and string instruments.

Each channel offers custom-wound Focusrite input transformers, switchable phantom power, phase reverse, an easily read illuminated VU meter, and a handy scribble disc for denoting channels. Mic gain is switched in 6 dB steps over a 66 dB range for accurate, precise channel matching and recall.

The many benefits of the Focusrite mic amp topology include superb common-mode rejection, a good overload margin, and, with its shared gain structure (20 dB from transformer and up to 40 dB from the amplifier), a very low noise floor with the signature wide bandwidth (10 Hz to 150 kHz). It also maintains this level of performance with a very wide range of impedances across the inputs.

The output stages of the Red 8, using its custom transformers, will easily drive very long cable runs of up to several kilometers without significant loss of quality, making them ideal for remote recordings.

MY IMPRESSIONS

I use the Red 8 mostly for recording Hammond organ, Fender Rhodes, and Wurlitzer electric pianos, electric guitars, and electric basses. The Red 8's transformer colorations add lots of 'warmth' and 'depth' to the sound of these instruments.

Figure 8.17
Focusrite Red 8
Stereo Mic Pre

Grace Design Model 201

During its 12 years in production, the original model 201 mic preamplifier found its way into countless recording facilities around the world, gathering a considerable amount of critical acclaim along the way. This classic two-channel mic preamplifier has been completely redesigned as the m201.

The new m201 (see Figure 8.18) delivers massive headroom and ultra-wide bandwidth and has a very open, musical character. The signal path is fully balanced from start to finish, resulting in a wider dynamic range, while new high-current output drivers enable even longer cable runs without signal loss. A ribbon mic mode is also included, which shifts the gain range up 10 dB while deactivating 48V phantom power, bypassing the decoupling capacitors and optimizing the input impedance.

Each channel has an 'input mode' rotary switch that selects between a standard 48V phantom input, ribbon mic mode, a front panel DI input, or optional DPA high-voltage inputs (130V or 190V). Newly designed M+S (mid-side) decoder circuitry with a built-in width control is also included. The front panel width control is wired with a precision 12-position rotary switch that provides a range from 100 percent mid (mono) to 30 percent mid/70 percent side.

The ultra-precision summing and difference amplifiers feed a set of dedicated outputs, which allows simultaneous recording of discrete mid+side signals as well as the stereo matrix.

The front-panel-mounted HI-Z inputs are designed to accommodate a wide variety of high-impedance input sources, making the m201 an excellent choice

Figure 8.18
Grace m201

as a DI box. The m201is also offered with a state of the art 24-bit/192kHz A/D converter module—factory or retrofitable.

MY IMPRESSIONS

The Grace Design m201 mic pre offers a wonderful balance of frequencies from low to high and produces a very wide, satisfying, three-dimensional stereo image. It sounds much more natural and 'bigger' than the Focusrite Red 8, with a 'tight' bottom end, a lovely top end, no added warmth, and with fantastic stereo definition and character. This is the closest competitor to the GML 8302 that I have found in terms of transparency.

Universal Audio 2–610 and 2–1176

The Universal Audio 2–610 is UA's top-of-the-range mic pre—see Figure 8.19. This is a two-channel tube microphone preamplifier based on the legendary Universal Audio 610 modular console. Designed by Bill Putnam, the Universal Audio 610 was among the first modular recording consoles. Early Universal Audio consoles were used in all of Putnam's studios including Universal Recording in Chicago and United Western in Los Angeles. Many prominent engineers such as Bruce Swedien began their careers in these studios with this classic rotary knob console. The UA 610 was particularly well noted for its preamplifier, and has left its mark on countless classic recordings from Sinatra to Pet Sounds to Van Halen to LA Woman.

Carefully modeled after the microphone amplifier section of the 610 console, the 2–610 is true to the design of its predecessor. UA has taken great care to preserve the quality and character of the original while adding functionality with more boost/cut settings, phantom power, direct inputs, and impedance controls.

The Universal Audio 2–610 is often paired with a 2–1176 Limiting amplifier—see Figure 8.20. Based on the classic 1176 models, the front panel controls

Figure 8.19
Universal Audio
2–610

385

include the usual attack, threshold, and ratio controls with input and output gain and a VU meter that can be switched to show either gain reduction or signal level. The two 1176 compressors can be linked for true stereo operation using a convenient front-panel switch, or you can apply different settings to each channel if you have different audio material in each channel.

Figure 8.20
Universal Audio
2–1176

MY IMPRESSIONS

I particularly value the Universal Audio 2–610 for its colored vintage sound, especially when paired with a 2–1176 Limiting amplifier. This combination is especially useful for recording lead vocals using two different microphones, or whenever you would like to use similar EQ and compression types. I often use this combination to record pianos in stereo, especially on popular music recordings where the valve circuitry adds interest. The EQ can be used to create a brighter or darker sound, according to your taste, and compression can be applied as necessary.

Manley VoxBox

The single-channel Manley VoxBox (see Figure 8.21) is primarily a voice processor that features an all-tube, transformerless input, transformer-output design with EQ, compressor, de-esser, and peak limiter sections.

With bags of headroom available and using super clean Class A valve circuitry, you can output directly from the opto-compressor and mic pre section. Alternatively, you can take your output from the Pultec-style mid-frequency EQ section, with or without the de-esser and peak limiter being switched in.

Figure 8.21
Manley VoxBox
Tube Mic Pre +
Opto compressor,
Pultec EQ, De-esser
and Peak limiter

The VoxBox combines a mic preamplifier, electro-optical limiter, a Pultec-style equalizer, and a compressor before the mic preamplifier—a unique feature that can substantially reduce distortion while leaving no more sonic footprint than the mic pre on its own.

The compressor can simulate an LA2A and has other settings for drums and gain riding. The Pultec EQ works across the audio bandwidth and includes a de-esser and a limiter that are completely independent of the compressor and that act post-EQ.

The front panel has an instrument input jack for use with bass guitars and keyboards. The associated phase switch may be used to disconnect the mic input transformer in this configuration. This phase switch is normally used with the mic input to allow polarity reversal, for example, when using a mic on the bottom of a snare (as well as one on the top)—or with vocalists if the mic or headphones are out of phase. If the sound in the singers' headphones is out of phase with the sound of their voices, it can be very off-putting!

A low filter associated with the mic, line, and instrument inputs provides a gentle 6 dB/octave high-pass filter.

The main input level control is placed before the compressor and preamplifier in the signal chain to allow 'hot' signals to be attenuated to prevent distortion occurring within the input circuitry. A gain switch is also provided. This lets you set the amount of negative feedback, which has an effect on gain, transient accuracy, noise, and clipping characteristics—it is not a pad. It is intended for use as a tone control and/or to optimize noise, and can be used to vary how 'forward' or 'aggressive' you want the vocal or instrument to sound.

The meter select switch lets you adjust the VU meter to show each of the three isolated audio inputs, and two other positions indicate the amount of gain reduction from the compressor and de-esser/limiter. This is a slow-responding meter that won't reveal the faster peak reductions. Instead it gives a reasonable indication of the audibility of any gain changes—depending on the amount and rate of needle swing.

The compressor section has threshold, attack, and release controls along with a bypass switch and a link feature for use with a second VoxBox for stereo applications. The bypass switch can be used even while music is being recorded—no clicks or pops will be heard and the change of level occurs smoothly.

The EQ section has a low peak frequency switch control to set the center frequency for this fixed-bandwidth filter. There are 11 frequency positions between 20 Hz and 1 kHz—extending the range of the original Pultec design, which had only six frequency settings. The low peak potentiometer provides up to 10 dB of gain boost at the selected frequency band. The mid dip frequency switch offers 11 frequency bands between 200 Hz and 7 kHz. The mid dip cut control allows the gain in the selected frequency band to be reduced by up to 10 dB. Dipping frequencies between 200 Hz and 1.5 kHz can reduce phase shift problems while improving general clarity and 'punch', for example. The high peak frequency switch again has 11 positions, this time ranging from 1.5 kHz up to 20 kHz. The high peak control allows up to 10 dB of gain boost at the selected frequency range.

A de-ess bypass switch is provided along with a de-ess select switch to allow selection of the de-essing frequencies. A further switch position turns the de-esser into a limiter that works as a 'flat' electro-optical limiter similar to an LA2A. A threshold control is provided for the de-esser to set how loud an 'ess' (or a transient when limiting) has to be before it will be reduced.

MY IMPRESSIONS

I particularly appreciate the versatility that this unit offers, giving you all the tools you need to record great vocals! The VoxBox also delivers tremendous depth and clarity of sound with instruments such as DI'ed bass guitar—it sounded amazing with my 1962 Fender Precision Bass, bringing out all the harmonics very clearly!

Focusrite Liquid Channel

For anyone who finds it hard to decide which combination to choose, the Focusrite Liquid Channel provides a solution: according to Focusrite, the Liquid Channel can precisely replicate the sound of any classic mic pre and compressor and has the ability to change its impedance and vary its signal to either transformer or electronic types. The Liquid Channel comes with 40 classic mic pre's and 40 classic compressors, and more can be downloaded from Focusrite's Web site via a USB port.

The Liquid Channel (see Figure 8.22) uses convolution technology to resynthesize the modeled compressors and mic pre's combined with an analog front end. This features a preamplifier with the ability to change its impedance and vary its signal path to either transformer or electronic, perfectly replicating the interaction characteristics of the original and allowing the Liquid Channel to replicate precisely the sound of any classic mic pre and compressor. A newly designed digital EQ is also included, providing a comprehensive and truly 'liquid' channel strip!

Figure 8.22
Focusrite Liquid Channel

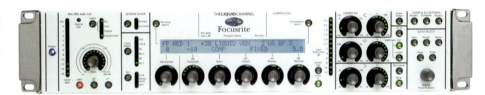

A press-button switch to the left of and just above the main display window lets you choose the preamp and compressor from the internal memories. Pressing this causes its LED to flash and engages the data knob at the bottom right corner of the front panel. Simply turn the data knob to select the mic pre/compressor replica then press the data knob to load the selected replica.

To account for variances in the amounts of harmonic distortion from one preamp to the next, a harmonics control provides harmonic distortion, which is usually perceived as adding 'warmth'. The amount of second-, third-, fourth-, or fifth-order harmonics will depend on the type of preamp chosen and the amount of mic pre gain. It allows users to create an overdrive sound without necessarily overdriving the preamp. The amount of harmonic distortion is indicated by the LED surrounding the dial, with a value from 0–15 displayed above it. '+ODD' appears next to the value when third- and fifth-order harmonics are present.

The compressor section offers controls for threshold, ratio, attack, release, and makeup gain and has an associated gain reduction bar graph meter.

The three-band EQ section offers high-shelving variable between 200 Hz and 20 kHz, mid-band parametric variable between 100 Hz and 10 kHz, and low-frequency shelving variable between 10 Hz and 1 kHz—each with a gain control allowing boost/cut of +/−18 dB.

> **MY IMPRESSIONS**
>
> The Liquid Channel is my go-to processor when I'm looking for just that right combination of mic pre and compressor to suit a new vocalist. The Liquid Channel lets me try lots of options much more quickly than if I were to try swapping lots of hardware combinations!

Vocal Processing Using Plug-ins

Some engineers like to apply de-essers, compressors, and equalizers as they are recording, while others prefer to capture the basic sound and process this as necessary later on. Whichever approach you take can lead to successful results—you just need to choose the one that works best for you.

Many engineers, given the choice, prefer to use hardware versions of the classic equalizers and compressors. But there are disadvantages to using hardware, not least of which is the cost of buying classic units. And you can do many things with software versions, such as using automation to process even individual words or syllables within a vocal performance, that would be difficult or impossible to achieve using the hardware.

Another major advantage of using plug-ins is that it is easy to recall the settings that you have used for a particular vocal, instrument, or mix. You can always make notes about the settings you have used on analog hardware units, but this can get very tedious and time-consuming.

So the $64,000 question is—'Do the plug-ins sound as good as the hardware versions?' I have tried running comparisons and have spoken at length with several plug-in designers while looking for the answer to this question. What it boils down to is this: some of the plug-ins are not direct equivalents for the hardware, by deliberate design. So although these sound similar, and sometimes even better, they do sound slightly different compared with the originals. In other cases, the design philosophy has been to get as close to the sound of the original hardware by modeling all of the typical flaws and imperfections found in all hardware units that exist in the real world. With these, many designers believe that they have got so close that no one can tell the virtual devices from the real devices.

My advice is simply to use your own ears with whichever plug-ins you are working with. Make your adjustments with these and then ask yourself if you like what you are hearing. If you do—then no one can argue with this!

Equalization

These days, the chances are that your favorite EQ is actually available as a plug-in. Two of my current favorites from Universal Audio are the Maag Audio EQ4 and Manley Massive Passive EQ.

Lots of high-profile producers are using the Maag EQ4—see Figure 8.23. This has a special high-frequency band called the Air Band. It has six bands of EQ, all fixed apart from top band, which goes from 2.5 kHz all the way up to 40 kHz! First there is a low sub, then 40 Hz, 160 Hz, and 650 Hz band-pass filters, and, finally, the 2.5 kHz shelving band with its associated Air Band and Air Gain controls. The final control is a level trim that can be used to compensate for level changes in the overall signal when you boost or cut the EQ bands. One of the key features is that this EQ does not introduce phase shifts, so it won't mess your sounds up. It sounds particularly great on vocals, whether adding color in the bottom end at 40 or 160 Hz, or boosting the Air Band at 10 kHz or higher.

Figure 8.23
Maag Audio EQ4

The Massive Passive is a two-channel, four-band equalizer with additional high-pass and low-pass filters—see Figure 8.24. It uses simple passive components and exploits their natural qualities to provide the most radical EQ settings needed for tracking as well as the more subtle shadings required for vocals and mastering. The Massive Passive incorporates the best strengths of Pultecs, choice console EQs, parametrics, and graphics—but lets you use twice as much EQ with half the coloration, allowing huge HF boosts without sibilance problems and unbelievable fatness without mud!

Figure 8.24
Manley Massive
Passive EQ

Compression

You can use a compressor to help make each word of the vocal audible by reducing the peaks in level then using the makeup gain to raise the overall level in the mix. Another way is to use the volume automation in Pro Tools to 'ride the gain' on the vocal tracks to control the levels, or you can adjust the clip gain controls on each clip. A third way is to use a combination of both these methods.

Softube offers emulations of Summit Audio's classic EQF-100 Full Range Equalizer and TLA-100A Tube Leveling Amplifier as individual plug-ins and combined as the Grand Channel plug-in—see Figure 8.25.

Figure 8.25
Softube Summit
Audio Grand
Channel

Softube also offers its emulations of the Tube-Tech PE 1C Program Equalizer, the ME 1B Mid Equalizer, and the CL 1B Compressor as individual plug-ins and combined as the Tube-Tech Classic Channel—see Figure 8.26.

Figure 8.26
Softube Tube-Tech
Classic Channel

Both the Summit and Tube-tech plug-ins come with plenty of presets that demonstrate how they can be used in different recording situations.

De-Essing

If the 'esses' are sounding distorted, then you can use a de-esser, such as Universal Audio's Precision De-esser—see Figure 8.27, to alleviate the problem.

Figure 8.27
Universal Audio
Precision de-esser

MORE INFO

See Chapter 9 for more about equalizers, compressors, and de-essers.

Pitch Correction and Effects Plug-ins for Pro Tools

When you are working on songs, the voice is the most important element of the recording. After all, if the singing is out of tune, the chances of the public wanting to buy the record are going to be greatly reduced.

One of the most useful processors to use with vocals, first introduced by Antares in 1997, is Auto-Tune. This was the first commercially successful software that made it easy to tune up vocals that were a little sharp or flat and it is now widely used in music production around the world.

It can also be used to create unusual effects—as on Cher's hit recording 'Believe', which featured Auto-Tune with the pitch correction speed set to produce strange effects. Recorded and released in 1998, 'Believe' hit the top of the Billboard chart in March 1999 and went on to reach #1 in 23 countries around the world. It was the most successful song in Cher's long career, and was all the more surprising as she was also the oldest female artist at that time (at the age of 52) to hit the #1 spot.

Today, it is difficult to avoid hearing Auto-Tune in use if you listen to Top 40 pop and rap songs, where its popularity seems to be increasing each year that passes! But even on records where you cannot hear Auto-Tune in use, it is more than possible that it has actually been used to fine-tune the vocal pitches in the seemingly never-ending search for perfection. It is even possible to use Auto-Tune to increase or decrease the vibrato on individual notes—or to remove this completely!

While several alternatives are available, Auto-Tune's biggest rival today is probably Celemony's Melodyne, covered later in this chapter.

Antares Auto-Tune

First introduced in the mid-90s, Antares Auto-Tune was designed by Andy Hildebrand, originally to correct out-of-tune vocals and instruments. Auto-Tune not only allows you to control pitch, but also rhythm and articulation. Its extremely high-quality time-shifting algorithm together with its intuitive user interface combine to make it quick and easy to correct timing errors—or to get creative with your sounds.

The software can detect and correct pitches up to the pitch C6, or as low as 25 Hz when the bass input type is selected—allowing intonation correction to be performed on virtually all vocals and instruments. Auto-Tune also lets you adjust the depth of any natural vibrato in real time—and you can even create your own styles of vibrato to add to notes that don't have any!

Using Auto-Tune with Pro Tools

Auto-Tune has two modes of operation—Automatic and Graphical. Using the default Automatic mode, you can simply insert the Auto-Tune plug-in on any track, such as a vocal, adjust the settings, and let Auto-Tune correct the pitch of that track in real time as you play back.

You can also select an audio clip in the Pro Tools Edit window, open the AudioSuite version of Auto-Tune, adjust the settings using either Automatic or Graphical mode, then render this to disk and play the rendered version back in place of the original audio.

> ### TIP
> You can sing 'live' into Auto-Tune. To set this up, simply insert Auto-Tune on an auxiliary input and route a microphone through to this. This can be very useful when you are first learning how to use Auto-Tune because you can hear exactly how Auto-Tune corrects what you are singing while you adjust the retune speed and other parameters in real time.

The way I prefer to work most of the time is using Graphical mode, especially when I am working on an important lead vocal or instrumental part. In this scenario, I prefer to set Pro Tools up so that I preserve the original vocal track at all times, while creating an auto-tuned version of this on an adjacent track.

This way I can easily redo any section that needs further work, for example, if the vocalist needs to resing part of the vocal that can only be improved by singing this again, or because it has been decided to change the lyrics or the melody or the vocal phrasing for that section—or for whatever reason.

Here's how I set this up:

Step 1 Insert Auto-Tune on the vocal track.

Step 2 Create a new auxiliary input next to the vocal track and name this A-T Monitor (or something equally descriptive).

Step 3 Create a new audio track next to this and name it Auto-Tuned Vocal (or something equally descriptive).

Step 4 Open the I/O dialog from the Setups menu in Pro Tools, click the Bus tab, identify a suitable available bus to use, name this A-T Out (or something equally appropriate), then okay this.

Step 5 Back in the Mix window, use this bus to route the output of the vocal track to the input of the auxiliary input that you will use to monitor the audio from the vocal track. Also route the output of the vocal track to the input of the audio track onto which you will record the auto-tuned vocal. See Figure 8.28.

Step 6 When you are ready, select a vocal clip in the Edit window and use Loop Playback mode in Pro Tools to play this back.

Step 7 Adjust the settings in Auto-Tune until you are happy with the results, then put the auto-tuned track into Record mode and record your auto-tuned clip onto the auto-tuned track.

Figure 8.28
Pro Tools set up to monitor and simultaneously record an auto-tuned vocal track to another audio track named "Auto Tuned Vocal"

Step 8 Mute the original clip and play back the auto-tuned version to make sure this has recorded okay.

Step 9 Move to each clip that needs auto-tuning and repeat the process.

TIP

When you have retuned all the clips that need attention, you can insert a third audio track and make a comp that contains original unprocessed clips together with processed clips to use in your mix session if you like. At this point it also makes sense to make the other two tracks inactive to conserve CPU resources.

Automatic Mode

You can use the Automatic mode to shift the pitch of an entire recording without any particular effort or expertise being required—see Figure 8.29. You can compare the pitches in the recording to any scale you define, including microtonal and ethnic scales, and you can even remove poorly executed vibrato and add accurate vibrato if necessary.

Figure 8.29
Auto-Tune
Automatic mode

Key and Scale

Auto-Tune lets you choose from major, minor, chromatic, or various historical, ethnic, and micro-tonal scales.

Individual scale notes can be bypassed, resulting in no pitch correction when the input is near those notes. Individual scale notes can also be removed, allowing a wider range of pitch correction for neighboring pitches.

The scale can also be detuned, allowing pitch correction to any pitch center—see Figure 8.30.

Figure 8.30
Auto-Tune's Key and Scale pop-up selectors and the scale detune control

Alternatively, you can learn the scale from MIDI or select the target pitches in real time via MIDI from a MIDI keyboard or using a prerecorded sequencer track.

Edit Scale Display

The edit scale display (see Figure 8.31) can be used to create custom scales or to modify any of the preset scales selected in the Scale pop-up. When any edits are made using this display, each scale retains its own edits independent of the other scales.

For example, if you were to select C Major in the Key and Scale pop-ups and remove or bypass certain notes, then you changed to C Minor and made different edits, if you then return to C Major, the previous edits that you made to the C Major scale will still be active there.

Figure 8.31
The Edit Scale display

NOTE

In most cases, you will probably tell Auto-Tune which notes are valid scale notes using the Key and Scale pop-ups, the edit scale display, and/or the virtual keyboard. However, there may be occasions when it is not clear exactly what key a melody line is in, or where the melody line has too many accidentals to fit comfortably into a conventional scale. For those occasions, the 'Learn Scale From MIDI' function allows you to simply play the melody into Auto-Tune from a MIDI keyboard or sequencer track and let Auto-Tune construct a custom scale containing only those notes that appear in the melody.

MIDI Control

At the lower left of the user interface, just above the virtual keyboard, are controls for the 'Learn Scale' and 'Target Notes' MIDI functions—see Figure 8.32.

Figure 8.32
Auto-Tune MIDI
parameters

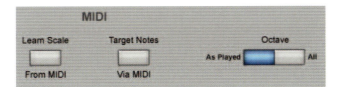

To use the 'Learn Scale From MIDI' function, make sure that the desired MIDI source is routed to Auto-Tune and then click the Learn Scale From MIDI button. Its color will change to blue and the edit scale display will automatically be set to a chromatic scale with all of the notes set to 'Remove'.

Now simply play the melody to be corrected from your keyboard or sequencer. Tempo and rhythm don't matter, so take your time and make sure you don't play any wrong notes. As each note is played, the corresponding Remove button in the edit scale display is turned off (adding that note to the scale as a scale note).

When you have played the entire melody, press the Learn Scale From MIDI button again to end the process. The edit scale display will now contain a scale containing only those notes that appeared in your melody. If you happen to have made an error during note entry, or want to try again for any other reason, simply click the Learn Scale From MIDI button and start the process again.

To use the 'Target Notes Via MIDI' function, ensure that the desired MIDI source is routed to Auto-Tune, then click the Target Notes Via MIDI button. Its color will change to blue and the edit scale display will automatically be set to a chromatic scale with all of the notes set to 'Remove'. While in this mode, Auto-Tune continuously monitors its MIDI input for Note On messages. At any instant, the scale used for correction is defined by all MIDI notes that are on. For example, if MIDI notes A, C, and E are held, Auto-Tune's input will be retuned to an A, C, or E, whichever is closest to the input pitch. The source of the MIDI input would typically be a MIDI keyboard or sequencer track, and could consist of chords, scales, or, most powerfully, the exact melody that the input should be corrected to.

> **NOTE**
>
> 'Target Notes Via MIDI' is used to specify target pitches in real time while pitch correction is occurring, while 'Learn Scale From MIDI' is used in advance of correction to create a custom scale.

Octave As Played/All Octaves: For both of the MIDI functions ('Learn Scale from MIDI' and 'Target Notes via MIDI'), you can choose whether you want incoming MIDI notes to affect all octaves or just the notes in the specific octaves in which they are played. Simply click the desired button. The button will change color to blue to indicate your choice.

> **TIP**
>
> Because Auto-Tune can accept target pitch information from MIDI notes, and can also quantize the pitch of incoming audio notes to a set scale, you can set it up to act as though it were a monophonic vocoder. Here's how to do this: make sure that the Target Notes Via MIDI button is enabled, then either set up Auto-Tune on a MIDI channel as a MIDI-controlled effect and set your vocal as the side-chain input, or insert Auto-Tune across the vocal track and route the output of a new MIDI track to this.

Virtual Keyboard

The virtual keyboard (see Figure 8.33) displays Auto-Tune's pitch detection range and acts as a real-time display of the currently detected pitch, a display of the current scale settings, and a tool for setting target note behaviors in specific octaves.

Figure 8.33
Auto-Tune virtual
keyboard

Pitch Correction Controls

The main pitch correction controls are for retune speed, humanize amount, and natural vibrato amount—see Figure 8.34.

Figure 8.34
Pitch correction
control

Retune speed controls the rate of change toward the desired pitch. Adjusting the natural vibrato lets you increase or decrease the amount of any vibrato that exists, and you can use the humanize control to preserve the original shape and character of the vibrato as much as possible.

The retune speed control lets you adjust how quickly the pitch is moved toward the scale tone. Use fast speed settings for short duration notes and for instruments such as oboe or clarinet, whose pitch typically changes almost instantly. Use slow speed settings for longer notes where you want to preserve the vibrato at the end of notes, or with vocal or instrumental styles that slide gradually between pitches.

One situation that can be problematic in Automatic mode is a performance that includes both very short notes and longer sustained notes. The problem is that to get the short notes in tune, you'd have to set a fast retune speed, which would then make any sustained notes sound unnaturally static. Luckily, the humanize function easily solves this problem. The humanize function differentiates between short and sustained notes and lets you apply a slower retune speed just to the sustained notes. Thus, the short notes are in tune and the sustained notes still allow the natural variations of the original performance. The higher the humanize setting, the more the retune speed is slowed for sustained notes.

The natural vibrato function lets you increase or decrease the depth of any vibrato present in the input audio while preserving the original shape and character of the vibrato. If the original performance does not contain vibrato, this control will have no audible effect.

The Targeting Ignores Vibrato feature is designed to help with target note identification when the performance includes vibrato so wide that it approaches adjacent scale notes. When this function is on, Auto-Tune attempts to recognize vibrato and differentiate between it and any intended note changes. Whether this will work depends very much on the actual performance, so don't expect successful results every time and turn it off if it doesn't improve the situation.

> **TIP**
>
> Using a fast retune speed, you can even remove a performer's own vibrato and replace it with Auto-Tune's programmed vibrato.

Create Vibrato

Auto-Tune lets you create your own vibrato style and apply this to any note. Using the pop-up shape selector, you can choose from sine wave, square, or sawtooth waveforms for the low-frequency oscillator. The rate control lets you set a rate in the range of 0.1 Hz to 10 Hz—see Figure 8.35. The variation control lets you set an amount of random variation that will be applied to the rate and amount parameters for each different note. This helps to humanize the vibrato by adding random 'errors'.

You can also specify an onset delay—the amount of time (in msec) between the beginning of a note and the beginning of the onset of vibrato—and an onset rate. This onset rate is the amount of time (in msec) between the end of the onset delay and the point at which the vibrato reaches the full amounts set in the pitch, amplitude, and formant amount settings.

Figure 8.35

Create vibrato controls

Controls are also provided to let you set individual vibrato depths for pitch, amplitude (loudness), and formant (resonant frequencies)—see Figure 8.36.

402

Figure 8.36
Pitch, amplitude,
and formant
amounts

NOTE

The create vibrato controls function completely independently of the natural vibrato function. Changes in that function have no direct effect on the depth of any vibrato resulting from the create vibrato controls. However, because both functions can operate simultaneously, they can interact in ways that may or may not be useful, depending on your intent. In most cases, you should probably use one or the other.

Transpose Controls

Three associated controls are provided for transposition—the transpose amount, the throat length, and a formant correction enable button—see Figure 8.37.

Figure 8.37
The transposition
controls

The transpose control lets you shift the overall pitch of your performance over a two-octave range (+/− one octave) using semitone increments, and this is in addition to any pitch correction applied by either Automatic or Graphical mode.

NOTE

In Automatic mode, this transposition takes place in real time. In Graphical mode, this function provides overall transposition on top of any pitch shifting applied using the graphical editing tools—and it does not affect the pitch edit display.

> **TIP**
> Although it is possible to transpose in Graphical mode by selecting all the correction objects in your track and manually moving them up or down, in most cases, using the transpose function will provide superior results.

The throat length control lets you modify a singer's vocal quality by varying the shape of Auto-Tune's modeled throat—which largely determines the singer's vocal character.

This control ranges from 50 to 180, with values above 100 representing a lengthening of the throat, while values below 100 represent a shortening of the throat. The actual values represent the percentage change in the throat length. For example, a value of 120 represents a 20 percent increase in throat length, while a value of 70 represents a 30 percent decrease in throat length.

The throat length control sets the overall throat length for your entire track. Additionally, throat length can be adjusted on a note-by-note basis in Graphical mode.

> **NOTE**
> The throat length control is only active when formant correction is engaged. When this is not engaged, it is disabled and is greyed out.

> **TIP**
> If you are transposing a female vocal performance down to make it sound more male, for example, then increasing the throat length will help to make this sound more convincing. On the other hand, if you are transposing a male vocal up to make it sound more female or childlike, decreasing the throat length will help.

Engaging the Formant Correction button (it turns blue) prevents the voice's resonant frequencies (also known as *formant frequencies*) from shifting during transposition—ensuring that the vocal characteristics are preserved over the pitch shift range.

The resonant frequencies of the vocal tract in men and women are usually different, so if you are transposing a female voice down or a male voice up, for example, and you still want it to sound of the same gender, then you need to make sure that the formant correction is preserving the frequencies of the tell-tale formants. If, on the other hand, you want to make a female voice sound more male, you should disengage the Formant Correction button (it turns pale grey) so that Auto-Tune will lower the formant frequencies together with the rest of the vocal.

Input Type and Tracking

The Input Type pop-up selector lets you choose from a selection of processing algorithms optimized for the most commonly encountered scenarios: soprano voice, alto/tenor voice, low male voice, instrument, and bass instrument. Matching the appropriate algorithm to the input results in much better tracking—with even faster and more accurate pitch detection and correction.

To accurately identify the pitch of the input, Auto-Tune requires this to be a periodically repeating waveform, characteristic of a voice or solo instrument.

Figure 8.38
The input type selector and the tracking control

The tracking control sets how much variation is allowed in the incoming waveform for Auto-Tune to still consider it periodic. Normally, you will leave this set to the default value of 50—see Figure 8.38.

Setting Retune Speed, Humanize, and Natural Vibrato

A fast enough retune speed setting will minimize or completely remove a vibrato, as well as produce the iconic Auto-Tune vocal effect.

For more natural sound, set the retune speed to, say 20, or even slower, so that Auto-Tune just tunes the sustained notes, not the faster notes.

> **TIP**
> To set up the widely used Cher/T-Pain, robot/vocoder effect: choose the input type, leave the tracking at 50, set the key to the key of your song, leave 'Scale Detune' and 'Transpose' unaltered, set retune speed to fast—then take a listen.

The natural vibrato control allows you to adjust the amount of vibrato present in the input audio. Vibrato is modulation in pitch as well as amplitude, so, with a wide vibrato, this pitch variation can make Auto-Tune think that the note is jumping between different pitches. The most common symptom of this problem is a pronounced warbling as the input is alternately tuned to each of the upper and lower adjacent notes. Enabling 'Targeting Ignores Vibrato' may help. If not, turn it off.

Once you get the short notes in tune with the retune speed, you can increase the humanize value until you hear more natural sustained notes. I usually make any adjustments to the vibrato first, or decide these are not needed, then I adjust the humanize control to allow some of the pitch changes on sustained notes, such as natural vibrato, to come through.

> ## TIP
>
> Start by setting 'Humanize' to zero and adjusting the retune speed until the shortest problem notes in the performance are in tune. If the sustained notes then sound unnaturally static, start advancing the humanize control. The higher the humanize setting, the more the retune speed is slowed for sustained notes. Your goal is to find the point where the sustained notes are also in tune and just enough of the natural variation in the performance is present in the sustained notes to sound natural and realistic. Beware that if you set 'Humanize' too high, any problematic sustained notes may not be fully corrected.

Graphical Mode

If you are good at judging pitch by ear, and if you are prepared to spend the extra time needed, you can use Auto-Tune's Graphical mode to analyze and correct even the smallest pitch deviations—see Figure 8.39.

Figure 8.39
Auto-Tune
Graphical mode

Figure 8.40
The Correction
mode buttons

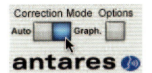

Figure 8.40
The Correction
mode buttons

When you first open the plug-in, this defaults to Auto Correction mode, so you need to switch to Graphical mode using the button near the top right of the window—see Figure 8.40.

Graphical mode features a large pitch graph display on which the vertical axis represents pitch (highest notes at the top, lowest at the bottom) while the horizontal axis represents time.

> **NOTE**
> There is also an envelope graph display below the pitch graph. This can be used with Auto-Tune's recently developed time control feature that lets you edit a note, word, or phrase to correct its timing. The envelope graph displays the amplitude (loudness) envelope of the sound whose pitch is shown in the pitch graph. Additionally, its central horizontal axis will display red in any range in which time has been tracked. When time control is enabled, the envelope graph will display two envelopes, one above the other; the original envelope on the bottom and the (potentially) time-shifted envelope on the top.

The pitch graph display's default mode displays horizontal lines that represent each pitch. This is probably the most useful mode with curve and line correction objects. However, for note objects, Antares added a lanes display mode that, as the name implies, displays horizontal lanes that extend from the left-hand 'keys' and are tinted to differentiate the sharps and/or flats. Note objects snap neatly into these lanes, so they are particularly useful when you are using note objects to shift the pitch of individual notes.

Track Pitch

The Track Pitch function is used to detect the pitch of the audio to be processed so that it can be displayed on the pitch graph display. With this button enabled, when you play back the audio a graphic representation of the pitch and its amplitude envelope will be drawn in red in the display as the audio plays. When all of the audio you want to correct has played, stop playback and exit Track Pitch mode.

TIP

You should always start tracking pitch in an area of silence before the audio you want to correct, or from the very start of the track. Starting tracking in the middle of audio will typically result in an artifact.

NOTE

If you are only correcting the pitch of your vocal and will not be editing the timing, then go ahead and use the Track Pitch function. However, if you will also be editing the timing, you should use the Track Pitch + Time function instead.

Draw Lines and Curves

When you have created a pitch contour for the audio you want to correct, you can draw lines or curves alongside these pitches, referred to as *pitch correction objects*. Then you can drag these pitch correction objects up or down in the graphical display to make the pitches higher or lower.

You can use the line tool to draw multi-segment straight lines on the pitch graph. Start the process by selecting the line tool and clicking anywhere on the pitch graph to set an anchor point. As you move the cursor, a line will extend from the anchor point to the cursor position. Click again to set a second anchor point and define the first segment of your pitch contour. Continue clicking and defining lines until your contour is complete. End the process by double-clicking on the final anchor point or pressing the Escape key on your computer's keyboard.

You can use the curve tool to draw arbitrary curves on the pitch graph. As with the line tool, start the process by selecting the curve tool and clicking anywhere on the pitch graph to set an anchor point. Hold down your mouse button and move the cursor to draw the desired pitch contour curve. End the process by releasing your mouse button. Unlike the line tool, the pitch graph will not scroll if you attempt to move the curve tool cursor outside the current display area.

TIP

If, while the line or curve tool is selected, you move the cursor onto the envelope graph display, it will temporarily change to the magnifying glass tool, allowing you to quickly and easily move to any other point in your audio and then resume editing without needing to manually change tools.

Make Curve

The Make Curve button is enabled whenever there is any red input pitch contour data present in the pitch graph (whether it is displayed in the current pitch graph view or not). Pressing the Make Curve button causes blue target pitch curves to be created from the input pitch contour data. These curve objects can then be dragged and stretched for very precise pitch correction.

Additionally, green output pitch curves are created that represent the exact pitches output at the currently selected retune speed. If you select the correction curve, move it, and adjust the retune speed, you will see the green output curve change in real time to reflect the changing retune speed.

Whenever you create correction curves with the Make Curve function, those curves will initially be assigned the default curves retune speed set in the Options dialog. If that default value is zero (as it may well be), the green output curve will be positioned exactly on top of the blue correction curve, effectively hiding it.

While all new curves are created with the default curves retune speed, you can then select individual curves (or cut up single curves to create multiple curves) and assign a custom retune speed to each one.

Retune Speed

The retune speed control allows you to specify how quickly Auto-Tune will change the pitch of the input to that of the target pitch curve or note object pitch. A value of zero will cause the output pitch to precisely track the target pitch of a curve line or be locked to the pitch of a note object. Slower values will have the effect of smoothing out the target pitch curve. As ever, you should let your ears be your guide to selecting the proper value for each note in a particular performance.

The retune speed setting is used only during the pitch correction process. It's similar in function but separate from the retune speed control in Automatic mode. In Graphical mode, the target pitch is not the scale tone nearest to the input, but rather the blue target pitch object (for curves and lines) or the exact note represented by a note object.

The ability to assign independent retune speeds to individual correction objects (curves, lines, or notes) is the key feature that allows Auto-Tune to achieve exactly the desired result for every note of a performance.

NOTE

A correction object's green output pitch curve is defined both by the object's retune speed and the setting of the Adjust Vibrato function.

Because each correction object (curve, line, or note) can have its own independent retune speed, the retune speed control is only active when at least one correction object is selected.

Graphical Mode Quickstart

Step 1 Set Auto-Tune to track the pitch of the incoming audio by clicking on the Track Pitch button near the bottom left of the user interface—see Figure 8.41—then play a short vocal clip that is out of tune in places.

Figure 8.41

The Track Pitch button

Step 2 Auto-Tune produces a red curve showing how the pitch of the incoming audio varies overlaid on a representation of the waveform in white, as in Figure 8.42.

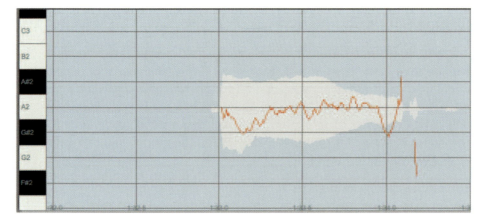

Figure 8.42

The pitch graph

Step 3 Click the Make Curve button, which is located just to the right of the Track Pitch button. An editable curve, colored green, is created and overlaid on top of the red curve—see Figure 8.43.

410

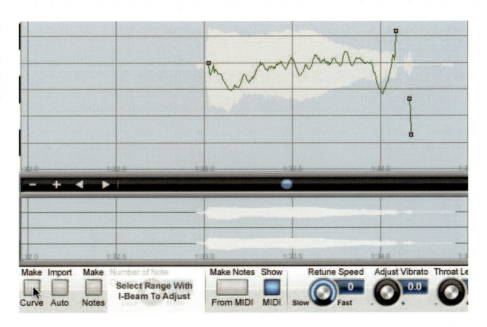

Figure 8.43
An editable curve,
colored green, lies
on top of the red
curve

Now you can use the various tools at the top of the graphical display to make adjustments to the vocal until it sounds the way you want it to sound.

Step 4 Using the scissors tool, for example, you can cut the curve into separate segments that encompass the out-of-tune parts and the in-tune parts. In my example, the leftmost segment at the beginning of this note falls flat, the middle segment is approximately in tune, and the end of the note falls then rises and generally sounds sharp in pitch—see Figure 8.44.

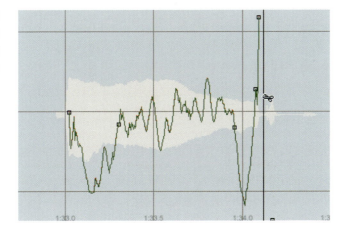

Figure 8.44
Using the scissors
tool to cut the curve
into segments

411

Step 5 Using the arrow cursor tool, first click away from the curve to deselect all the segments. Then click on the first segment to select this and drag it until the pitch of the note is centered around the correct pitch—see Figure 8.45. Audition this and adjust the position of the curve segment again if necessary.

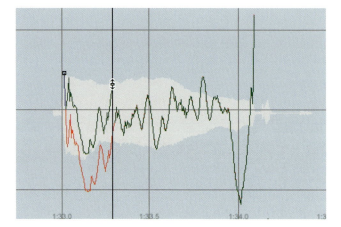

Figure 8.45
Dragging the first pitch curve segment to a new pitch

Step 6 Now repeat this process for any other out-of-tune segments, such as the end segment in my example—see Figure 8.46.

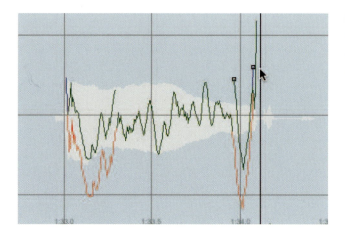

Figure 8.46
After moving the end segment to adjust its pitch

Step 7 Enable the audio track that you have set up for this purpose and record this auto-tuned clip to disk.

Make Notes

Auto-Tune 7 includes the Make Notes function—an entirely new range of functions for correcting the pitches of individual notes or phrases with a minimum of tweaking.

Pressing the Make Notes button causes Auto-Tune to analyze the input pitch and create target notes, each of which is centered on a horizontal pitch graph line. These notes represent the pitches that Auto-Tune 'sees' as the performer's target notes.

> **TIP**
>
> You can also draw notes when the note tool is selected by clicking and dragging near the desired horizontal graph line or lane to create a new note.

When Auto-Tune analyzes the input pitch for the purpose of creating note objects, it must make decisions about what constitutes a note and what constitutes a transition between notes, as well as differentiating between a single note with wide vibrato and a series of separate notes of alternating pitch. Often, the right choice depends on the style and technique of a specific performance. The number of note objects control lets you give Auto-Tune some guidance in making these decisions.

These note objects are much easier to manipulate than lines or curves. Once created, note objects can easily be dragged up or down to change their pitches, or have their beginning and/or end positions moved forward or backward, or be cut into multiple shorter notes for individual processing.

> **TIP**
>
> In selecting target pitches, the Make Notes function considers only the notes in the currently selected key and scale. If the melody includes many accidentals, it may be more convenient to select the chromatic scale.

Using Make Notes

In the example here, I used the same vocal note that was flat at the beginning and a little sharp at the end. With the default settings, Make Notes created two notes and left the vibrato alone—see Figure 8.47.

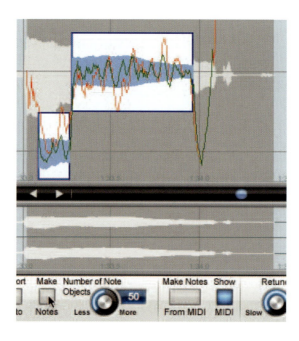

Figure 8.47
Make Notes created two separate notes

Of course, I already knew that the note goes flat and then sharp at the end, so I increased the 'Number of Note Objects' value until more note objects appeared—in this example, both at the end and at the beginning (which was also a little flat)—see Figure 8.48.

I clicked away from the note boxes, which were selected by default, to deselect these in the display. Then it was trivially easy to drag the two lower note boxes

Figure 8.48
More Notes are
created when the
number of note
objects value is
increased

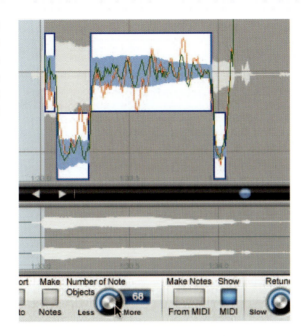

to line up with the other two, correcting all the off-pitch sections within this sung note—see Figure 8.49.

Figure 8.49
Dragging a note to
the correct pitch

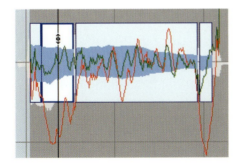

Notice that the green output pitch curves within each note box have automatically centered the pitches of each note around the horizontal (theoretically correct) pitch graph lines. The actual pitches of the recorded notes will be moved toward these target pitches at a rate determined by the retune speed settings for each note (which default to 50, but can be individually set for each note).

> **NOTE**
>
> Always make sure that you adjust the 'Number of Note Objects' value first, before proceeding with any pitch correction or note-based (or curve-based) pitch shifting. Otherwise, things can get confused.

Correcting this out-of-tune word using the notes correction objects was quite a lot faster and easier to set up than using the lines or curves correction objects. However, the lines and curves correction objects do allow more detailed editing, which can be useful with more complex material.

Having finished my editing session, I recorded the pitch-corrected vocal clip onto the audio track set up for this purpose—see Figure 8.50.

Figure 8.50
Recording the corrected clip to disk

Melodyne

Melodyne can be used for correcting the intonation and timing of vocal or instrumental performances; for audio quantization; for creating harmonies; or for remixing and restructuring the melody, tempo, or timing of existing recordings.

Melodyne also lets you edit and optimize any single-voice instrumental recordings, drum, or percussion tracks. With a drum loop, for example, you can alter the position, length, pitch, and volume of individual hits, straighten up the timing, make the whole loop swing, or even give it an entirely new rhythm by quantizing to new values.

Originally designed to work with monophonic source material, Melodyne versions 3 (and later) can handle polyphonic guitar or piano tracks just as easily as monophonic lines using Celemony's Direct Note Access (DNA) technology.

Melodyne's excellent time-stretching and pitch-shifting capabilities allow you to correct audio material unobtrusively or to create extraordinary transformations and effects, yet the results remain impressively natural. You can always manually edit the pitch, vibrato, drift, timing, volume, and formants of every note in the recording individually—but, for the most common edits, you can save a lot of time by using the correct pitch and quantize time macros, which allow you to optimize an entire recording with a single mouse click. With a few more mouse clicks, you can thicken your singing through doublings or allow a second voice to run with it at the octave, adding slight random variations of pitch and timing so that it sounds natural. Even more 'magically', Melodyne automatically knows the key and the scale so you can use its Scale Snap feature to automatically place notes correctly!

For choral or instrumental ensemble work, you can use the multitrack Melodyne studio. This allows you to create a complete choir arrangement from a single lead vocal: just copy the vocal to the other tracks, select a key, and move the additional voices into pitch. Melodyne will automatically adjust the new tracks to the correct key and give you authentic-sounding backing voices within a few minutes—adding slight deviations in pitch, rhythm, and color to make them sound even more realistic if you wish. In the same way you can create a string quartet from a solo violin or make your brass section sound a little fatter!

Almost all of Melodyne's parameters can be controlled remotely via MIDI and automated. You can use any MIDI controller you like for this and Melodyne comes with ready-made adaptations for lots of popular models. Virtual knobs that can be automated in the host and controlled via MIDI offer real-time control of pitch, formants, and volume.

Melodyne also allows you to export audio notes using the 'Save As MIDI . . .' command from the Settings menu to create a Standard MIDI file and save this to your hard disk. This file can then be loaded back into Pro Tools and used, for example, to double your vocals using a software synthesizer. Each MIDI note is created with the same position, length, and pitch as each audio note and the velocity of each MIDI note is derived from the amplitude of the audio note it represents. When you save rhythmic material, all the MIDI notes will share the same pitch but take their position, length, and amplitude from their audio equivalents in the rhythm track. So you can use this technique to derive a quantization reference from a drum loop for other MIDI tracks in your DAW, for example.

Melodyne Versions

Melodyne is available for both Windows XP and Mac OS X, with versions available for ReWire, AAX, RTAS, VST, AU, and ARA, and comes in four configurations—the basic Melodyne essential, the mid-range Melodyne assistant, the top-of the range Melodyne editor, and the Melodyne studio bundle—which includes the multitrack Melodyne studio 3 together with Melodyne editor. Melodyne essential, assistant, and studio do not offer DNA Direct Note Access—this is only available with Melodyne editor.

Melodyne Studio

Melodyne studio is a multitrack audio editing application with an unlimited number of tracks—depending on your computer system—and supports audio resolutions of up to 32-bit and 192 kHz. Geared more toward stand-alone use and less toward interaction with a DAW, Melodyne studio is not limited to editing monophonic melodies, but can also handle complex audio textures and even polyphonic material, such as chords or song mixes. Harmonic material such as rhythm guitar or piano parts can be transposed without altering the tempo, slowed down or sped up without altering the pitch, and even quantized. All this can be done in real time at the highest sound quality—and even extreme alterations can still sound natural and convincing. Melodyne studio also allows you to transfer timing or pitch data from one track to another.

Priced at 699 euros, Melodyne studio works as a stand-alone application and also includes the Melodyne Bridge plug-in that lets you link the stand-alone version to your DAW using ReWire. Melodyne studio can be used either as a ReWire master or slave. However, at the time of writing, Melodyne studio is only available as a 32-bit application and the Melodyne Bridge plug-in only works in RTAS, VST, or AU formats.

> **NOTE**
> You can link Melodyne studio to your DAW either using ReWire or using Celemony's MelodyneBridge. In most cases, the MelodyneBridge is far more convenient to work with than a ReWire connection.

Melodyne Editor

Melodyne editor offers the most extensive range of functions for editing vocals and instruments, including an extended range of timing tools that allow you to control the attack and internal time path of individual notes, allowing you to edit vocal phrasing in microscopic detail. Celemony's Direct Note Access (DNA) technology lets you edit recordings of polyphonic instruments such as pianos and guitars—even individual notes within chords! Polyphony, in this case, denotes more than simply the presence of chords. Polyphony exists whenever notes overlap, even though they may have been played one after the other. Direct Note Access also lets you output recordings of polyphonic instruments in the form of MIDI notes. Another unique feature, Scale detective, even lets you extract scales from audio, edit, and apply them to other recordings.

Priced at 399 euros, Melodyne editor can be used as a stand-alone application or as a plug-in with your DAW in AAX, RTAS, AU, VST, or ARA formats. It is both 32- and 64-bit compatible and will also work as a ReWire slave.

Melodyne Assistant

Melodyne assistant is the mid-range, more easily affordable, alternative to Melodyne editor. Melodyne assistant does not include Melodyne editor's extended timing tools and doesn't include the polyphonic DNA feature.

Priced at 199 euros, Melodyne assistant can be used alone or as a plug-in with your DAW in AAX, RTAS, AU, VST, or ARA formats. It is both 32- and 64-bit compatible and will work as a ReWire slave.

Melodyne Essential

Melodyne essential is the entry-level version and only allows you to modify the pitch center, position, and duration of notes. Melodyne essential works fine for basic pitch and timing correction, but does not provide the more detailed editing capabilities of the more expensive versions.

Priced at 99 euros, Melodyne essential can be used as a stand-alone application or as a plug-in with your DAW in AAX, RTAS, AU, VST, or ARA formats. It is both 32- and 64-bit compatible and will also work as a ReWire slave.

Using the Melodyne Plug-in

Step 1 Insert the Melodyne plug-in into the first slot on the audio track that contains the vocal or instrument that you want to work with. This version of Melodyne is listed in the Other category—don't confuse it with the Melodyne ReWire plug-in, which is listed in the Instrument category.

NOTE

Always use the first slot—before any compressor, EQ, or other effects. During the transfer, Melodyne records the audio that you intend to edit, so any effects ahead of it in the signal path would also be recorded. Normally, you will not want to do this, so make sure any effects are inserted after Melodyne.

Step 2 Select a clip in Pro Tools that needs tuning, open the Melodyne plug-in, click on the Transfer button, then play the clip. During transfers, Melodyne records the audio material from the Pro Tools track, creating its own audio files, which it stores on your hard disk. Melodyne then analyzes the audio and displays a representation of this—see Figure 8.51.

Figure 8.51
Transferring data to Melodyne

Step 3 In this example, I sang the word 'you' with two notes—E and D that were both slightly sharp—using a fairly wide vibrato on the D, and inadvertently falling flat at the end of the note—see Figure 8.52. To fix this, I separated the last part of the D using the note separation tool so that I could move this to the correct pitch manually, as the automatic correction would move this to C#, not D.

Step 4 The two notes that now exist are selected—highlighted in dark red— see Figure 8.53. Immediately after cutting the note, I clicked away from the notes in the display to deselect these.

Figure 8.52
Transferred vocal

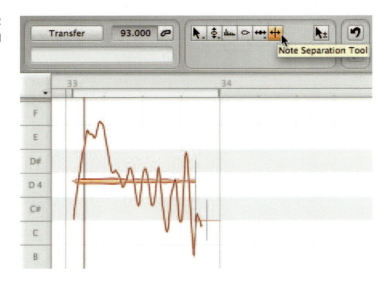

Figure 8.53
Separating the
notes

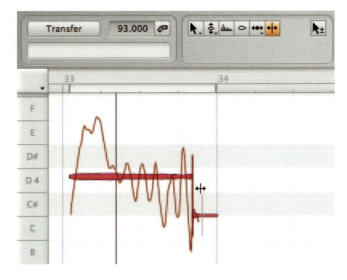

Step 5 To move the last note to the correct pitch, I selected the main pointer tool and dragged the C# up to D—see Figure 8.54.

Step 6 The quickest way to correct out-of-tune notes is to use the correct pitch macro, which you open by clicking on the button provided near the top right-hand corner of the toolbar—see Figure 8.55.

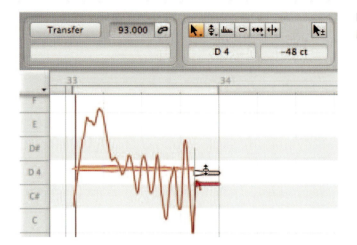

Figure 8.54
Moving the note

Figure 8.55
Opening the correct
pitch macro

Step 7 Moving the correct pitch center slider toward 100 percent centers the
pitches of all the notes (the E and D contained in the first segment and the D
in the separated note segment at the end of the word) around the correct D
pitch—see Figure 8.56.

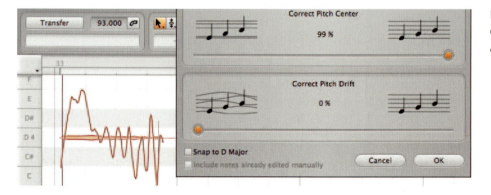

Figure 8.56
Correcting the pitch
center

Step 8 The vibrato was a little too much and there were places where the pitch still drifted a little sharp or flat. In this example, I found that a pitch drift correction of 25 percent improved this a lot—see Figure 8.57.

Figure 8.57
Correcting the pitch drift

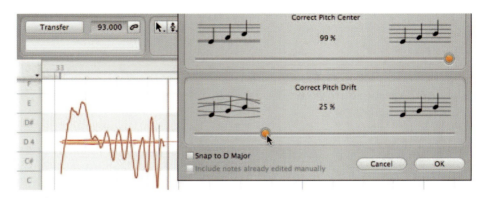

NOTE

Melodyne allows much more detailed editing than this 'quick fix' methodology that I have outlined here—which only scratches the surface of Melodyne's powerful capabilities.

Using Melodyne with ReWire

Generally you will want to use Melodyne as a plug-in in Pro Tools. This is the most convenient way of working and keeps all the Melodyne data stored with your Pro Tools session, making archiving and passing on projects easier.

Occasionally, however, you may wish to integrate the stand-alone version of Melodyne into Pro Tools as a Rewire client. For example, this can be very useful if you want to bring in audio samples or files to your Pro Tools session and have these adjust to the project tempo. When the program is integrated via Rewire, this happens automatically as soon as you drag an audio file from Mac's Finder or from Windows Explorer and drop it into the Melodyne window.

Here's how to set this up:

Step 1 Launch Pro Tools first. Add an auxiliary input and choose 'Melodyne' from the Instrument category. This will launch Melodyne as a stand-alone application that will be integrated via Rewire, with the transport functions and tempos of the two programs synchronized.

Figure 8.58
Selecting
ReWire outputs
in Melodyne
singletrack

Step 2 Launch Melodyne and choose a pair of outputs to route the audio into Pro Tools from the Rewire pane in the Melodyne user interface—see Figure 8.58.

Step 3 Open the Pro Tools Rewire window and choose the same pair of outputs if you are working in stereo or just, say, the Left output if you are working in mono—see Figure 8.59.

> **NOTE**
> You can open multiple Melodyne single-track documents and transfer audio from each of these to separate auxiliary inputs in Pro Tools using the eight channels available.

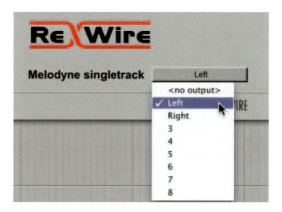

Figure 8.59
Selecting Melodyne
output as track
input to Pro Tools

Step 4 Now, when you drag and drop your audio samples—or complete tracks—into Melodyne these will be analyzed and adapted to the project tempo. When you have finished editing your audio in Melodyne, you can transfer this via Rewire into Pro Tools, route the Pro Tools auxiliary input to an audio track, and record the audio from Melodyne to disk—see Figure 8.60.

Figure 8.60
Recording audio
from Melodyne
singletrack via
ReWire into Pro
Tools

Step 5 If you want to preserve the data that you have edited in Melodyne, you must save the Melodyne document manually as a Melodyne project document. To save this in your Pro Tools project session folder, make sure that the 'Save Audio File(s) in Copy' option is ticked—see Figure 8.61.

Figure 8.61
Saving the
Melodyne
singletrack project
document

Working with Polyphonic Material

Step 1 Insert Melodyne on a track that plays polyphonic material, click on the Transfer button, and play an audio clip that you want to work with. The audio is recorded into Melodyne, analyzed, and displayed. For this example, I chose a recording of two four-note guitar chords with the top two notes common to both chords—see Figure 8.62.

The notes I had actually played were A#-G#-D-F# and A-G-D-F#. The notes that actually showed up were G and octave below, the correct A#, D, and octave below the one I played, and the correct F# at the top. The second chord was analyzed correctly as A but with a greyed out G below this, G in the correct octave but with a greyed out A above this, and it correctly found the D above this, with the sustained F# at the top.

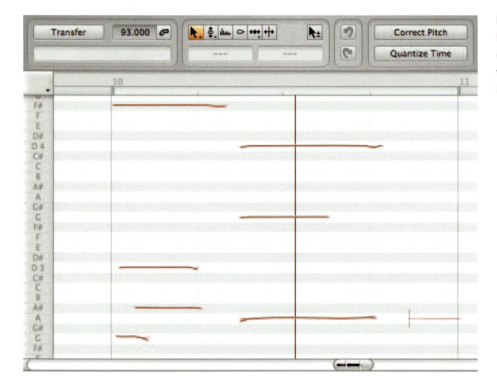

Figure 8.62
Polyphonic material (two guitar chords) transferred to Melodyne

Step 2 After transferring or loading your audio, the procedure is to select the ± tool to switch to Note Assignment mode. The color of the editing background changes to remind you that in Note Assignment mode no 'audible' editing is possible—see Figure 8.63.

Figure 8.63
Note Assignment
mode

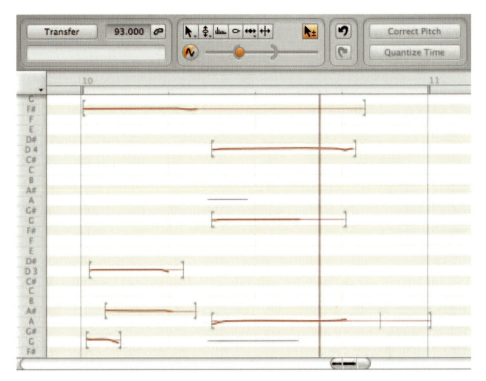

As the manual explains: 'You will use this mode to check Melodyne's interpretation of the audio material and correct it where necessary. Notes that have been "swallowed" (where a fundamental has been mistaken for an overtone) can be activated, which makes it possible later to edit them. Conversely, overtones that have been mistaken for fundamentals can be deactivated.' This was more or less exactly the situation that I encountered with my guitar chords.

The manual goes on to explain how to sort this out: 'In this mode, the outline of active blobs is filled in (i.e. they are solid) whereas, with inactive blobs, only the hollow outline is seen. When you click on a blob, you will hear the pitch of the corresponding note. Where a solid blob has been assigned to what is, in fact, merely one of the overtones of some other note, you can deactivate it by double-clicking on it. Now only the hollow outline of the blob will be seen and its energy in the frequency spectrum will be attributed to the note of which it can most plausibly be assumed to be an overtone. Conversely, by double-clicking on a hollow blob, you can turn a potential note currently interpreted by Melodyne as an overtone into an active one. Only active notes can be edited later using the tools in the Melodyne toolbar, which is why all the notes played and only those notes should be represented by solid blobs.'

Step 3 In my example, I double-clicked on the low G in the first chord, and left the other notes alone—see Figure 8.64.

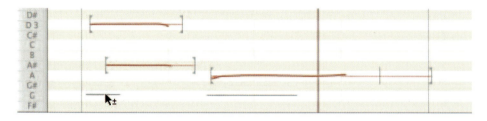

Figure 8.64
Blob deactivated in
Note Assignment
mode

This still left me wondering what to do about the greyed out G and the greyed out A in the second chord.

As the manual explains: 'The double-bracketed Note Assignment slider that appears beneath the toolbox in note assignment mode allows you to control the number of potential notes displayed and the number of active notes derived from them. If you move the large right bracket (or "crescent") in the slider to the left, fewer potential notes will be displayed. If you drag it to the right, more potential notes will appear. Now drag the orange knob on the slider to the left and right. As you drag it to the left, you reduce the probability of the potential notes displayed becoming active notes, thereby reducing the number of active notes. As you drag it to the right, you increase that probability, thereby creating more active notes from the potential notes displayed. Adjust the Note Assignment sliders until the number of active notes displayed is as close as you can get to the number of notes that were actually played.'

So, set the brackets on either side of the orange knob, and adjust the position of the orange knob until just the notes you played are visible.

Step 4 In this example, I adjusted the note assignment sliders and the orange knob to display just four notes in each of the two chords—see Figure 8.65.

Figure 8.65
Two four-note
chords displayed

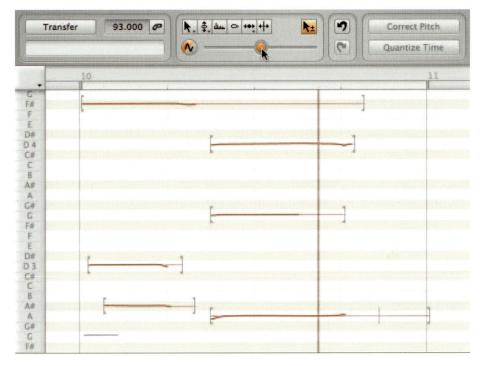

Step 5 When you have finished checking the detection of the notes in your polyphonic material, quit Note Assignment mode by selecting any of the other tools. Then you can manually correct the individual notes.

Bouncing/Printing Melodyne Edits

When you have finished using Melodyne in your project, you have two choices: you can either keep Melodyne active until you have finished your final mixdown, which lets you make changes whenever you like but uses CPU resources, or you can record the audio you've edited using Melodyne onto disk then deactivate the plug-in to free up resources.

The fastest way to do this is to use the Bounce function, selecting the 'Offline' option. To set this up, solo the Melodyne track, deactivate any other plug-ins and any automation that would otherwise be applied to this track, then select 'Bounce to Disk' from the File menu. In the Bounce dialog, choose the correct bounce source, name the file, then choose a directory to save the new file to. Check the 'Import After Bounce' option to make sure that this new file will automatically be imported to your session. When the bounce is complete, choose 'New Track' from the menu that follows. A new track will be created containing the bounced material and you can deactivate the original Melodyne track.

Otherwise, you can record to a Pro Tools track. Again, deactivate any effects such as EQ or compression and any automation that you don't want to include in the recording on the new track. Select 'New Track' from the track's Output menu to create a new track and Pro Tools will automatically route the output signal from the Melodyne track to this new track. When you finish recording this, you can choose 'Hide and Make Inactive' from the Track menu to remove the Melodyne track from Mix and Edit windows and conserve your CPU resources.

> **TIP**
> Recording your Melodyne edits as audio tracks has the further advantage of allowing you to pass projects on to colleagues who don't have Melodyne.

Technical Notes

Taken directly from the Melodyne manual, the following technical notes are particularly relevant for Pro Tools users.

Saving Melodyne Plug-in Settings

Just as with an effects plug-in you can store different settings as presets, in Melodyne you can save different edits. You may wish to do this in order, for example, to allow a performer or artist to hear and choose between different edits of the same take. To save and reload Melodyne settings, follow the same procedures as for all other plug-ins: click in the upper part of the current Melodyne Plug-in window on 'Preset' and select 'Save Settings As . . .'. Then assign a name to the current Melodyne edit. You can store alternative edits as additional presets and switch between them using the preset selector.

H/W Buffer Size

Under 'Setup > Playback Engine' set the H/W buffer size to 1,024 samples. Smaller values lead to a significant increase in the CPU load. Should you require a smaller buffer, for example when adjusting the headphone mix directly in your computer and not via an external channel strip or mixer, switch all instances of Melodyne during the recording to bypass. Reactivate Melodyne as soon as you begin editing your new track.

Running Pro Tools 10 and 11 Simultaneously

Melodyne is integrated into Pro Tools 11 as a 64-bit AAX plug-in but into Pro Tools 10 as a 32-bit RTAS plug-in. Version 2.1.2 of the Melodyne installation program installs both formats, RTAS und AAX, side by side. Under OS X, you can run both Pro Tools 10 and Pro Tools 11 on the same computer. Please note, however, that, according to Avid, Pro Tools 11 can only be co-installed with Pro Tools 10.3.7 or higher (under Mac OS X 10.8).

Session Compatibility between Pro Tools 10 and 11

Older sessions using Melodyne can be opened in both Pro Tools 10 and Pro Tools 11, regardless of whether they were saved by Pro Tools 10 or Pro Tools 11. Putting it another way, a '10' session can be imported into Pro Tools 11 and an '11' session into Pro Tools 10. This bidirectional compatibility is subject, however, to one condition: the same version of Melodyne (v.2.1.2) must be installed on both systems (Pro Tools 10 and 11).

iZotope Nectar 2

Available in AAX, RTAS, AudioSuite, VST, VST 3, and Audio Unit (AU) formats, Nectar 2 is a single plug-in that incorporates just about everything you might need to process a vocal.

Nectar has a total of 11 audio processing tools, including automatic pitch correction; a noise gate; a 12-voice harmonizer; an eight-band equalizer; saturation controls with five types available including analog, retro, tape, tube, and warm; a versatile compressor with digital, vintage, optical, and solid state options; a de-esser; digital, tape, and analog-style delays; an EMT 140 plate reverb emulation; distortion, modulation, and other FX; and a limiter.

The user interface is compact and efficient with lots of visual feedback—see Figure 8.66.

More than 150 presets are provided so you can immediately conjure up sophisticated vocal sounds from dry vocal recordings. These can be accessed

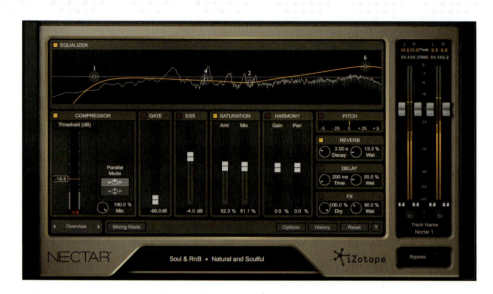

Figure 8.66
Nectar 2 plug-in

using the preset manager where they are subdivided into folders for specific genres such as pop, rock, soul, and R & B—see Figure 8.67. These presets are very useful to get you started with vocal processing—suggesting lots of practical applications and providing excellent starting points for these.

Figure 8.67
Nectar 2 preset
manager

If you want to get even more serious about your vocal processing, Nectar 2 also features an Advanced mode which can be accessed by clicking where it says 'Overview' at the bottom left of the window. The Advanced window has all the modules listed at the left, and these can be activated or deactivated by clicking on the small square to the left of each module's name.

Pitch provides Auto-Tune-style tuning for real-time pitch correction and the harmony module can automatically generate up to 12 voices to harmonize with your vocal or instrumental track—or you can choose the harmony notes using a MIDI controller. You can change the order of these modules by dragging these up or down in the list—although the pitch module must always remain at the top of the signal flow order. As you can see in the screenshot in Figure 8.68, comprehensive and detailed controls are provided for each module, such as the EQ shown here—which has up to eight filters and more than 10 filter types.

Figure 8.68
Nectar 2 Advanced window showing the EQ controls

The Nectar 2 Production Suite also includes breath control and pitch editor plug-ins.

The breath control plug-in—see Figure 8.69—targets breath sounds and adds/reduces gain as needed to make all breaths the same consistent volume level in Target mode or attenuates them by a user-defined amount in Gain mode.

433

Figure 8.69
Nectar 2 breath
control

The breath control module automatically analyzes the incoming vocal and distinguishes breaths from sung vocals based on the harmonic structure. If any piece of the incoming audio matches a harmonic profile similar to a breath, the module will suppress that portion of the audio until sung vocals are detected.

Different from a threshold-based process in which the module is only engaged once the audio has risen to a certain volume, breath control will perform its analysis regardless of level. This allows for accurate breath recognition with a multitude of quiet or loud vocal styles with minimal adjustment of the module's controls.

The sensitivity fader controls how sensitive the breath control module is when detecting the harmonic structure of breaths in your incoming audio.

When in the default Target mode, the reduction amount set using the target slider represents the level that you wish all detected breaths to be reduced to. So if a singer breathes in quietly and then loudly, both the loud and quiet breaths will be reduced only by as much as is necessary to have both breaths reach the desired level. This results in the most natural-sounding breath reduction as the detected breaths in your audio are only reduced as necessary. Loud and annoying breaths will be reduced heavily while quiet, natural-sounding breaths will be left at the same volume.

When a breath is detected in Gain mode, the breath control module will reduce the gain of that breath, regardless of the level of that particular portion of audio. So, for example, with the gain slider set to a reduction of −30 dB, if the singer breathes in very quietly and then very loudly, both loud and quiet breaths will be reduced in gain by −30 dB. This can be more effective with particularly troublesome breathing sounds, or as a way of removing all breaths from a particular spoken or sung vocal. However, this can sound more unnatural, as very quiet breaths may be inaudible while loud breaths will be reduced to a normal level.

The graphical pitch editor plug-in—see Figure 8.70—lets you correct pitches manually. Logically laid out, the user interface is easy to learn and to use and the window can be easily resized by dragging the bottom right-hand corner, which really helps with detailed editing.

The pitch editor works similarly to Auto-Tune and Melodyne: you capture some audio from your DAW first, then the notes are displayed in the main note workspace area together with pitch contours for each note. By default, these contours will represent pitch after correction. The piano roll on the left side of the note workspace represents the note scale. After capture, the waveform of the input audio will be represented in a display above the note workspace. This display may also be used for navigation and zooming within the timeline.

Below the note workspace are five note editing tools. Looking from the left:

- The cursor is the default tool used for selecting notes and for dragging note regions vertically within the piano roll to change their pitches. Clicking and dragging with this tool allows you to lasso select multiple notes at once. To correct notes, you may either drag a note region to the correct note lane or simply double-click any note region to lock it to the nearest note lane. Double-click the same note region again to return it to its original pitch.

- The split tool allows you to divide a single detected note into two individual note regions that you may then edit independently.

- The merge tool allows you to heal a separation between two detected notes, turning them back into one unified note region.

- The zoom to selection control allows you to easily zoom to a specific location in the piano roll/timeline. Simply click the zoom to selection icon, then click and drag across a specific location in the note workspace to zoom to fit that selection.

- The zoom to show all captured notes control allows you to instantly zoom in or out to show all the detected note regions within the audio you have captured.

A set of transport controls lets you capture and play audio within the plug-in.

Below the note editing tools, controls are provided for vocal range, correction speed and correction strength, formant preservation, and formant shift, together with a vibrato control that lets you increase or reduce vibrato in captured notes.

After capture, you may set the key/scale of your vocals to force them into specific note lanes. After choosing the root note and scale settings, clicking 'Snap To Scale' forces all note regions into note lanes available in the specified key/scale.

Finally, at the bottom right of the pitch editor plug-in's interface, you will find the I/O meters and a toolbar with Bypass, undo History, and Options menu buttons.

Figure 8.70
Nectar 2 pitch editor

The Nectar 2 Production Suite does succeed in providing all the tools you need for vocal production in one easy-to-use interface—and at an attractive price. The pitch editor is not as feature-packed as its rivals, Auto-tune and Melodyne, but it will get most basic jobs done! The breath control plug-in is the icing on the cake here—really helping to create natural-sounding vocals. The main Nectar 2 plug-in with its well-designed presets provides lots of creative options for vocal processing using its comprehensive set of signal-processing tools. If you are on a budget, buying the Nectar 2 Production Suite is the most affordable solution for vocal processing and will work perfectly well in most situations.

Summary

So the secrets for recording great vocals are, first, to work with a great vocalist, then use a great microphone with a great preamplifier—and, second, to make sure you don't mess things up by setting levels too high and causing overloads anywhere in the signal chain or by applying too much EQ, compression, or other effects!

Digital technology allows us to emulate just about any analog (or digital) hardware, including equalizers and dynamics processors, microphone preamplifiers, and even microphones themselves! Nevertheless, to capture the sound of any singer still requires a real microphone, which still uses analog technology, even if an A/D converter is built into the microphone casing and the audio is transferred to the digital audio workstation via a digital AES, S/PDIF, or USB connection.

Many engineers and producers still prefer to use analog hardware—classic microphones and mic preamplifiers at least. However, once the audio has been captured digitally, there are many excellent processors available as plug-ins to use 'in-the-box', as we have seen something of in this chapter (and will learn much more about in the next chapter).

There has been a lot of comment and discussion about the pros and cons of using digital technology to help recordings of singers and musicians to sound pitch-perfect and timing-perfect. My view is that the goal should always be for the producer to help the singers and musicians to achieve the very best performances that they are capable of delivering and then to use the technology as subtly and imperceptibly as possible to fix things that would really spoil the recordings if they were left unrepaired. There are lots of circumstances in which it is not a viable option to get a performer back into the studio to rerecord, and this is where Auto-Tune and Melodyne can play very important roles when producing 'in-the-box'.

And there are many new creative possibilities with these new tools—let's not forget that lots of people loved the Cher record and paid good money to buy

437

a copy, others like T-Pain, and many more enjoy the latest R & B chanteuse or those boy-band heartthrobs with their so obviously (and mostly deliberately) Auto-Tuned vocals. I am pretty sure that Miles Davis (a musician who always moved with the times) would have approved . . . even though I normally don't like to hear the 'Auto-Tune effect' used on vocals myself.

Nevertheless, as a busy producer working mostly 'in-the-box', I am extremely grateful for the new level of control these tools give me over the fine detail of my recordings—even allowing me to correct a wrong note within a guitar chord using Melodyne! O Brave New Digital World!

in this chapter

CHAPTER 9

Audio Signal Processing and Effects Creation

Introduction

Although balancing the relative levels of your audio tracks is the most crucially important aspect of mixing, signal processing probably comes a close second.

So what do we mean by *signal processing* anyway? This is a general term for applying various types of change to the audio—which is considered to be a signal just like any other kind of information that is communicated in some way. In music production, the term *signal processing* usually refers to applying processes such as EQ or compression, or adding reverberation, delays, or other special effects.

In analog studios, the mixing console would typically have EQ built into each channel, while more advanced mixing consoles might also have compression and possibly other effects built in. Reverberation, delays, and more sophisticated EQs, compressors, and other effects such as harmonizers, phasers, and flangers were typically available in external units, often referred to as *outboard gear*.

Pro Tools has a basic complement of signal processors provided as plug-ins. These include all the common types of signal processors and work well enough for straightforward tweaking of your audio. Various additional plug-ins are bundled together with the different Pro Tools systems and these can include both Avid-owned and Avid-distributed plug-ins, along with some third-party plug-ins.

These additional plug-ins are more akin to the traditional studio outboard gear that you would expect to be of higher quality and would expect to pay a premium for. But how good are they? Well, the best emulations of classic analog hardware devices are getting very close to the real thing, and with digital hardware devices, it is possible in some cases to get virtually identical versions in software these days. When he developed his first TDM plug-ins, Ray Maxwell at Eventide told me, 'We take the algorithms that we develop for our hardware processors (which happen to use Motorola DSP) and we use a special re-compilation engine that we created to convert the code to work within a Pro Tools plug-in. So when you buy one of these plug-ins you are getting exactly the same effects that we put into our hardware units.'

And there are distinct advantages to having all this signal processing available for you to use 'inside the box'—you don't need to use potentially noisy connectors and cables to hook everything up. So plug-ins typically provide much better noise performance than their hardware equivalents, and because they are implemented as software, they also provide virtually instant storage and recall of all the settings.

> **TIP**
>
> One thing I have come to realize is that there is something of a learning curve with what can seem to be the easiest pieces of kit to operate—even something with just one or two knobs and a couple of switches. Do not underestimate the length of time it may take you to become familiar with how to operate and get the best out of the signal processing plug-ins— even the apparently simplest of these may need your full attention for quite some time.

Equalization (EQ)

So-called *equalizers* were originally used in telephony applications to correct for frequency losses over distance on telephone lines—to 'equalize' the frequencies.

In music production, equalizers are normally used creatively to alter the tonal quality of the audio—and are typically called *tone controls* on consumer equipment (think of the treble, middle, bass, and presence controls found on many popular guitar amplifiers or hi-fi amplifiers).

You might also use EQ to remove particular frequencies: in other words, to 'filter out' certain frequencies from the audio signals. So you will sometimes see equalizers referred to as *filters*.

For example, a singer or musician might accidentally touch a microphone stand during a performance, which can produce a loud low-frequency 'bump' that is very difficult to remove. Or if your studio is not effectively soundproofed, the low-frequency rumble from a passing truck or a door slam from somewhere in the building can find its way onto your recording. Ideally you will record this performance again. But sometimes in the heat of the moment such things can get overlooked and it may not be possible or practical to rerecord the performance. In this case you will have to use all the tricks of the trade to try to minimize the problem without adversely affecting the rest of the recording. A high-pass filter may work, but you can also try various other types of EQ—and possibly even a multiband EQ or compressor.

You can also use automation to apply corrective changes solely to the affected clip of audio. Be careful, though: sudden changes can sometimes be too noticeable—so it can be better to apply an overall change. You always have to judge the situation for yourself by careful auditioning.

Bandwidth

Often, you will want to increase (boost) or decrease (cut or attenuate) a range (or band) of frequencies. The width of the band (i.e., the range) of frequencies that you boost or cut within the overall frequency spectrum is defined as the bandwidth of the EQ filter.

Some filter designs let you alter the bandwidth (sometimes referred to as the *resonance* or 'Q') of the filter that you are using. Make the Q 'narrower' and the range of frequencies included becomes less. Make the Q 'wider' or 'broader' and the filter's bandwidth is increased to encompass more frequencies.

> ### TIP
> You would use a narrower bandwidth EQ filter to reduce or remove some specific frequencies, to take out 'boominess', for example. You would use a broader bandwidth EQ filter to shape the sound more generally.

Basic Two-Band and Three-Band EQ

The most basic EQ that you would use with a mixing desk might just have a couple of bands of EQ—one for the lower frequencies and one for the higher ones.

Slightly more sophisticated EQ designs provide a frequency selector control that you can use to 'sweep' through a range of 'center' frequencies, either side of which the filter's bandwidth extends. Using this control you can focus on the exact range of frequencies that you want to modify. Typically you will find controls for three separate bands covering low, mid, and high frequencies. This type of EQ is sometimes called *semi-parametric EQ*, although it is more often called *sweepable EQ* for obvious reasons.

Parametric EQ/Bandpass Filter

Getting more sophisticated again, you will come across so-called *parametric equalizers*. These usually have three or more sweepable bands, each of which also has a resonance or 'Q' control to let you individually adjust the bandwidth of the EQ filters.

These filters are often called *peak* or *peaking filters* in reference to the shape of the frequency response curves for this type of filter that typically have a peak at the center frequency.

Another name for this type of filter is a *bandpass filter*, so called because you can alter the width of the band of frequencies that will pass through and therefore be affected by the filter. The other important parameter here is the center frequency of the band that is allowed to pass.

These filters are used to generally shape the tonal qualities of your audio recordings, especially when recording and mixing popular music. Orchestral, jazz, and most acoustic music genres rely much more on using good microphone techniques to capture the desired sounds directly from the sources, so there is less need for creative applications of EQ in these genres.

Notch Filter

An extreme version of a bandpass filter with a very narrow bandwidth is often referred to as a *notch filter*. You may also see this type referred to as a *band-stop filter* or *band-rejection filter*—names that reflect the function of the filter. This type of filter typically lets you cut (stop or reject) the frequencies within the narrow band, effectively cutting out a 'notch' in the frequency spectrum—hence the names.

The main application of notch filters is to cut out problem frequencies while leaving the rest of the frequency spectrum untouched.

High-Pass and Low-Pass Filters

You will also come across high-pass and low-pass filters. These do what they say—they pass high or low frequencies above or below the frequency that you set, which is called the *corner* or *cutoff frequency*. A high-pass filter cuts low frequencies while a low-pass filter cuts high frequencies, so you sometimes see them labeled as *low-cut* and *high-cut*, respectively, which is a more intuitive way of describing what you hear when you apply these filters.

These filters typically 'roll off' (decrease) the amplitude of signals below the corner frequency at a rate of 12 or 18 dB per octave—a parameter called the *slope* of the filter that can sometimes be selected by the user. Various corner frequencies, switchable between 40 or 80 Hz for low-cut filters or 12 or 16 kHz for high-cut filters, may be provided—or the corner frequencies may be continuously variable over much wider ranges.

Typically, you will use these filters to reduce or eliminate frequencies that you don't want, such as low-frequency rumbles or high-frequency noises or hiss.

Shelving EQ Filters

Shelving filters affect a broad range of frequencies either below a certain frequency in the case of bass controls or above a certain frequency in the case of treble controls. Rather than cutting or boosting frequencies progressively, this type of filter changes the gain equally at all frequencies below or above the cutoff frequency—so the frequency response graph looks as though it has a shelf at a higher or lower level than the zero gain-change (0 dB) level at the low-frequency or high-frequency end of the spectrum.

Two important parameters for shelving filters are the corner frequency, which is defined as the frequency at which the response is 3 dB above or below 0 dB, and the frequency at which the filter curve levels out to form a shelf, which is called the *stop frequency* (i.e., the frequency at which the filter's frequency response curve stops changing). The maximum slope of the changing part of the curve is typically 6 dB per octave.

> **NOTE**
> One disadvantage of shelving filters is that it is possible to boost frequencies above or below the audible range, which can lead to problems such as wasted amplifier power and even loudspeaker damage when used with public address and sound reinforcement systems.

Graphic EQ

Another EQ type that you might come across is the so-called *graphic equalizer*. This uses a simple but effective design. Typically, it will have a set of 10 (or even as many as 31) EQ filters each controlling a band of frequencies an octave (with 10 bands) or one-third of an octave (with 31 bands) wide.

Each frequency band has a vertically mounted fader that you can use to boost or cut the level and each fader is controlled using a button attached to a slider. When no boost or cut is applied, the buttons all line up to form a straight horizontal line on the front of the equalizer. When you boost or cut at various frequencies, the line becomes a curve that provides a graphical representation of the effect that the equalizer filters are having on the frequency response of the device.

So, you can see the effect of the changes that you are making to the audio passing through the graphic equalizer by looking at the shape formed by the buttons that you use to boost or cut the levels at the different frequencies and regarding this as a curve on a graph showing the frequency response of the set of filters.

Although perhaps the easiest type of EQ to use, graphic equalizers are not very popular in music production because they usually suffer from very poor phase response, which can adversely affect the sound quality.

EQ Plug-ins for Pro Tools

The plug-ins supplied with Pro Tools, although fairly basic, are actually very good and should not be overlooked. Nevertheless, there are lots of equalizer designs that engineers favor because they have used these in studios for many years—and plenty of these are now available as plug-ins from companies such as Universal Audio. New designs are also being developed that take advantage of the digital domain to do things that their analog counterparts could never do—and new user interfaces are also appearing from developers such as FabFilter.

Sonnox Oxford EQ and Filters

If you are looking for just one extremely capable EQ plug-in, that would have to be the Sonnox Oxford EQ and Filters—see Figure 9.1.

Figure 9.1
Sonnox Oxford EQ
and filters

Using the same design criteria as the EQ found in Sony's top-of-the-range Oxford OXF-R3 recording console, the Sonnox Oxford Equalizer and Filters plug-in provides just about all the control you would ever want over the EQ types together with a large frequency response graph that helps a lot when you want to visualize how the various EQ filters are interacting. It has five bands of EQ, with selectable shelf settings on LF and HF sections, plus high-pass and low-pass filters to filter the extremes of the audio frequency range. It also has four different EQ types that cover most of the EQ styles currently popular among professional users, including several important legacy styles.

You can also buy a GML 8200 option for the Oxford EQ plug-in. This adds a highly accurate emulation of the GML 8200 EQ designed in collaboration with GML for the OXF-R3 console. This emulation has all the finer characteristics of the classic analog outboard unit, faithfully reproducing all the control ranges and responses of the original EQ, even to the point of producing center frequencies up to 26 kHz whilst running at 44.1 kHz or 48 kHz.

449

McDSP 4020 Vintage EQ

McDSP's 4020 Vintage EQ features a 'retro'-styled user interface that will immediately feel familiar to most users—see Figure 9.2.

There are four EQ bands—low-frequency LF shelf, low-mid frequency LMF parametric, high-mid frequency HMF parametric, and high-frequency HF shelf—together with high-pass and low-pass filters. Other controls include input level and output level, and the 4020 has phase invert switches for each channel.

What's unique about this plug-in is that the slope (Q) of the shelving and parametric equalizers varies as the gain is adjusted. The filter section uses a new design that varies the slope from a gradual 6 dB/octave at the selected frequency to a maximum of 24 dB/octave. Also, both the EQ and filter section frequency controls are calibrated in octaves (instead of decades). The EQ section frequency ranges do not completely overlap; however, they are designed for the optimum range their dynamic shape can cover in the audio spectrum.

Figure 9.2
McDSP 4020
Vintage EQ

The 4020 Vintage EQ is one of my go-to plug-ins whenever I want to tweak a drum kit, brighten up a guitar part, or quickly tame some vocals.

McDSP AE400 Active EQ

Now let's get serious! As most engineers and producers know, there will be situations where the audio in a track changes in tone quality during the course of the performance because the singer or player is performing more softly or more loudly. If you get the EQ just right in one part, it will be wrong in other parts. An 'active' equalizer—which can vary its gain based on the incoming signal levels—is the best tool to use to deal with such situations.

McDSP's AE400 has four bands of fixed and active equalization—see Figure 9.3. Each band is completely overlapping with all other bands and has its own Q (bandwidth), fixed gain, and active gain controls. A key filter for each band allows the active gain response to be as selective or as broad as needed, based on the Q (bandwidth) control. Input and output controls, individual band bypass, band control linking, and band key signal monitoring round out the features of the AE400.

Figure 9.3
McDSP AE400
Active EQ

Active parametric equalization, as implemented in the AE400, has the same three main controls—gain, freq, and Q—and operates similarly to a fixed equalizer, with the useful exception being that the gain varies from 0 dB (no effect) up to the selected gain as the signal rises above the selected threshold and at a rate determined by the attack (time needed to reach selected gain) and release (time needed to return to 0 dB). Additional adjustment in overall sensitivity is controlled by the ratio control, which lets you set the input signal level sensitivity to further control how the active EQ will reach maximum active equalization—a feature unique to the AE400.

To get the AE400 set up, you can start by using it like a standard fixed parametric EQ. Using the output, freq, and Q control text fields in any of the four bands, or the 'dot on the plot', the fixed EQ response can be adjusted easily. Or you can just drag the mouse over the output, freq, and Q text fields to update them.

Once the fixed EQ response is created, then the active EQ response gain control can be dialed in via the vertical slider next to the IO curve plot in each band. When you do this, a transparent curve representing the maximum equalizer response appears behind the fixed equalizer response. The total (maximum) equalizer response then becomes the sum of the fixed equalizer response and the maximum active equalizer response.

When you start playback and slowly decrease the threshold control, the peak meter text readout will show the peak values of the key signal for each band. When the threshold control is below the values shown in the peak meter text readout, the solid line representing the current (active) equalizer response will start to move toward the maximum active equalizer response. Adjustments to active equalization sensitivity (ratio) and response time (attack/release) can also be observed in the movement of the current (active) equalizer response curve.

The AE400 also offers an inverted threshold mode in which the active gain is 0 dB (has no effect) until the signal level drops below the threshold—reversing the behavior of the active equalization. When the 'INV' button located underneath the threshold control slider is pressed, the current (active) equalizer response will start to move toward the maximum equalizer response when the signal level is BELOW the threshold level.

The key signal for each band can be the original audio input or a side chain input—and this key signal can be auditioned by pressing the speaker icon button at the bottom right of each band.

TIP

A typical application would be to help the lead vocal to 'sit' properly in your mix. You can make sure that the vocal always has space in the mix using the AE400's side chain features. Here's how to do this: Route the main mix through the AE400, and the vocal into the side chain input. Sweep the key filters around using the 'Key' plot display button until you find which frequencies the vocal signal levels are at. You can use the key peak text readouts to assist in this effort, along with the key monitoring button—the one with the speaker icon at the bottom right of each band of controls. When you have identified the vocal frequencies, reduce the gain of these by 2 or 3 dB using a Q setting of, say, one (for one octave) so that the side chain will cause the AE400 to actively EQ the mix using the vocal's own frequency response. Now when you play your mix through the AE400 you should hear the vocal much more clearly despite competition from the other mix elements!

FabFilter Pro-Q

FabFilter Pro-Q (see Figure 9.4) is an extremely high-quality EQ plug-in with an innovative user interface. Key features include up to 24 separate EQ bands (bell, low cut, low shelf, high shelf, and high cut), state-of-the-art filter algorithms with precise analog modeling and unlimited internal headroom, zero-latency or linear-phase processing, mid/side support, customizable stereo placement for every band (stereo, left/mid, or right/side), 6 dB, 12 dB, and 30 dB display ranges, and one of the easiest, yet most powerful, EQ interfaces I have encountered.

At the top right of the user interface you can browse through the factory presets or save your own settings using the preset buttons. The Undo, Redo, A/B, and Copy buttons enable you to undo your changes and switch between different states of the plug-in. The Help menu provides access to help and version information. One of the 'friendliest' design features is the way that lots of pop-up help messages appear as you hover the mouse over different parts of the user interface—helping you to learn how to use Pro-Q.

The selection controls below the main display let you change the settings of the currently selected EQ bands.

453

The bottom bar offers advanced features such as the processing mode (zero latency or linear phase), channel mode (stereo or mid/side), a spectrum analyzer, a global bypass, and output level and panning.

Figure 9.4
FabFilter Pro-Q

The main part of the user interface presents the interactive EQ display that lets you add and edit EQ bands. To add a new EQ band, simply click on the yellow overall curve and drag it up or down. If you Right-click on the dot for an EQ band, a pop-up menu will appear with various options that you can tweak.

When you move the mouse cursor near to an EQ dot, a parameter value display pops up showing the current parameter values for the corresponding EQ band. Click and hold the solo button (with the headphones icon) to enter Solo mode for the current EQ band. The other EQ bands will dim, just like the yellow overall curve—as can be seen in Figure 9.5. In Solo mode, you won't hear the effect of the EQ band itself, but instead you will hear the part of the frequency spectrum that is being affected by that band. When using Solo mode with low-cut or high-cut bands, for example, you will hear the frequencies that are being cut away instead of the frequencies that pass, which helps you to determine whether you are cutting the right frequencies. Generally, Solo mode aims to expose the parts of the incoming audio that matter to the current EQ band, but that you can't hear just by listening to the regular EQ sound.

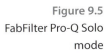

Figure 9.5
FabFilter Pro-Q Solo
mode

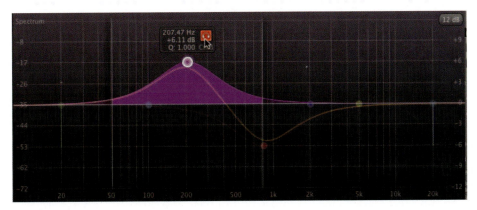

One of FabFilter Pro-Q's best features is that it's very easy to equalize both stereo channels in a different way. This is a great way to surgically remove unwanted sound artifacts, or even to add stereo effects. To make this even more powerful, Pro-Q offers both Left/Right and Mid/Side Channel modes. In the default Left/Right mode, each EQ band works either on both stereo channels or on the left or right channel only. This is controlled by the stereo options at the left-hand side of the selection controls:

- Click the L or R button to let the selected bands affect only the left or right channel.

- Click the Stereo button (in the middle) to let the selected bands affect both stereo channels.

- Click the split button underneath the buttons to duplicate the selected band, making two identical copies, one operating only on the left channel and one operating only on the right channel. This makes it very easy to slightly adjust one of the channels.

As soon as one or more of the EQ bands are operating on a single channel, the EQ display switches to Per-Channel mode, where it shows two overall frequency response curves: a white one for the left channel, and a red one for the right channel.

The Channel mode pop-up selector in the bottom bar lets you switch between left/right and mid/side operation. In Mid/Side mode, the incoming stereo signal is converted into mid (mono) and side parts, which you can then easily filter independently. This is often an even better way to fix artifacts or modify stereo information because it represents the stereo signal in a more natural way. In Mid/Side mode, the display shows the two overall frequency response curves in white (Mid) and light blue (Side) so you know at a glance in which mode Pro-Q is currently operating.

TECHNIQUES

Independent channel equalization is very useful when dealing with stereo audio containing unbalanced frequency content over the stereo field. Let's say you want to combine a stereo drum recording with a stereo acoustic guitar recording. The drum recording contains more low-mid frequencies in the left channel (for example, a low tom-tom), and more high frequencies in the right channel (like cymbals or a hi-hat). The guitar sound, recorded with a mic capturing the soundboard/hole panned left and one capturing the fretboard/neck panned right, might have similar frequencies as the drum recording, making it hard to combine them in a balanced way. By using independent left/right channel EQ-ing, it is possible to balance these elements so that they do not fight each other. Instead of EQ-ing the whole stereo track of the drums and guitars one can simply EQ where it is necessary to get the two elements to complement each other.

Mid/side EQ is perhaps most commonly used to bring some stereo elements further up within a recording, either by cutting certain freq-uencies in the mid channel or by boosting the wanted frequency range in the side channel. It is great for adding a bit of depth to typical hard-panned rock or heavy guitar recordings where you boost the 'bite' frequency range of the guitars (around 2–4 kHz) with quite a narrow EQ. Combining this with cutting some of the 'mud' away from the side channels will give the illusion of huge guitars that still sit well within the mix.

Independent mid/side equalization is also often used during mastering. For example, raising high frequencies in the side channel can freshen up the sound, while a low-cut filter in the mid channel can work very well to clear up the low end.

Consider using linear-phase processing when filtering both stereo channels (either in Left/Right or Mid/Side mode) differently to avoid introducing unwanted phase changes.

To help you judge the effect of the combined EQ bands on the incoming audio signal, FabFilter Pro-Q also includes a powerful real-time frequency analyzer. You can switch this on and off using the pop-up selector at the bottom of the user interface window. Here you can choose whether to display the spectrum of the incoming audio, or of the audio that has been processed by the EQ, or of both. These spectra are then displayed in the background of the main EQ display—see Figure 9.6.

Figure 9.6
FabFilter Pro-Q
frequency analyzer
display

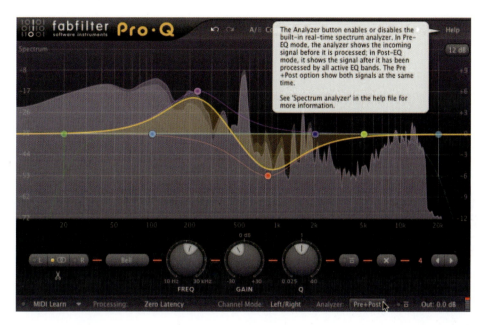

Figure 9.6
FabFilter Pro-Q
frequency analyzer
display

Two other pop-up selectors at the bottom of the user interface window allow you to activate the MIDI Learn mode, and to switch between the default Zero Latency mode and the four available linear phase modes. Linear phase filters change the phase in the same way for all frequencies, ensuring that no phase cancellations occur as a result of mixing unprocessed audio with audio that has been EQ-ed using Pro-Q.

Pro-Q is undoubtedly one of the finest EQ plug-ins available for Pro Tools today. It is a pleasure to use and from a technical standpoint it is absolutely first-rate!

Flux Epure V3

Flux Epure V3 (see Figure 9.7) is a state-of-the-art five-band equalizer that uses a state-space implementation to produce the best possible signal-to-noise ratio whatever the settings. According to Flux, 'Usually digital equalizers damage this ratio by increasing the noise level for negative gain values. Epure's algorithm ensures the best quality possible for a digital equalizer.'

The user interface features an effective frequency response display that shows the filter curves in different colors with shaded areas that reveal the overall response. The resolution of this display changes automatically when you boost or cut beyond the resolution currently being displayed, making it easier to see what you are doing.

A flexible signal routing matrix is hidden away beneath a small Setup button just below the master gain control. This allows you to use mid/side encoding on input and output, and, by splitting the controls of the two channels via the group matrix, you can EQ the mid (sum) signal differently from the side (difference) signal.

As with all the Flux plug-ins, Epure provides two preset slots referred to as slot A and slot B positioned at the lower left and lower right of the user interface window, just above the preset Save, Recall, and Copy buttons. These slots provide access to two sets of parameter settings that can be recalled for each slot individually so you can alternate between these settings. A morphing slider is also provided that allows you to morph between the two slots with their corresponding settings—allowing for some very creative tweaking. To open the preset manager, just click on either of these slots. To enable automation for read and write, click on the Automation button below the morphing slider.

Figure 9.7

Flux Epure

SoundToys FilterFreak

FilterFreak is not a technical tool to use for corrective equalization—it's a creative tool that you can use to process your sounds in myriads of ways. Two separate plug-ins are provided—FilterFreak 1 (see Figure 9.8) and FilterFreak 2. Both of these come with hundreds of presets that can be used in an extremely wide range of contexts.

Figure 9.8
SoundToys
FilterFreak 1

A mix control lets you set the balance between the 'wet' filtered sound and the 'dry' original audio—very useful when using this as an insert effect.

You can also add distortion using the Analog mode toggle switch and its associated 'flavor' pop-up selector button to choose how FilterFreak will distort or 'saturate' as the signal input increases. In 'analog' mode, FilterFreak saturates in a similar way to the way real analog gear responds, adding a certain amount of distortion at all signal levels but particularly at high levels. With the analog mode switch off, FilterFreak is in 'digital' mode, so higher signal levels will clip in the typically nasty, crunchy digital way—which can also be a desirable effect at times.

FilterFreak actually has a number of different modulation modes that can be selected using the button positioned at the lower right of the window. Using this pop-up mode selector, you can choose between the normal LFO mode and five more sophisticated modes: Envelope, Random, Random Step, ADSR, and Rhythm.

The FilterFreak LFO has a wider range than most—up to 100 Hz. Modulating the filter with a repeating LFO produces an 'Auto-Wah' effect. This can also be used to create many other effects depending on the waveshape, filter, poles, mode, resonance, and so forth.

The Rhythm mode is a much more sophisticated version of the LFO that lets you sync the LFO to the tempo of your Pro Tools session, producing complex filter modulations that can be programmed in musical and rhythmic ways.

> **TIP**
>
> One of FilterFreak's coolest features is its ability to synchronize its filter sweeps to your music's tempo—and to the downbeat (the '1' of 1, 2, 3, and 4) of your music using MIDI clocks.

The Envelope mode employs an 'envelope follower' that 'follows' the volume level of the input signal and dynamically controls the amount of filter modulation based on volume changes in the input signal. Envelope mode works great on audio that dynamically changes in volume in rhythmic ways—such as guitar or drums.

The Random S/H (Sample and Hold) mode produces a waveform that jumps from one value to another at each cycle—good for special effects. The Random Step mode combines the sample and hold effect with an envelope follower. Instead of changing to a new value at a specific set rate, a new random value is triggered when you press the Trigger button, in response to a MIDI note trigger, or when the input signal exceeds the threshold setting. This works really great on drums and other highly percussive signals, and can be used to create a dynamic effect that varies with each audio event.

ADSR mode is a recreation of the standard envelope generator found on most synthesizers. ADSR stands for Attack, Decay, Sustain, and Release. With the ADSR you can define a specific envelope shape that will be used to modulate the filter each time it receives a trigger based on the level of input signal. This is quite a bit different than the Envelope mode, whose shape changes and responds dynamically to the input signal. In FilterFreak, the ADSR is triggered by pressing the Trigger button, by receiving a MIDI note event, or when the input signal exceeds the threshold setting.

FilterFreak 2 (see Figure 9.9) is identical to FilterFreak 1 except that it provides TWO separate filters that can be used together to create an even greater set of filtering effects! The filters themselves are identical but can be set totally independently of each other. You can combine a low pass with a band pass, a notch with a high pass, a low pass with a high pass, and so forth to create all sorts of different filter sounds. A serial/parallel switch determines how the incoming audio signal is fed through FilterFreak's two filters. Switching on the link switch 'links' the frequency, resonance, and gain controls of the two filters together.

Figure 9.9
SoundToys
FilterFreak 2

Dynamics Processors

Automatic control of audio signal levels, also known as *dynamics processing*, started out as a means to prevent levels exceeding available dynamic range—to avoid waveforms becoming 'clipped', as in a 'limiter'. This was also developed as a means of controlling the level from a singer's microphone or from a bass guitar or other instrument, as in a 'compressor'. Today, dynamic signal level control has grown in complexity and popularity to become an extremely important part of the sound production process. Compressors are combined with limiters or with gates to keep out unwanted noise while expanders can be used to bring up the level of quieter signals.

Compressor

A compressor is used to reduce the highest level peaks in the audio to make it sound more even in level and to allow the average level to be raised without the peaks exceeding the maximum level that the system can handle without distortion.

When the average level is increased in this way, any low-level audio (a softly sung word, or a quietly played note, or whatever) becomes much more audible. This can be very important in popular music to make sure that the vocals can be heard properly and to make sure that all the other instruments work together smoothly and supportively in the mix.

Compression is used very sparingly in classical and orchestral recording, and should be used very carefully when recording jazz and acoustic music. In these types of music the full dynamic ranges of the instruments are used to create the effects that the composers and performers intend, so compression can ruin these recordings if applied too liberally.

Compressors have a threshold control that is used to set the level beyond which compression is applied to the audio. Most of the audio should pass through the compressor without being affected (unless you are trying to create special effects).

Audio that exceeds the chosen threshold level is reduced in level according to the compression ratio. With a 2:1 compression ratio, if the input level increases above the threshold by 2 dB, the output level only increases by 1 dB. With a 10:1 compression ratio, if the input level increases above the threshold by 10 dB, the output level only increases by 1 dB, and so on.

If the compressor produces an abrupt change as soon as the threshold is reached, it is said to have a 'hard knee'. This is a reference to the shape of the curve on the compressor's transfer characteristic (a graph of input against output level). If the change is smoother, and looks more rounded on the graph, this is said to be a 'soft knee'. Some compressors let you switch between these two types.

Soft knee settings are best used with high compression ratios, especially with vocals, or when using the compressor as a limiter. Hard knee settings are useful for extreme limiting when there are lots of transient peaks to deal with.

Other control parameters include the 'attack time', that is, the time it takes for the compression to start working after the threshold has been reached, and the 'release' time, that is, the time it takes for the compressor to stop affecting the audio after the input level has fallen back below the threshold.

You should set the attack time in relation to the type of audio you want to compress. If you make this too fast, it can take away the impact of the sound. On the other hand, if it is too long, it may not act quickly enough to control the dynamics properly.

You can soften drums and other percussive sounds by using a short attack time to apply compression to the transient at the beginning of a percussive sound. Or you can shorten sounds using longer release times to make the decays quieter. You can also bring sounds forward or backward in the mix by carefully adjusting the compressor's attack and release times. It is even possible to completely alter the amplitude envelope, to make the audio sound as if it is going backward, for example, by using a very short attack time with a very long release time.

These can be useful techniques for pop music, but they do make the instruments sound unnatural, so you would usually want to avoid this with most other genres of music. If the attack or release times are set too short, this can produce a click at the beginning or end of the sound. Again, this can be used as a creative effect in pop music but is unlikely to be what you are looking for in other genres. You should also avoid using too short a release time with too high a ratio because you will hear the audio level changing if the compression effect is released too quickly. This sounds like the audio is 'breathing' or 'pumping'. If the audio levels are fluctuating a lot, a longer release time will help to smooth out these fluctuations.

Another unwanted side effect would be low frequency distortion. This can occur especially with bass instruments, when the attack or release times correspond with the time taken for half a cycle of the waveform to take place. At low frequencies, which have long wavelengths, this can happen, in which case the compressor acts on each positive and negative cycle of the waveform, distorting the shape of the waveform and consequently the sound. Some compressors (e.g., Sonnox Oxford Dynamics) have a hold control that prevents the compressor from acting for a short period of time so that the attack and release times can be set to short without affecting the waveform in this way.

With slow, legato music and especially with instruments such as sustained strings, organ, or synthesizers, you can use slower attack and release times very successfully. With faster, more percussive music, you will find that faster attack times with shorter release times work best.

Some compressors let you choose whether to use peak detection or RMS averaging detection. Peak detection works best with percussive sounds while RMS averaging, which corresponds more closely with how we perceive loudness, works best with vocals and provides more transparency for acoustic instruments.

When you have adjusted all these parameters correctly, preventing the peaks from rising as high as previously, the overall level of the audio will often sound too quiet. Most compressors provide a 'makeup gain' control that lets you increase the output level from the compressor to make up for this loss of 'gain' or reduction in apparent level.

So, generally speaking, a compressor compresses the dynamic range of the audio by reducing peaks so that it sounds smoother and allows the average level to be raised accordingly, making it sound louder. It can also be used for a variety of creative effects, making sounds 'fatter' or emphasizing inner detail, especially when used to alter the amplitude envelopes of elements within the mix.

Limiter

A limiter is an extreme version of a compressor that uses a ratio setting somewhere between 10:1 and infinity:1 (or, perhaps more typically, between 20:1 and 40:1) to prevent the output from exceeding the chosen threshold level by any significant amount.

It is possible to use any compressor that offers ratios above 10:1 with relatively short attack and release times as a limiter. However, you will often encounter limiters that use preset ratio, attack, and release parameters.

Extremely short attack and release times are used to make it more difficult to hear the gain changes as the peaks are reduced in level. If gain changes are applied to successive peaks and the audio has lots of peaks, these gain changes are likely to become audible. So limiting should only be used to remove occasional peaks. If there are lots of peaks, the gain should be reduced until fewer peaks cross the threshold and need to be limited.

> **TIP**
> A typical application of limiting in music production would be to control the loud 'snaps' that a bass guitarist might play in between fingered notes played at a lower volume. The threshold would be set so that the limiter would only act to limit the snaps without affecting the rest of the bass notes.

Side Chain Processing

The way dynamics processors work is to first detect the amplitude of the input signal to check whether this exceeds the threshold or not, then, if it does, to use this to trigger the gain change.

This detected signal can be replaced in many dynamics processors by an external 'side chain' key signal. The external key input lets you trigger the dynamics processing from another audio track or audio source.

Side chain filters can also be used to make the dynamics processing react to particular frequency ranges. So, for example, you could use the low frequencies from a bass drum to trigger the dynamics processing.

De-esser

When you are working on vocals, you will often encounter problems with S's, and F's, T's, P's, and K's and similar sounds, which have high-frequency peaks that can cause distortion and sound annoying. These problematic frequencies are called *sibilant frequencies*.

A de-esser is designed specifically to alleviate problems with S's and other sibilant sounds by applying fast-acting compression just to the sibilant frequencies. A frequency control sets the frequency band in which the de-esser operates (typically 4–10 kHz) and a threshold or range control sets the level above which compression is applied to the sibilant frequencies without affecting the rest.

NOTE

You may already be using a compressor on the vocal with a relatively slow attack time that works well for most of the material but seems to make the problem with the sibilant frequencies even worse. This happens because the sibilant sounds have very fast transient attacks that even a reasonably fast attack time on the compressor will not catch.

TIP

If you are already using a compressor or limiter on the vocal, you should apply the de-essing after these processes in your effects chain, inserting the de-esser after any compressor or limiter plug-ins.

Expander

An expander is the opposite of a compressor—it expands dynamic range rather than compressing it. Expanders can be used to reduce signal levels when the input signal falls below the selected threshold—a process typically referred to as *downward expansion*.

Expanders can also increase signal levels when the input signal falls below the selected threshold—typically referred to as *upward expansion*. This can be useful in bringing out quieter details in a track without increasing the signal levels of other parts.

Expanders have threshold, ratio, attack, and release controls, just like compressors, and usually feature a range control that lets you set the lowest level to which the signal will be reduced once it falls below the threshold.

The ratios on an expander are the inverse of those on a compressor but typically lower numbers such as 1:2, 1.5:2, or 1:3. With a ratio of 1:3, for example, when the input drops 1 dB below the threshold the output drops 3 dB below the threshold. In other words, when the input signal falls below the threshold, the expander lowers the level even further, expanding the dynamic range and making low-level sounds even quieter.

> ### TIP
> Expanders can be used creatively to alter the balance between louder and softer sounds. For example, a recording of a rhythm played on a drum or cymbal with a pattern containing both strong and weak notes can be modified using an expander so that the weaker hits get softer in relation to the stronger hits. So now you know how to make those hi-hats really 'groove'!

Noise Gate

An expander can be used as a noise gate. When the input signal falls below the threshold, the range control can be set such that the expander makes the quiet sounds so quiet that they cannot be heard, or even such that it will not pass any audio from the input through to the output—that is, the gate closes. When the audio input signal rises above the threshold again, the gate opens.

> ### NOTE
> Noise gates can be used to remove unwanted low-level sounds between passages of wanted higher-level sounds. For example, you could use a noise gate to remove any low-level sounds from other drums that are spilling into the snare microphone and becoming audible in between the snare hits.

A gate has similar controls to an expander. The attack time is the length of time that it takes for the output to change from full on to full off once the gate has been triggered, so attack times are typically as short as one millisecond. The release time is the length of time that it takes for the gate to open after the threshold has been exceeded. The range control lets you set the level to which the signal is reduced when the gate is triggered to be off.

> ### TIP
> A gate can also be used to affect the envelope of the input signal, using the attack and release controls. If you have a snare that is not snappy enough, for example, you can set the gate threshold level high enough to be roughly equal to the level of the initial 'crack' of the snare sound. Playing around with the gate's attack and release parameters then affects the way you hear the attack of the snare.

Dynamics Plug-ins for Pro Tools

Dynamics control is an essential part of the mixing process for most types of music. As with equalizers, Pro Tools comes with a basic set of dynamics plug-ins that work fine—especially in straightforward situations. Experienced engineers and producers will, of course, have their favorites from the physical world, and more of these are becoming available as plug-ins. However, the digital world does allow for much more sophisticated designs than were available in the analog world and companies such as McDSP, Sonnox, Flux, and even Avid itself have developed tremendously powerful new dynamics processors that can give you far more control over your audio than was possible previously.

McDSP DE555 De-Esser

The DE555 de-esser (see Figure 9.10) uses a look-ahead design to analyze the incoming audio and automatically adjusts the compressor's threshold to provide consistent, transparent, natural-sounding de-essing at any signal level. For example, if a portion of dialog is at −24 dB (a whisper), and later at −2 dB (a shout), the DE555 would still de-ess by the amount specified using its range and ratio controls. With rival de-essers that do not have this feature, you will inevitably spend more time adjusting (or automating) the manual threshold controls found on these.

The DE555 also has a high-frequency HF-only mode that will only reduce the signal levels of the sibilant frequencies and not those of the main vocal or speech frequencies (which can help to produce a more natural-sounding result).

Figure 9.10
McDSP DE555 de-esser

Sonnox Oxford Supresser

Designed to be the last word in de-essing, the Sonnox Oxford Supresser (see Figure 9.11) goes a long way toward achieving that goal. Although primarily a de-esser that you can use to reduce sibilance and fricatives in vocals and the 'spirant' artifacts associated with wind instruments, the Supresser can also be used to remove low-end plosives and thuds from vocals without affecting nearby components in the frequency spectrum, keeping the low end intact.

Figure 9.11
Sonnox Oxford
Supresser in Easy
View mode

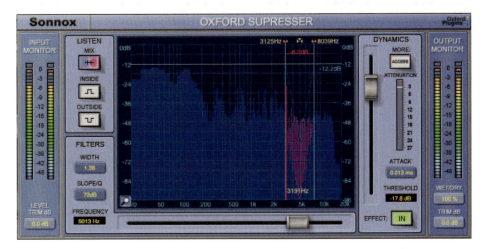

The central graphical display provides excellent visual feedback, allowing quick identification of troublesome frequencies and where to set the threshold. The threshold level and peak hold level of the user-definable frequency band that will be processed are shown on the graph, together with a Fast Fourier Transform (FFT) display of the signal that falls within this band. The peak level and the frequency containing the most energy within this band are retained visually within this display.

NOTE

When you first insert the plug-in, the display defaults to Easy Access View with a zoomed-out graphical display, so you can see the overall picture of where the center band is within the overall spectrum. Whenever you move the center band in any way, the center frequency of the band is clearly displayed in yellow at both the top and bottom of the display. So if you are aiming for a particular frequency, you can always see the center frequency clearly as you move toward this. To view additional controls, click on the button at the top right of the dynamics section to switch to More Access View.

Additional controls in More Access View include mode button switches for trigger and audio, auto level tracking controls, reaction envelope controls, and compression ratio and knee controls—see Figure 9.12.

Figure 9.12
Sonnox Oxford
Supresser in More
Access View

The Supresser uses an enhanced version of the compressor section from the Oxford dynamics plug-in together with a pair of linear-phase crossover filters taken from the Oxford EQ plug-in to make the compressor react only to the defined frequency band. Three listening modes allow the user to listen to the mix (Mix), the output of the bandpass filter (Inside), or the output of the band-reject filter (Outside).

The Supresser is very easy to use: once the frequency band of interest has been defined, you simply lower the threshold fader until the gain reduction meter starts to kick in. The plug-in uses its Auto-Level-Tracking mode to automatically track the general signal level so the threshold follows accordingly and the same relative amount of gain reduction is applied as the signal level rises and falls—perfect for vocals!

By default, the Supresser feeds the band signal to the compressor so that only the in-band frequencies are affected when the threshold is exceeded. 'Wide' band operation can be used instead, allowing full-band compression reacting only to specific frequencies. Or any combination of band and wide can be used to define when the compressor reacts and what it compresses.

Another excellent feature is the wet/dry control at the output stage that lets you mix some of the original signal back into the output together with the compressed signal.

The Supresser comes with plenty of presets that serve as good starting points—see Figure 9.13. Alternatively, the advanced section provides access to many more controls to allow precise corrective treatment or creative effects.

Figure 9.13
Sonnox Oxford
Supresser presets

FabFilter Pro-DS

FabFilter's Pro-DS (see Figure 9.14) is not only an excellent de-esser, but also incorporates an 'Allround' mode that works well for high-frequency limiting of any material, including drums and full mixes.

Key features include highly intelligent single vocal detection to detect sibilance in vocals, transparent program-dependent compression/limiting, high-quality 64-bit internal processing, adjustable threshold, range and detection HP and LP filtering, wide band or linear-phase split band processing, optional look-ahead time up to 15 ms, adjustable stereo linking with optional mid-only or side-only processing, and up to four times linear phase oversampling.

As with all FabFilter plug-ins, the user interface is a joy to use. At the top right of the user interface you can browse through the factory presets or save your own settings using the preset buttons. The Undo, Redo, A/B, and Copy buttons enable you to undo your changes and switch between different states of the plug-in. The Help menu provides access to help and version information.

In the upper part of the plug-in, the real-time moving level display and the level meters show you at a glance what's happening to your audio. The unaffected output level is shown in transparent grey, while affected parts (where gain reduction is applied) are shown in light green. At the right of the interface are meters for output level meter and gain reduction.

At the left are two large yellow knobs to control threshold and range with a pair of sliding controls underneath these to set the high-pass and low-pass filter corner frequencies. These controls may be used to set the amount of gain reduction and the frequency range over which the de-esser will trigger.

To the right of the large knobs, a pair of buttons allows you to select either the Single Vocal or the Allround mode.

> **NOTE**
>
> In Allround mode, triggering only depends on the frequency range, specified by the HP and LP filtering sliders, in combination with the threshold setting. This is intended for processing entire mixes, for example.

Another pair of buttons below these lets you select either wide band or split band processing. When working with single vocals, the wide band option often gives the best results, but when de-essing full mixes or more complex audio, it's often best to use the split band mode and leave the lower frequencies untouched.

> **TIP**
>
> To brighten up your sound during mastering, you can use Pro-DS as a high-frequency limiter. Choose the Allround mode and 'Split Band Processing', and limit the transients of the high frequency range. Then, bring up that frequency range again using a high shelving EQ filter (from FabFilter Pro-Q, for example).

At the bottom of the window, two smaller buttons let you select Internal or External Side Chain mode.

> **TIP**
>
> You can use the side chain buttons to choose whether to use the normal input signal or an external side chain input signal to trigger the de-esser. For example, if you decide that a heavily processed, distorted vocal track needs de-essing, and you still have access to the unprocessed vocal track, you could use this original 'clean' vocal as a side chain input. This would achieve better results than attempting to directly de-ess the distorted vocal track.

Next to the level meters, at the right of the interface, you can control the amount of stereo linking (with optional mid-only or side-only processing) and look-ahead time. At the bottom of the window, a pop-up oversampling selector lets you set either two times or four times internal oversampling, which reduces possible aliasing for fast/aggressive de-essing at the cost of additional CPU usage. To the left of this, a MIDI Learn button lets you associate any MIDI controller with any plug-in parameter.

Figure 9.14
FabFilter Pro-DS

With the look-ahead knob, Pro-DS can be set to start de-essing up to 15 ms before the trigger audio level actually exceeds the threshold. This is an excellent way to catch transients and/or the start of sibilance. You can enable or disable look-ahead with the Lookahead Enabled button at the top right of the look-ahead knob.

When choosing a look-ahead value, it can be useful to enable the Audition Triggering mode using the round button with the small headphone icon at the top left of the threshold knob. This way, you can hear right away whether you're catching all of the detected sibilance.

All in all, Pro-DS is a very capable de-esser that is flexible enough to be used for a variety of other specialized audio signal processing applications—adding significantly to its appeal.

FabFilter Pro-G

FabFilter's Pro-G (see Figure 9.15) is an extremely versatile expander/gate plug-in. Key features include five program-dependent expander/gate algorithms including an upward expansion style; high-quality 64-bit internal processing; adjustable threshold, ratio and range, attack, and release settings; adjustable hold time up to 250 ms; optional look-ahead time up to 10 ms; mid/side processing; up to four times linear phase oversampling; and an expert mode that offers highly customizable side chain options.

Figure 9.15
FabFilter Pro-G

At the top right of the user interface you can browse through the factory presets or save your own settings using the preset buttons. The Undo, Redo, A/B, and Copy buttons enable you to undo your changes and switch between different states of the plug-in. The Help menu provides access to help and version information.

In the middle section of the plug-in, the real-time moving level display and the level meter show you at a glance what's happening to your audio. They show the output level in light blue on top of the input level in dark blue, with a 60 dB scale. In the foreground of the level display is an input/output transfer curve with thin dashed horizontal and vertical lines overlaid to show the threshold levels—making it easy to set up the gate/expander.

At the right of this display, a meter shows the output level versus the input level.

To the left of the level display are the controls that determine the range and amount of gating/expansion: threshold, ratio, and range. To the right of the level display are controls that affect the reaction speed, hold time, and look-ahead time of the gate/expander algorithm, together with a style selection button and a custom knee setting.

NOTE

With the look-ahead knob (often also called *pre-open*), the gate/expander can be set to open up to 10 ms before the audio level actually exceeds the threshold. This is an excellent way to preserve transients while avoiding ultra-fast attack times that might cause distortion or aliasing. You can enable or disable look-ahead with the Lookahead Enabled button at the top right of the look-ahead knob.

At the left of the bottom bar, the MIDI Learn button lets you easily associate any MIDI controller with any plug-in parameter. The oversampling setting sets the amount of internal oversampling, which reduces possible aliasing for fast attack or release settings at the cost of additional CPU usage. On the far right of the bottom bar, you can bypass the entire plug-in and adjust the initial input and final output levels.

TIP

The gate/expander in FabFilter Pro-G can also be also triggered by MIDI notes. If at least one note is on, the gate/expander behaves as if a 0 dB signal is entering the side chain at this moment, so it is fully triggered. The 'Disable/Enable MIDI' option in the MIDI Learn menu turns this off or on as well.

One of the best things about Pro-G is the excellent range of presets provided. These include settings for bass, drums, guitar, vocal, bus, and FX—see Figure 9.16.

Figure 9.16
FabFilter Pro-G
presets

Bass ▶	
Bus ▶	
✓ Drums ▶	Added Dynamic Range bM
FX ▶	Beats
Guitar ▶	Beats on the side
Vocal ▶	Fractured Beat MTK
Default Setting	Heavy Pumping
	High Tom (Adjust Filters)
Save As...	Kick Focuses Middle bM
	Kick Long
Options ▶	Kick Me MTK
	Kick Metal MTK
	Kick Rock MTK
	Kick Short
	Kick You MTK
	Low Tom (Adjust Filters)
	✓ Overheads – Expressive Sizzle bM
	Snare – Hihat leakage suppression bM
	Snare Long
	Snare Short
	Snare–it MTK
	Soft Snare MTK
	Solidify Center bM
	Straight Perc MTK
	Subtle Beats
	Subtle Beats 2
	Subtle Stereo Punch bM
	Top End MTK

STYLES

With the Style Selection button, you can choose among various gate/
expander styles, all carefully designed to meet specific needs or offer a
certain character:

Classic offers the style of gating and expansion often found in vintage,
high-end mixer channel strips. It can be quite aggressive, but also subtle
when needed. It's a great all-around style for mixing purposes and works
especially well on drums. Clean style is great for transparent gating and
expansion—minimizing flutter and distortion. Vocal style retains the
natural feel of the vocal, opening the gate gently when the singer breathes
in and releasing gently, yet fast enough to reduce unwanted noise or bleed.

Another common application of a gate/expander is on electric guitar before distortion, to reduce or minimize rumble. With Guitar style, especially when used in the lower ratio range (2:1 to 5:1), Pro-G gently follows the natural decay of the guitar sound, ensuring that even after distortion, the result still sounds very natural and lively.

An upward expansion algorithm is also included. When you choose this style, separate threshold and ratio parameters are used with custom, smaller ranges. In Upward mode, the expander will amplify signals above the threshold instead of reducing them below threshold. When used moderately and with care, you can achieve very natural and transparent-sounding expansion effects.

Finally, Pro-G also features a dedicated ducking mode, as found in many classic gates. A typical application of ducking is to automatically lower the level of a musical background track when a voice-over starts, and to automatically bring the level up again when the voice-over stops (in movies and on radio broadcasts). It is similar to compression with a side chain, but it can sound quite different because it's processed as 'inverted' gating instead of compression. Ducking can also be used to achieve the well-known and very popular 'pumping' effect, much used in modern dance music.

All styles have built-in, carefully tuned hysteresis where needed. This will cause the gate to close at a slightly lower threshold level than the threshold at which it will open (as set by the threshold knob), avoiding flutter when the incoming audio signal hovers around the threshold.

Using the Expert button under the level display, you can enable or disable expert mode. When enabled, the interface will resize itself to offer additional options to customize the side chain signal that triggers the gate/expander and to adjust the wet/dry levels individually—see Figure 9.17.

NOTE
If Expert mode is off, FabFilter Pro-G does not use any of the expert controls, and always triggers on the main input signal, fully stereo linked, without filtering, and with 100 percent wet signal and 0 percent dry signal.

Figure 9.17
FabFilter Pro-G
Expert mode

Normally, when 'Side Chain' is set to 'In', the side chain receives input from the regular plug-in input signal. Alternatively, when set to 'Ext', the side chain can receive its signal from any other track from your host. You can use the Audition button to listen to the side chain signal exclusively.

To narrow the frequency range on which FabFilter Pro-G will trigger, you can filter the side chain signal with steep 48 dB/octave low-pass and high-pass filters. To adjust the filters, drag the triangular buttons in the middle of the side chain section to the left or right. Alternatively, double-click a button to type a value directly. The filters will be bypassed completely when they are at the far left and right sides of the filter controller. To drag both filters at once, either hold down the Alt/Option key or drag the highlighted area between the buttons. Again, use the Audition button to listen to the filtered side chain signal.

You can use the stereo version of Pro-G 100 percent stereo linked, but it is also possible to fully unlink the side chain channels and best of all, you can have anything in between. By using the combined gain/mix knobs at the left side of the expert section, it is possible to precisely determine the trigger signal that will be sent to each level detector of each channel.

The combined left knob is connected to the left gate/expander channel, the right knob to the right channel. The inner knobs set the volume of the signal that is sent to the channels; the pan rings control where the signal is coming from: the left or right side chain input or a mix of both.

For example, if you put both volume levels at zero dB and both mix rings in center position, Pro-G will be fully stereo linked because both channel level detectors use the same audio signal. This keeps the stereo image from wandering from left to right when expanding/gating a stereo signal. Alternatively, if you turn the mix ring for the left signal to the full left position and the right mix ring to the full right position, Pro-G's side chain is fully unlinked. Now expansion on both channels will work independently. (You can hold down the Alt/Option key to link both panning rings.)

When Channel mode is set to 'Mid/Side', the mix rings let you specify a trigger signal as a mix of the mid and side signals, instead of a mix of left and right signals. This way, you can, for example, choose to trigger on the mid signal only or to expand mono and stereo information independently.

All other settings will result in anything between linked and unlinked. You can even cross-link the channels, so that the side channel is gated when the mid signal drops below the threshold!

Finally, the wet and dry combined knobs on the right of the expert section offer individual control of the wet (processed) and dry (unprocessed) signal. This makes it possible to slightly 'dilute' the gating/expansion effect and to control panning.

Avid Pro Compressor

Avid's Pro Compressor (see Figure 9.18) uses algorithms based on those used in the acclaimed Euphonix System 5 console. The user interface is clearly and attractively laid out, with input and output meters at the left and right, respectively, and a dynamics graph in the upper central position with a frequency graph for side chain filtering in the lower central position. Between these two graphs are four rotary controls for knee, attack, release, and depth. To the left of the dynamics graph is a threshold control and to the right a ratio control. To the left of the frequency graph is a dry mix control and to the right a makeup gain control.

The meters show both sample peaks and averaged values for input and output. The peak hold value is displayed numerically at the top of each meter and the peak hold indicator appears as a thin orange line within the meter that retains its position until a higher peak comes along, while a thicker orange bar shows the instantaneous sample peak.

Figure 9.18
Avid Pro compressor

The main part of the meters display average levels with an integration time of about 400 ms. Color coding is used for these levels, with dark blue indicating nominal levels from −90 to −20 dB, light blue indicating pre-clipping levels from −20 to 0 dB, and yellow indicating full scale levels from 0 to +6 dB.

Both input and output meters have level controls at the bottom of each meter. You can click once on the displayed value to activate the text entry field then type a numerical value and press Return to register this value. Or you can click on the displayed value, keep the mouse button held down, and drag this upward or downward to adjust the value, releasing the mouse when you reach the value you wish to enter.

The input meters allow you to enter values in the range −36 to +36 dB. The output level control lets you set the output level after processing, either to

Figure 9.19
Avid Pro
compressor
showing output
attenuation meters

make up gain or to prevent clipping on the channel where the Pro Compressor plug-in is being used. You can enter values in the range -INF dB to +12 dB.

If you click the Output/Attenuation button at the top right of the output meter section, the meter will be switched to display attenuation for the processed signal from 0 dB to −6 dB—see Figure 9.19.

The dynamics graph display shows a curve that represents the level of the input signal (on the horizontal x-axis) and the amount of gain reduction applied (on the vertical y-axis). The display shows a vertical line representing the threshold setting for the compressor.

The dynamics graph display also features an animated red ball in the gain transfer curve display. This ball shows the amount of input gain (x-axis) and gain reduction (y-axis) being applied to the incoming signal at any given moment. To indicate overshoots (when an incoming signal peak is too fast for the current compression setting), the ball temporarily leaves the gain transfer curve.

You can use this graph as a visual guideline to see how much dynamics processing you are applying to the incoming audio signal. You can also drag in the dynamics graph display to adjust the corresponding compressor controls—threshold, ratio, knee, or depth.

At the top of the dynamics graph are five buttons to let you select the compressor's detection mode. The default option is smart detection, which analyzes the incoming signal and interpolates between the various detection modes—RMS, Average, Peak, or Fast—as necessary.

To the right of these is a button with a small loudspeaker icon. If you click on this, it will flash to indicate that you have entered the Attenuation Listen mode. Attenuation Listen mode lets you isolate the gain reduction part of the processed audio signal. This can help you hear what parts of the input signal are triggering compression, which, in turn, can help you better understand the characteristics of the compressor with the current settings.

The Pro Compressor also has comprehensive side chain filtering features, with a Side-Chain Listen mode.

NOTE

The compression ratio ranges from 1.0:1 to 20.0:1. Once the ratio control hits 21.0:1, it displays LMTR. The LMTR setting marks the highest 'normal' compression mode before the onset of negative compression values (from −20.0:1 to 0:1). At the LMTR setting, for every decibel that the incoming signal goes over the set threshold, 1 dB of gain reduction is applied. Once the ratio control passes the LMTR setting, it provides negative ratio settings from -20.0:1 to 0:1.

With these settings, for every decibel that the incoming signal goes over the set threshold, more than 1 dB of gain reduction is applied according to the negative ratio setting. For example, at the setting of -1.0:1, for each decibel over the set threshold, 2 dB of gain reduction is applied. Consequently, the output signal is both compressed and made softer. You can use this as a creative effect or as a kind of ducking effect when used with an external key input.

The depth control sets the maximum amount of gain reduction applied regardless of the input signal. For example, if ratio is set to LMTR (between 20.0:1 and -20.0:1) and depth is set to 'Off', up to 20 dB of gain reduction is applied to the incoming signal (at 0 dB). If you set depth to −10 dB, no more than 10 dB of gain reduction is applied to the incoming signal.

Overall, Avid's Pro Compressor sets a new standard for user interface design together with excellent audio quality.

Avid Pro Expander

Avid's Pro Expander (see Figure 9.20) offers a wide range of easy-to-use features and improvements that go way beyond what you can do with the Avid Channel Strip plug-in, plus some unique features not found in Avid's Pro Compressor. For example, it has a look-ahead mode that prevents the gate from cutting into the attack of upcoming audio signals, with up to 15 ms of look-ahead time.

Typically, you will use its upward mode to increase the dynamics of the audio signal and you can use this creatively in combination with steep Q side chain filter settings or an external key input to selectively modify parts of the frequency spectrum or reemphasize the rhythm of the music. The Pro Expander has four expansion detectors—average, peak, fast peak, and

RMS—and also has a SMART mode that analyzes the incoming audio and morphs among the four detectors to prevent distortion or pumping effects. The Pro Expander also has an advanced ducker mode.

Figure 9.20

Avid Pro expander

The Pro Expander controls and displays are extremely easy to operate and view on-screen—in contrast to other designs that are confusingly laid out with text that is too small to read with comfort. Pro Expander's logical design, together with its excellent audio quality, makes this a very desirable addition to any set of dynamics tools.

Avid Pro Limiter

The Avid Pro Limiter provides 'brick-wall' limiting and metering designed to prevent output clipping and distortion without necessarily coloring the sound. You can use the Pro Limiter to ensure that your mix output never exceeds digital 0 dB when hitting the digital-to-analog converters on your audio interface.

The user interface is quite similar to those used for the Pro Compressor and Pro Expander, with input and output meters, various rotary controls, information displays, buttons and pop-up selectors, and a histogram that shows loudness over time—see Figure 9.21.

Figure 9.21
Avid Pro limiter

The gain reduction meters are interleaved with the input meters and show the amount of gain reduction applied to the input signal. The gain reduction meter scaling is shown on the right side of the input meters (from 0 dB to −36 dB). The gain reduction amount varies depending on the level of the input signal, as well as the threshold and character settings.

In the upper left corner of the input section is a Dim Input Meter button that dims the input meters, but not the attenuation meters—see Figure 9.22. This lets you visually focus on the gain reduction applied rather than on the incoming signal levels.

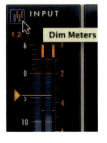

Figure 9.22
Avid Pro Limiter Dim Input Meter button

The input meters show peak signal levels before processing with the input meter scaling shown on the left side of the input meters, running from −90 dB to +6 dB.

Pro Limiter displays the input sample peak hold value (in dB) at the upper left corner of the input meters and the gain reduction peak hold value (in dB) at the upper right corner of the input meters.

There is a horizontal threshold control, colored yellow, that you can drag up and down to set the threshold level and, just below the meters, is an input trim control that you can either type a value into or drag up or down to adjust the input level in the range −30 dB to +30 dB.

The output meters show peak signal levels after processing, with the output meter scale shown on the right side of the output meters, running from −90 dB to +6 dB. An output sample peak hold value appears just above the meters, and you can drag the horizontal, yellow-colored control bar up or down to set the output ceiling level.

Pro Limiter provides controls for setting the threshold, ceiling, character, and release time. The character control adds soft-saturation processing with no additional gain before applying limiting.

To the right of these controls, Pro Limiter provides numeric displays that show the current loudness or peak level of the processed signal. Pro Limiter conforms to the 'EBU R128' and 'ITU-R BS.1770–3' standards for loudness metering.

- Integrated: Displays the current integrated level of the processed signal level in LUFS.

- Range: Displays the range of the processed signal level over time in LU.

- True Peak: Displays the true peak hold value of the output signal in dB.

- Short Term: Displays the short-term output signal level in LUFS.

At the top right of the user interface a pop-up selector lets you select various channel linking options for multichannel operation. Below this is a button that lets you enable the Auto Release function, which overrides the release setting and automatically adjusts the limiter release time based on changes in the program material. Next to this is a Listen button that lets you isolate the processed part of the audio signal so that you can hear what parts of the input signal are triggering the limiting while you are adjusting the threshold, character, and release settings.

In the lower section of the user interface is a histogram that shows a graphic representation of loudness over time within a window of 60 seconds. The graph displays true peak levels as a yellow line and the range of loudness over time as a blue shadow around the peak level line.

The loudness meters to the right of the histogram show the level of the summed output of Pro Limiter. The meters range from 0 LUFS down to −50 dB LUFS. Negative 23 LUFS is a common standard loudness reference level.

The first meter, immediately to the right of the histogram, is the momentary loudness K-meter. This provides an alternative display of the loudness range (as in the histogram). The current peak level is shown as a yellow line using Bob Katz's K-scale metering.

To the right of this are three meters marked M, S, and I: the Momentary Loudness Meter (M) displays the current true peak level; the Short-Term Loudness Meter (S) displays the current short-term output level; and the Integrated Loudness K-Meter (I) displays the current integrated level of the processed signal using K-scale metering.

In practice, the Pro Limiter is very straightforward to use and does exactly 'what it says on the tin'—provide transparent limiting with effective controls and feedback to let you apply this to any type of audio.

McDSP ML4000 Multiband Dynamics

McDSP's ML4000 Multiband Dynamics mastering limiter is primarily aimed at mastering applications, but can also be used creatively and as a problem solver during tracking sessions.

The ML4000 (see Figure 9.23) has four bands, each with a gate, an expander (upward or downward), and a compressor. The output from this is fed into a limiter. Each band's gate, expander, and compressor can be configured separately or linked together and each band has an output gain fader with associated Solo, Master, and Link buttons to the left of this and meters to the right.

At the far left of the user interface is a set of meters showing the input into the mastering limiter to the left, the amount of gain reduction applied to signal peaks in the center, and the master output level to the right. Next to the meters are controls for output ceiling, threshold, knee, and release together with a mode selector that lets you choose the type of limiting: clean, soft, smart, dynamic, loud, or crush. Below this mode selector are bypass switches for the gate, expander, compressor, and limiter sections.

The controls and displays for the gate, expander, and compressor appear in the large area to the right of the limiter section. Along the top of this display is a pair of tabs that allow you to select either the main or the crossover display pages.

The main display page shows the IO response curves for the gate, expander, and compressor in each band, and switchable meters let you view either the input or output signals in each band. Below the response curves, meters for each band are positioned to the right of the band gain sliders. To the right of these meters, the combined gain reduction from the gate and expander and the compressor gain reduction are displayed.

The X-Over (crossover) page also shows the IO response curves for the gate, expander, and compressor in the upper part of the display area, but has the band gain sliders repositioned to the right of each band's IO curves. A large frequency response display is shown in the central part of the window—replacing the band gain sliders and meters. The In, Out, and Dyn buttons, positioned to the right of this, let you switch the display to show the overall frequency response in real time of the input, or of the output, or of the total dynamics effect, displayed on top of a graphic illustration of the crossover frequencies. You can adjust these crossover frequencies by simply clicking and dragging on the graph.

In the lower half of the display are controls for the dynamics processors in each of the four frequency bands. Arranged in columns are sliders that can be used to adjust the threshold, the range, and the hold for the gate (or the ratio and range for the expander, or the ratio and knee for the compressor), the attack, and the release times for each dynamics processor. Beneath these sliders are three buttons that let you select which dynamics processor's controls to display—gate, expander, or compressor—with a small bypass button for each processor located underneath these selector buttons.

TIP

The multiband gates in the ML4000 can be used to remove background noise from instruments in close proximity to each other—such as with a drum kit. The different drums are all usually mic'ed separately, but sound from adjacent drums often leaks into the other tracks. A conventional single-band gate is typically used to only 'open', say, the snare track when the

(Continued)

snare drum is actually being played. However, because different drums can have overlapping frequencies, a multiband gate can solve many previously unmanageable problems. For example, using a multiband gate, the kick drum could have a less sensitive gate in the low bands (with a lower threshold) and the other bands could have more sensitive (higher threshold) gates to mute background noises that would otherwise be impossible to remove.

Figure 9.23
McDSP ML4000
Multiband
Dynamics in
expander mode

Drawmer Dynamics

One of the first third-party designed plug-ins for Pro Tools TDM systems was Drawmer Dynamics. This offers similar features to those available in the Drawmer hardware units that have been used in many professional studios.

Although not available in the AAX format, older Pro Tools systems very often included Drawmer Dynamics with the bundles of plug-ins, so we will look at this here.

Drawmer Dynamics is a good example of a dynamics processor to look at in some detail, as it has a compressor, an expander, and a noise gate with a side chain keying section.

Drawmer Dynamics Compressor

The Drawmer Dynamics compressor section (see Figure 9.24) appears in both the expander and noise gate plug-ins. The compressor features a 'soft-knee' design to provide subtle level control that works particularly well for vocals, strings, and wind instruments. It has lots of user-friendly features including auto-adaptive attack and release and automatic gain makeup, which calculates the most suitable amount of output gain to give the maximum dynamic range. A peak limiter is also included to catch any peaks that a slow compressor attack might miss, which is particularly useful for digital recording.

Figure 9.24
Drawmer Dynamics
compressor

Drawmer Dynamics Expander

The Drawmer Dynamics expander (see Figure 9.25) has controls for threshold, ratio, attack, release, and range. The threshold control sets the level below which expansion starts to take place. The ratio control sets the amount of attenuation applied to the signal as it decreases below the threshold level. The attack control sets the rate at which the expander opens from a closed state and the release control determines the speed at which the expander closes to the range setting once the input signal has fallen below the threshold level.

> **TIP**
> One application is to reexpand overcompressed material. The threshold setting can range from −70 dB to +12 dB. If you set the threshold above 0 dB this produces upward expansion, which can be used to reexpand overcompressed material.

A unique feature of the expander is the Auto-Adaptive Expansion mode in which the expander will automatically vary ratio and release times depending on the dynamics of the input signal—making it easier to avoid cutting off quiet word endings, for example.

The expander's ratio control ranges from 1.1:1 up to 50:1 and then has two further settings—soft and softer. These both cause the ratio and release times to be automatically varied according to the dynamics of the signal being processed. The way this works is that low-level signals are processed with a lower expansion ratio while any residual noise during pauses is processed with a higher expansion ratio—making this even quieter. The aim is to preserve any wanted audio signals that are only slightly above the residual noise floor—by expanding the residual noise floor downward.

Figure 9.25
Drawmer Dynamics
expander

Drawmer Dynamics Noise Gate

The Drawmer Dynamics Noise Gate controls are separated into two sections: the key section and the noise gate section—see Figure 9.26.

The key section has controls for the HF and LF key filters and a switch to enable the side chain. The side chain has an external 'key' input into which you can feed any mono signal. The dynamics of the input signal, which could be a guitar or a bass drum or whatever, can be used to control a mono or stereo signal passing through the gate. The idea here is that you set the key filters to define a specific frequency range in the key input signal to trigger the gating action.

The noise gate section has all the usual controls for threshold, attack, hold, release, and range, and switches are provided to enable the ducking mode and to bypass the gate section. In Gate mode, a signal above the threshold will cause the gate to open. In Duck mode, the audio passes unattenuated until the signal exceeds the threshold. Ducking is mainly used for ducking the music behind voice-overs, although it can also be used to remove pops and clicks. The range control lets you set the level to which the gate closes after the input signal has fallen below the threshold setting. When ducking is enabled, the range control sets the level to which the signal will drop when the ducker is triggered from the key input.

Figure 9.26
Drawmer Dynamics
noise gate

Avid Smack!

Avid's interestingly named 'Smack!' compressor/limiter (see Figure 9.27) supports the full range of Pro Tools sample rates up to 192 kHz, is available in all Pro Tools plug-in formats, and works with all Pro Tools multichannel track types.

The retro-style user interface incorporates all the controls you are likely to need and the range of compression ratios provided will allow users to apply anything from subtle compression all the way through to hard limiting. You can also change the way the compressor behaves by switching among its three compression modes, one of which is a special mode for emulating classic electro-optical limiters. Side chain support is included to allow external or internal side chain processing using a basic side chain EQ.

How is it in action? Well, this thing will not stand for any nonsense! I would definitely describe Smack! as 'aggressive'! There is nothing subtle about this compressor at all . . . it 'grabs' those peaks and 'wrestles' hard until it gets them under its control, then it is loath to let them go! Actually, that last bit is not entirely accurate—there is a release control, and you will need to use this, along with the attack and ratio controls, to adjust Smack! until it affects the sound in exactly the ways that you want it to. Overall, I have found Smack! to be most useful for bass guitars and percussion instruments, and occasionally as a mix bus compressor.

Figure 9.27
Avid Smack!
compressor/limiter

Avid Impact

What's good about Avid's Impact (see Figure 9.28) is the minimalist simplicity of its user interface, which seems to foster the idea that what we have here is a transparent-sounding high-quality compressor for everyday uses—which is exactly what Impact happens to be!

Figure 9.28
Avid Impact
compressor

Sonnox Oxford Dynamics

Now let's take a look at the Sonnox Oxford Dynamics plug-in—see Figure 9.29. Like the rest of the Sonnox range, this was originally developed for the Sony Oxford OXF-R3 console, and if I had to choose just one high-quality dynamics processor, this would probably be the one.

The Oxford Dynamics plug-in has six separate processing sections to the left, a display showing a graph of input versus output to the right, and various controls in the top half of the window. At the lower left side of the window are six 'Access' buttons with associated 'In' buttons that can be used to turn on each processing section. To display the appropriate controls for a section, just click on its Access button and the correct set of controls will appear in the controls section of the window.

The 'Oxford Dynamics Full' version of the plug-in incorporates all six sections—gate, expander, compressor, limiter, side chain EQ, and warmth. The two-band side chain EQ facilitates de-essing and other effects. The neat thing here is that this EQ can also be inserted into the main signal path—even while still being used in the side chain. The warmth section adds loudness, punch, and definition to the sound of the dynamics section. It works by increasing the density of higher-value samples within the audio—thus boosting average signal levels without increasing peak levels or risking digital clipping.

Oxford Dynamics is not overburdened with presets, but does give you a small selection of 10, including starting points for kick, bass, buss, master, and vocal compression.

Figure 9.29
Sonnox Oxford Dynamics full version accessing the compressor controls

Various alternative versions of the Oxford Dynamics plug-in are provided that exclude certain sections to lighten the load on the processor. For example, the standard Oxford Dynamics plug-in omits the side chain EQ, so its Access and In buttons are greyed out in this version. The Oxford Compressor/Gate, Oxford Gate/Expander, and Oxford Compressor/Limiter plug-ins also omit the side chain EQ, while including the compressor and gate, gate and expander, or compressor and limiter sections, respectively—and they all include the warmth section.

A special Buss Compressor plug-in is available for 5.1 surround work, again with just compressor, limiter, and warmth sections included. In this version, a 24dB/octave variable low-pass sub-filter replaces the side chain EQ to allow automatic generation of a sub-channel from the normal wide-band channels.

There is also a separate plug-in, the Oxford Limiter, that just does limiting—see Figure 9.30. This gives you much more control than most other limiters, allowing you to adjust the attack and release times and the shape of the 'knee', for example. The enhance function lets you add much more volume and punch to your audio than conventional limiters are able to. The enhance control ranges from 0 percent to 125 percent. In Normal mode, full sample value limiting occurs only with settings at or above 100 percent. In Safe mode, sample value limiting takes place permanently and the enhance slider controls the degree of dynamic loudness boost. A further function allows the

Figure 9.30
Sonnox Oxford
limiter 'Gentle
Master' settings

495

user to dynamically correct for reconstruction overloads in real time, thereby achieving maximum possible modulation levels without risk of producing the unwanted artifacts associated with compression and limiting. There are more than 40 presets to put you quickly into the right ballpark with your settings.

The user interface for all the Sonnox plug-ins is clearly laid out and most recording professionals will find them very intuitive in use. Most impressive, the Oxford Dynamics plug-in gives you all the control you need for corrective and creative applications while at the same time producing the 'warmth' with your digital audio that is normally associated with valve equipment when working with analog audio.

FabFilter Pro-C

FabFilter Pro-C (see Figure 9.31) features an innovative user interface design with a 'smart' display that shows the input waveform and the output waveform overlaid in grey and yellow colors respectively, with a red gain reduction indicator running along the top. Three small controls at the bottom of the window let you adjust the opacity of the different curves to help you distinguish these.

To the right of this display are three very accurate peak level meters, which also display the current input level, gain reduction level, and output level. To

Figure 9.31
FabFilter Pro C

the left of the 'smart' display (or overlaid in the foreground when in Expert mode) is a transfer curve displayed in white showing the relationship between the input (horizontal axis) and output (vertical axis) levels. These displays really help you to visualize how the compression is working—making it much easier to understand what is happening to your audio.

FabFilter offers three main types of compression including clean, which, as its name suggests, is very transparent; classic, which introduces analog characteristics and warmth; and opto, which models a vintage optical compressor and has a very soft knee and a generally slower response. A knee selector lets you switch between hard and soft knee types.

The output of the compressor and the dry (unprocessed) signal each have their own output knobs (and panning rings when Expert mode is activated) allowing for parallel compression. Expert mode also features mid/side compression and allows side chaining with built-in filters or with external sources.

What's appealing about FabFilter Pro-C is the innovative user interface design combined with cutting-edge technical capabilities.

Flux Elixir

Flux Elixir (see Figure 9.32) is a high-quality limiter that provides a true peak output level that conforms to the ITU-R-BS 1770 and EBU R128 standards. The significance of this 'true peak' style of operation is that Flux takes account of

Figure 9.32
Flux Elixir

the inter-sample peaks that can occur in the output from digital-to-analog converters that can lead to unexpected overloads. Using an algorithm that takes account of these 'true peaks' that occur after conversion back to analog audio, Flux Elixir prevents this from happening.

It couldn't be much easier to use—just set the input level, adjust threshold according to the amount of limiting you want, then enable the Make Up button to compensate the gain and to add loudness—and that's it: no need to worry about release time or any other conventional limiter settings!

Actually, there are a few other settings that you can make. You can set the algorithm to perform the limiting processing in up to five separate stages if you wish. This increases the processing quality—but requires more CPU power. Because the algorithm adapts itself to the audio material, using multiple stages allows for the processing to be even more precise and to achieve even more natural-sounding results. For example, if the threshold is set to −3 dB and the stages parameter is set to three, the first stage will limit at −1dB, the second stage will limit at −2dB, and the third stage will limit at −3dB—with analysis carried out separately for each stage.

You can also set a speed percentage. This determines how the algorithm will react to the audio material. The speed setting changes how the gain envelope will be generated—with more or less look-ahead, release, and curve smoothing. The default is 50 percent, which will be optimal for most cases. It is better to increase the number of stages before trying to reduce the speed, and be aware that settings from 50 percent to 100 percent can generate progressively more distortion.

A button marked 'Diff.' lets you listen to the difference between the input and output signal that can be helpful to check while you are setting up the parameters.

Elixir, like all Flux plug-ins, provides two preset slots referred to as slot A and slot B positioned at the lower left and lower right of the user interface window, just above the preset Save, Recall, and Copy buttons. These slots provide access to two sets of parameter settings that can be recalled for each slot individually so you can alternate between these settings. A morphing slider is also provided that allows you to morph between the two slots with their corresponding settings—allowing for some very creative tweaking. To open the preset manager, just click on either of these slots. To enable to automation for read and write, click on the Automation button below the morphing slider.

Flux Elixir manages to pack an enormous number of technical features into a relatively compact user interface without looking too daunting to use. This plug-in is definitely worthy of your attention if you need more dynamics processing for your Pro Tools system.

Mix Bus Compression

Some engineers and producers choose to insert a compressor and/or a limiter on a master fader in Pro Tools to apply this to the entire mix. Others feel that any final compression and/or limiting of the mix ought to be left to the mastering stage, where each track of an album is compared with the others to make sure that the final levels and overall EQ settings are appropriate, or where a single track is adjusted appropriately for the distribution medium.

Classic compressors that are used for mix bus compression in top studios include the SSL G-series, the Neve 33609 (see Figure 9.33), the Focusrite Red 3, the API 2500, the Universal Audio 1176, and the Cranesong STC-8. Plug-in versions of most of these are available for Pro Tools, along with unique software designs, such as the UAD Precision Buss Compressor—one of my favorite choices.

Figure 9.33
Universal Audio
Neve 33609
compressor

TIP

If you are going to use a mix bus compressor, you should make sure that this is in place as soon as you start working on your mix. Remember that any fader moves you make with the compressor in place will produce a different balance than without the compressor.

Generally, I don't recommend that you use multiband dynamics processors on your mix bus. These let you treat particular frequency bands differently and can be very useful in mastering when you don't normally have access to the individual tracks. However, when you are mixing a multitrack recording, you do have access to all the individual tracks, so it is better to EQ, compress, and balance these on a track-by-track basis rather than affecting the whole mix using a multiband processor.

If you do choose to use a mix bus compressor, then because you are compressing the whole mix, you will need to pay very careful attention to the compressor's attack and release times to make sure that these work appropriately in relation to the tempo and nature of the music. For instance, if the music is fast and the release time too slow, this will not sound good—and vice versa.

The 'auto-release' settings available on some compressors can be very helpful when the nature of the music changes throughout the recording. In this case, the release settings change automatically in response to the audio material.

You also need to be careful not to use too fast an attack time, or you will adversely affect the transients in your mix, especially from the drums and percussion instruments, making the sound 'duller'.

NOTE

If your aim is to prevent transients from clipping while you increase the average level of the mix so that it sounds louder, then this is best done using a look-ahead limiter or one with an extremely fast attack time—a process best done at the mastering stage.

TIP

If you want to make a 'listening CD' for yourself or a client that sounds comparable in loudness to typical commercial CDs, then use a mix bus compressor/limiter to create a special 'listening' mix after you have saved your main mix to disk without mix bus compression. This way, you can give the mastering engineer a much better mix to work with than one that has already been 'squashed to death'.

Delay and Modulation Effects

Some of the most interesting delay and modulation effects were developed during the analog era and, thanks to companies such as Universal Audio, many are now available as plug-ins. Of course, quite a few digital effects were developed during the 1980s and 1990s, many of which are no longer available in hardware. Fortunately, some of the original designers of these effects have made them available in their plug-ins. SoundToys, for example, specializes in this area.

Delays, Echoes, Phasing, Flanging, Harmonizing, Ring Modulation, Resonant Filtering

Delay and echo are the oldest and most commonly used effects in recording. From the classic double-tracked effect on Elvis's voice to the more subtle delays on the Beatles' records—they all used delay and echo extensively, as has most music recorded during the past 50 years. The first echo devices I came across during the '60s and '70s were the WEM Copycat, Maestro Echoplex (see plug-in version, Figure 9.34), and Roland RE-201 Space Echo (see plug-in version, Figure 9.35)—all of which used tape.

Figure 9.34
Universal Audio
(Maestro Echoplex)
EP-34 tape echo

The Binson Echorec, which used a spinning magnetic disc with a variable playback speed and head, was also quite popular. All of these produced a warm, analog sound in a relatively compact unit and provided control of echo time, feedback, mixing, and sometimes EQ. And some units, like the Space Echo, provided more than one playback head to create a variety of echo patterns.

Figure 9.35
Universal Audio
Roland Space Echo
RE-201

NOTE

Universal Audio has recreated several of these classics, most recently the Cooper Time Cube—see Figure 9.36. 'The original Cooper Time Cube was a UREI-branded, Bill Putman/Duane H. Cooper collaborative design that brought a garden-hose-based mechanical delay to the world in 1971. The limited feature set of 14ms, 16ms or a combined 30ms delay meant it was less flexible (and popular) than tape-based delays such as the Echoplex or later electronic units, and only 1,000 were ever made. However, the CTC was noted for its uncanny ability to always sit perfectly in the mix and was used on many hit records, such as "Tell Me Something Good" by Rufus and "Low Rider" by War, for its spectacular short delay and doubling effects. Since its early discontinuation, the CTC has grown a very strong cult following and finds a home in such prestigious studios as Blackbird and Sunset Sound. The Cooper Time Cube sends live audio through long pieces of tubing, not unlike a garden hose, to create a time delay. The 2' x 2' plywood housing is filled with coiled tubing, Shure mic capsules (of the SM57 variety) at both ends of each line are used as speakers and pickups, and a series of tooled aluminum blocks with 'tuning screws' are used in various places to tune the delay to a relatively flat response. The whole delay mechanism is suspended on springs within the housing to maintain 'acoustical isolation,' and is then filled with packing peanuts, which stay in the housing, even after shipping!

Universal Audio's Cooper Time Cube Mk II is a feature-enhanced improvement on the original delay system design. While the Cooper Time

Cube Mk II plug-in retains the all-important mechanical delay sound of the original hardware, it goes much further by offering all the necessary features expected from a modern delay device:

Delay, Decay, Pan, and Volume controls plus Tempo Sync and Automation for each of the two independent delay lines. The distinct sound of the single- or double-hose coil is preserved regardless of delay setting, and either sound is available at the flick of a switch. The CTC Mk II also incorporates other enhanced tone-shifting features such as the Color switch, which presents the user with the original (A) or 'leveled' (B) frequency response, plus tone controls and a 2-pole high pass filter. Lastly, a switch is presented for soloing the Wet signal, and the Send switch allows momentary interruption of the signal being sent into the delay system. As posted by David Crane to the Universal Audio blog at www.uaudio.com/blog/cooper-time-cube-power/, where you can also see pictures of the original device.

Figure 9.36
Cooper Time
Cube Mk II

In the late '70s, solid-state echo devices, such as the Marshall Time Modulator and the MXR Flanger Doubler, appeared. Other classic analog delays like the Electro-Harmonix Deluxe Memory Man, MXR Analog Delay, and various Boss and Ibanez units gave musicians affordable access to analog delays in compact pedals.

The 1980s saw the digital revolution get established with increasingly affordable rack-mountable devices such as the Lexicon Prime Time, PCM-60 and PCM-70, and the Yamaha Rev 7 and Rev 5, which all sounded extremely clean by comparison with analog devices.

In the '90s digital multi-effects were everywhere, with hardware units from Yamaha, Alesis, and others claiming to emulate just about any type of effect—never mind the quality, feel the width! Now that we are into the new millennium, we are seeing a return to individual designers and design teams offering high-quality, specially designed plug-ins for Pro Tools and other popular software.

Basic Delay Effects

Echo Delay

One of the simplest but often most powerful effects that can be used with audio is an echo. Sometimes referred to as *slapback echo*, which you might hear if you shout out loud in a canyon or in a hall with a reflective back wall, this is a single repeat of the original sound, delayed by anything between about 35 and 350 milliseconds.

A very short delay of less than 35 milliseconds produces an effect called *doubling*. This sounds like two people playing or singing the same thing at almost, but not exactly, the same time. A common music production technique is to ask a singer or instrumentalist to 'double' their part by singing it or playing it again as closely as possible to the first recording. It is never an exact match, and serves to widen out the sound. You can save time and get a more precise result by doing this using a short delay.

Echo Delay with Feedback

A simple slapback delay can be made more interesting by feeding some of the delayed signal back to the input of the delay unit so you hear a second echo. The feedback control may be labeled *regeneration* or *repeat*—terms that are descriptive of the result produced rather than the thing that causes this. Turn the feedback up some more and you will hear more repeats, and so on. A slapback delay with two or three repeats can really help to smooth out a vocal track or make a guitar figure sound much bigger and fuller. Often, the delay chosen will be in time with some subdivision of the beat, such as quarter notes, 8th note triplets, 16th, or whatever, which can have a significant effect on the 'groove' of the music. Line 6 Echo Farm, the Bomb Factory Tel-Ray Variable Delay, and moogerfooger Analog Delay all fall into this category.

Doubling

Sometimes, a slightly delayed copy of an original vocal or instrumental sound is mixed together with the original to create a 'fatter' sound. Delays of between 10 and 35 milliseconds are typically used to create this effect.

Modulation Delay Effects

The next enhancement to this basic effect is to add modulation. This involves using a low-frequency oscillator (LFO) to vary the amplitude or the pitch to create changing effects.

Tremolo

Tremolo is not a delay effect, but it does use modulation, so it seems appropriate to mention it briefly here. Tremolo is another name for amplitude modulation. Cyclically varying the amplitude by a little above and below the original signal level causes the sound to get a little louder then softer then louder, and so on. The neat thing about amplitude modulation is that it can be used creatively to add effects that change over time, developing in slow waves or fast ripples depending on the speed of the LFO.

SoundToys Tremolator

SoundToys Tremolator (see Figure 9.37) lets you create the widest range of tremolo effects that you will find anywhere. It will create the standard tremolo effect that you often hear on guitar or electric piano sounds, but it does a lot more than that. It can create some very ring-mod-like sounds when you speed it up into the audio range; it can create rhythmic beats with its built-in rhythm generator; and it lets you use lots of different modulation types.

Figure 9.37
SoundToys
Tremolator

The rate control lets you set the speed of the tremolo in BPM, although when the MIDI switch is on, incoming MIDI clock signals control the rate so that you can synchronize the LFO sweep from the first beat of the bar to the tempo of your music. A depth control lets you decide how much of the incoming signal's amplitude will be modulated by the LFO.

505

Tremolator also has a groove control that allows you to add a shuffle or swing groove feel to the tremolo and an accent control lets you add two more types of rhythmic feel to the tremolo effect: sync and max.

As you turn the accent control clockwise toward the max setting, the first beat of the rhythm pattern will be emphasized, and the other beats will be deemphasized. As you turn the control fully clockwise, the '2–3-4' beats will be virtually silent with no amplitude modulation happening on those beats. The only pulse in the tremolo effect will be on the '1', that is, on the first or down beat.

As you turn the accent control counterclockwise toward the sync setting, the pulse of the tremolo on the first beat will become less pronounced. When the knob is set fully counterclockwise the first beat will be almost fully off. So you won't hear any modulation of the signal on the '1', but you will hear the tremolo modulation on the '2–3-4' beats.

So you can use Tremolator to create anything from classic tremolo effects to completely 'wacky' new sounds, and it has a great selection of presets to prove the point—see Figure 9.38.

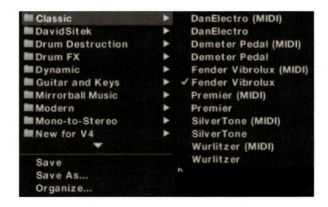

Figure 9.38
SoundToys
Tremolator presets

SoundToys PanMan

Panning is not a delay effect either, but SoundToys PanMan (see Figure 9.39) does use modulation, with all the different types available as in SoundToys FilterFreak.

Figure 9.39
SoundToys PanMan

SoundToys EchoBoy

Sound Toys EchoBoy (see Figure 9.40) has an LFO that can be used to vary the delay time above and below the normal setting, causing the pitch to be raised and lowered in a cyclical manner—that is, pitch modulation. The popular name for this effect is *pitch vibrato*.

EchoBoy can emulate virtually any classic or contemporary delay device—from the subtle warmth of high-end studio tape echo, to the 'low-fi' sound of classic tape echo boxes like the Echoplex and Roland Space Echo. EchoBoy even lets you introduce the glitches, distortions, and other anomalies that always existed in tape and analog devices. But it doesn't really sound quite as 'dirty' as my old Echoplex—although that is probably a good thing!

EchoBoy also features a Rhythm mode that has a delay line with up to 16 individual outputs (or 'delay taps'). This type of configuration, known as a *multi-tap delay*, can be found in classic delay units such as the Roland Space Echo, Eventide's H3000, and the Lexicon PCM-72.

Figure 9.40
SoundToys Echo Boy interface showing the various controls, switches, and displays neatly integrated into the main window

The selection of presets is truly comprehensive, with folders full of effects to suit vocals, guitars, and so forth, most of which hit the spot—and the drums presets are all killers! I tried each of these in turn on one of my recordings and used Pro Tools automation to switch between these in the different sections on playback—instant remix heaven! Two of my favorite presets are 'Vibrato', which is a true pitch vibrato that sounds great on guitars or keyboards, and 'CE-1 Chorus', which is a recreation of the classic Boss CE-1 chorus pedal. You will also find an excellent selection of special effects presets including 'Telephone', which produces the familiar narrow-bandwidth telephone sound, 'AM Radio', which models the compressed medium bandwidth of AM radio, 'FM Radio', 'Shortwave Exaggerated' radio, and even 'Transmitter', which is a kind of CB radio with distorted and resonant mid frequencies.

The user interface looks (and is) classy with all the controls you could ever wish for in one relatively compact design, winning my personal recommendation as one of the best delay effects processors that money can buy.

SoundToys Crystallizer

Based on an original effect created as a preset in the Eventide H3000 Harmonizer, SoundToys Crystallizer (see Figure 9.41) is a powerful creative sound-processing tool that offers unique delay effects to suit just about any type of audio input. Using the H-3000 Harmonizer's Reverse Shift algorithm with a lot of feedback and delay creates crystal-like echoes and shimmering effects that sound absolutely incredible, especially on guitar. A preset for this effect was developed called 'Crystal Echoes'. The Crystallizer plug-in builds on this concept but adds many more capabilities.

Figure 9.41
SoundToys
Crystallizer Granular
Echo Synthesis
preset

Crystallizer is basically a pitch shifter that grabs a slice of audio and plays it back at a different speed to create the alteration in pitch. Most pitch shifters automatically adjust the size of the slice to keep it constant so that the effected signal is played back with just a little bit of delay. Crystallizer, on the other hand, offers you the ability to take much larger slices of audio, shift the signal up or down by an incredible four octaves, and play the pitch-shifted signal back either in forward or reverse. It also allows you to feed the output of the shifted signal back into the input so that, as the signal repeats, it is reshifted with each successive repeat. And, as with other SoundToys plug-ins, Crystallizer can lock the playback to incoming MIDI timing clocks so that the repeating signal always starts playing back on the selected beat.

Phasing

Phasing is a more sophisticated effect than vibrato. This effect uses short delays of less than 10 milliseconds between the original signal and a copy of this—putting the waveforms out of phase with each other. An LFO varies the delay time, making it alternately longer and shorter, which alters the amount by which the waveforms move in and out of phase in a cyclical way. When the waveforms are combined at the mix output from the delay device, the various frequencies will sum together when they are in phase and will partially or totally cancel each other when partially or totally out of phase. A depth or intensity control defines the bandwidth of the frequencies affected. As the delay time is swept backward and forward, this produces the characteristic sweeping, swooshing kind of sound that we call *phasing*.

SoundToys Phase Mistress

The most advanced phaser plug-in available for Pro Tools is the SoundToys Phase Mistress—see Figure 9.42. This has a vast library of presets that emulate just about every existing type of phaser including those introduced in the late '60s and during the 1970s, such as the Univox Univibe, the Maestro Phase Shifter, the MXR Phase 90, the Electro-Harmonix Small-Stone, the Eventide Instant Phaser, and my personal favorite, the Mutron Bi-Phase, which combined two separate phasers in one box.

Analog phasers can have different numbers of phase shift 'stages' in their design. As you add stages (resulting in more notches), the overall strength of the phase shift effect increases and becomes more prominent. The number of stages also significantly affects the overall tonal character of the processed sound. For example, a two-stage phaser will be very 'washy' and wet (think Hendrix 'Machine Gun', Robin Trower's 'Bridge of Sighs', and the opening guitar on Dark Side of the Moon's 'Breath'). The phased drum sound on Led

Zeppelin's 'Kashmir' from Physical Graffiti was created with a four-stage phaser that provided a much more pronounced effect. As you might expect, with 6-, 8-, 10-, and 12-stage phasers the effect becomes progressively even more pronounced. Using an odd number of stages sounds totally different than using an even number of stages and virtually all phasers use only even numbers. Phase Mistress allows you to pick any number of stages between 2 and 24, including all the odd numbers, expanding its tonal palate exponentially beyond the sounds of those early phasers, so the presets also include an enormous number of creative new phasing effects.

The user interface includes many of the same features as the other SoundToys plug-ins. So, for example, it has a mix control that lets you set the balance between the 'wet' filtered sound and the 'dry' original audio. You can also add distortion using the Analog mode toggle switch and its associated 'flavor' pop-up selector button to choose how Phase Mistress will distort or 'saturate' as the signal input increases. Phase Mistress also has a number of different modulation modes that can be selected using the button positioned at the lower right of the window. Using this pop-up mode selector, you can choose between the normal LFO mode and five more sophisticated modes: envelope, random, random step, ADSR, and rhythm.

Figure 9.42
SoundToys Phase
Mistress

Flanging

Flanging is a more extreme version of the phasing effect that occurs with delays between, say, 10 and 20 milliseconds.

I can recommend two specialized plug-ins that offer flanging effects—the Universal Audio MXR Flanger-Doubler (see Figure 9.43) and Blue Cat's Flanger (see Figure 9.44).

Reverberation

Whenever you listen to musicians playing, you are also listening to the sounds of the acoustic space that they are playing in and that you are listening in.

Any sound you make is, by definition, in some sort of acoustic space. Unless this acoustic space is totally 'anechoic' (without echoes), then alongside the direct sound you are also hearing reflections of the sound from objects within the space, such as a wooden floor or partition, and from any boundaries of the space, such as the walls of a room or hall. This is what we call *reverberation*.

If you are in a hall with a reflective back wall, you will often hear a slapback echo of the sound bouncing off the back wall. And in extremely reflective conditions this echo may bounce back again from the front wall to regenerate the echo, possibly many times, building up into a very dense reverberation.

NOTE

Back in the 1960s, my first band regularly played in a public bathing pool with walls covered in glazed ceramic tiles that had been converted into a dancehall by putting a wooden floor over the pool (The Baths Hall at Ashton-in-Makerfield, if I remember correctly). Depending on the tempo (and volume) of the music, it often became almost impossible to play our

(Continued)

instruments without getting totally confused by the echoes from the back wall that would build up into a complete cacophony of sound with the echoes sounding almost as loud as the direct sound. Whoever thought it was a clever idea to book bands to play in this place must have been totally crazed in the head! But it would have made a heck of a good reverberation chamber . . .

There are various ways to create reverberation artificially in studios. One way is to play the music back through a loudspeaker at one end of a room that has lots of reflective surfaces to create natural reverberation and to return this reverberation to the mix using a microphone at the other end of the room. Known as a *reverb chamber*, this was popular in the leading recording studios from the 1950s until around the end of the 1970s, but it is very rare to find such chambers in studios today. Also, the only way to change the reverb time in a chamber is to use movable acoustic baffles, which is not very convenient. The main advantage is that chamber reverberation sounds very natural, as you are using a real room to create the reverb.

The next way is to use a reverberation plate such as the venerable EMT 140. These are still quite large, but more like the size of a large cupboard than the size of a room. With this device, the direct signal is made to vibrate a metal plate using an electro-mechanical driver. As with the air in a room, many modes of vibration of the plate become excited, with multiple reflections of the initial vibrations. One or two (for stereo) contact microphones pick up the resulting 'reverberant' vibrations of the plate and feed these back to the mix. Limited adjustments of the reverb time and damping are possible with a plate, but it is still a relatively inflexible device at heart. On the positive side, plate reverb can sound very, very good. Several software developers have produced plug-in versions of the EMT plates, most notably Universal Audio—see Figure 9.45.

Figure 9.45
Universal Audio
EMT 140 plug-in

A similar device is the reverberation spring. These are still used in guitar amplifiers to this day. Larger, higher-quality spring reverbs, such as those manufactured by AKG in the 1970s and 1980s, found their way into some professional studios as far back as the 1950s and 1960s, but are rarely used these days. However, you can now get good plug-in emulations of reverberation springs, such as Softube's Spring Reverb—see Figure 9.46.

Figure 9.46
Softube Spring
Reverb

The most common way to create reverberation today is by using some form of digital reverberation unit that models the various types of reverberation. A digital reverberation unit can create many different echoes with multiple repetitions that build up into dense reverberation. These devices can simulate most common types of reverberation, such as that of a room, a small or large hall, a cathedral, or many other types of real (or imaginary) acoustic space. Typically, they allow you to control the pre-delay, which is the time between the start of the original sound and the start of the first reflection. This helps to create realism because it mimics the way that in real rooms the early reflections arrive before the reverberation builds up. You can also usually control the attack time, decay time, and density of the reverberation, and sometimes the high-, middle-, and low-frequency decay times or various other parameters. Some designs give you separate control of various parameters for the early reflections and for the reverb 'tail'—that is, the decaying part of the reverberation.

Of course, many reverb units these days come with a good selection of presets that you may find very suitable, leaving you with little need to tweak the controls other than to adjust the reverb time or amount. And, as well as simulating room reverberation, you can often use digital reverberation devices to create unnatural but interesting effects such as reverse reverb and gated reverb.

First available in 1978, the original Lexicon 224 Digital Reverb (see plug-in version, Figure 9.47) was quickly adopted by the leading recording studios and was widely used throughout the 1980s, almost singlehandedly defining the sound of that era! You can hear the 224 in action on classics such as Talking Heads' 'Remain in Light', Grandmaster Flash and The Furious Five's 'The Message', Vangelis' incredible *Blade Runner* soundtrack, U2's 'Unforgettable Fire', and Peter Gabriel's 'So'.

Figure 9.47
Universal Audio
Lexicon 224 Digital
Reverb plug-in

Convolution Reverbs

An alternative to using reverberation-modeling algorithms is to add reverberation using convolution techniques. In this case, an impulse sound (like a gunshot, electrical spark, or other types of noise source) is created inside an acoustic space and the response of the acoustic space to this impulse (the impulse response) is recorded that contains the characteristic sound of the reverberation. This impulse response is then combined mathematically (convolved) with the sound that you wish to process—making this sound as though it is in the acoustic space that the impulse response was taken from. Convolution with high-quality impulse responses from desirable acoustic spaces usually produces more natural-sounding reverberation than devices that employ modeling techniques, although convolution reverbs typically have a much smaller set of parameters that can be controlled. The difference is similar to the difference between synthesizers and samplers—which sound much more realistic when it comes to reproducing the sounds of real instruments. In practice, impulse responses can also be obtained from any electro-acoustic or electronic device, such as a Lexicon 480 or an AMS reverb, so the better convolution reverbs come with a selection of these as well.

> **NOTE**
> The complete range of reverb effects available from equipment such as the Lexicon 480L digital reverb cannot be captured completely faithfully using convolution reverb because these effects are not all time invariant: some use chorusing, for example—an effect that varies over time. So convolution reverbs are best used for effects that don't vary over time, while algorithmic reverb and effects designs are more effective when it comes to emulating time-variant effects.

Reverberation Plug-ins for Pro Tools

Avid offers various high-quality reverb plug-ins for Pro Tools including Reverb One and ReVibe II, which are modeling reverbs, and Space, which is a convolution reverb. Modeling designs simulate reverb using algorithms and are not too demanding on your CPU—but don't sound as natural as convolution reverbs. Convolution reverbs use impulse response recordings taken from real acoustic spaces or devices and combine (or convolve) these with your audio, superimposing the reverb onto your sound. Convolution reverbs also require a lot more CPU processing than algorithmic reverbs.

Avid Reverb One

Avid Reverb One (see Figure 9.48) is much more advanced than the basic D-Verb that comes with all Pro Tools systems. It comes with a useful selection of presets and a well-designed graphical user interface that makes it easy to customize the settings to your taste. You can modify the reverb using standard controls for decay time, pre-delay, reverb dynamics (for long tails or gated effects), and early reflections. It also has chorus effects and a multiband equalizer.

Figure 9.48
Avid Reverb One

Avid ReVibe II

Avid ReVibe II (see Figure 9.49) is a full-featured mono, stereo, and surround reverb plug-in with an extensive set of room-modeling controls. It builds on the intuitive user interface of Reverb One, adding more functions. Reverb reflections and shape can be displayed independently for front and rear channels and Reverb EQ and Reverb Color have been combined into a single interactive graph that accurately displays the frequency and gain of the EQ and coloration. ReVibe also includes native 96 kHz processing, allowing its algorithm to process audio at 96 kHz without the need for sample rate conversion.

Figure 9.49
Avid ReVibe II

Oxford Reverb

The Sonnox Oxford Reverb (see Figure 9.50) is a modeling reverb that competes well with the Avid offerings. This has a large selection of presets, in categories ranging from halls and rooms to post and emulations.

The uppermost section of the user interface has display areas showing the early reflections and the reverb tail with input and output meters to the left and right of these, respectively. The lower sections have controls for the input, early reflections, equalizer, reverb tail, tail mix, and output.

Figure 9.50
Sonnox Oxford
Reverb

Flux Verb 3

One of the more recent reverb plug-ins available for Pro Tools comes from Flux, which offers its Ircam Tools Verb v3 algorithmic reverb processor plug-in—see Figure 9.51. This has an excellent user interface with a large time structure main display at top center and a row of seven time structure settings sliders below.

Using these sliders, you can set minimum times at which early and cluster reflections will appear and maximum times at which they will disappear. You can also control the distribution in time of these reflections.

> **NOTE**
> The early shape slider controls the amplitude rise or fall of the early reflections. The default setting of 0.5 corresponds to early reflections all having the same level. This mimics an acoustical space where reflective surfaces are all located at roughly the same distance to the listener. Below 0.5 early reflections decay with time; above 0.5 they rise with time. Early reflections of decreasing level would be typical of a space where most of the reflective surfaces are grouped at a range closest to the listener. Careful adjustment of these settings can help you to create a better sense of 'positioning the listener' appropriately.

To the right of the main display is a second display area that can be switched to show the three-band filter settings for the overall room, for the early reflections, for the cluster reflections settings, or for the late reverb tail. Cluster reflections are those that occur in the transition between early reflections and the late reverb tail.

To the immediate left of the main display is an options column with controls that let you adjust the simulated air absorption roll-off frequency or the diffuseness of the reverb. At the far left, a column of decay time controls lets you adjust the decay times at low, mid, and high frequencies, as well as the overall decay time of the reverb tail.

The input/output section is positioned in the lower right quadrant. This has slider controls for input and output gain together with associated meters, and a rotary dry/wet control.

A narrow section at the bottom of the window allows access to the presets. This has two slots—A and B—into which reverb presets can be loaded. Clicking on the preset name area in the center brings up the preset manager where you can copy presets into just one of the slots if you only want to work with one preset. Or you can copy presets into both slots—which lets you use the automatable morphing slider to morph between the parameters from slot A to slot B.

> **TIP**
>
> An attractive-sounding and technically very capable reverberation device, Verb v3 gets my vote as the best of the algorithmic modeling reverbs available for Pro Tools.

Figure 9.51

Flux Verb V3 ircam Tools

TSAR-1

Softube's design philosophy for the TSAR-1 (True Stereo Algorithmic Reverb Model 1) was to create the perfect reverb—and make it easy to use! After realizing this vision in the form of the TSAR-1 (see Figure 9.52), with just a handful of parameters available to tweak, Softube went even further and produced a version of this, the TSAR-1R, with the same sound qualities but just three parameters, time, pre-delay, and color!

Of course, to make sure that you have some sensible starting points before you start tweaking yourself, there are plenty of presets supplied, with everything from drum plates to guitar rooms, vocal chambers to scoring stages.

According to Softube, convolution reverbs work best when you use these to add reverb to completely 'dry' recordings. One of the strengths of an algorithmic reverb compared with a convolution reverb is that it can be more easily tailored to coexist with any already recorded ambience on a recording to which you wish to add more space or ambience.

The TSAR-1 top panel has controls for early reflections type, early reflections mix, diffusion amount, slow, fast, or random modulation, reverb mix, and output volume. On the lower panel, there are just five sliders to control pre-delay, reverb tail decay time, reverb density, reverb tone, and a high cut control that affects both the tail and early reflections.

Figure 9.52
Softube TSAR-1

Audio Ease Altiverb

My first choice of convolution reverb has to be Altiverb from Audio Ease—see Figure 9.53. This has the largest set of high-quality impulse responses taken from many of the world's finest rooms, halls, chambers, studios, and so forth, giving you access to some of the best acoustic spaces from around the world.

I use this as my primary reverb most of the time. I have compared all the other reverb plug-ins for Pro Tools systems and Altiverb is way ahead of everything else. Altiverb has been around since 2001, when it was the first real-time convolution reverb plug-in to become commercially available. It was the first convolution reverb available for Pro Tools systems and has reached a stage of development that none of its competitors have been able to match.

Figure 9.53
Altiverb
convolution reverb

Developer Arjen van der Schoot has posted a great video to YouTube in which he explains lots of interesting details about Altiverb, illustrating this with photographs and audio examples—www.youtube.com/user/audioease.

The whole point of using a convolution processor is that if you have the impulse response of a particular acoustic space or piece of equipment, then you can use the sound of that room or desirable vintage reverb device yourself. And you can do this without paying a fortune to fly off to Hansa Studios in Germany to use the same room where Bowie recorded 'Heroes', or to get the sound of your band playing onstage at Wembley Stadium or in the famous Paradiso venue in Amsterdam.

If you like the sound of vintage reverb devices you won't be disappointed either. Altiverb has impulse responses for just about every classic bit of gear that has been used in recording studios for the past 50 years, such as the Lexicon 224 digital reverb, a Fender Super Reverb spring reverb from this classic guitar amplifier, the Roland Space Echo, and even the spring reverb from the ARP 2600 analog synthesizer!

The user interface packs a lot of controls very efficiently into a relatively small window, using 'tabs' to switch parts of the display to show different features.

Often-used controls are all presented in the main window, while less-used controls are in 'drawers' that can be opened below these.

The big reverb time knob lets you lengthen or shorten the reverb tail, the smaller knob resizes the room, and the bright(ness) knob adds a synthetic type of brightness to the reverb that EQ could never create from a real-world room. So far, so simple!

I/O controls let you adjust in and out and mix volumes and you can test sounds in the 'drawer' below. Opening the EQ and Damping 'drawers' you can access two bands of parametric EQ controls and crossover settings for the damping— all very easy to understand and work with.

If you want to get more creative, the time controls let you set a pre-delay between dry or direct sound and reverb tail. You can lock this to the tempo of your song and adjust the attack control to control the onset of the reverb tail. In the 'drawer' below, you get reverse reverb and modulation. Add a bit of modulation to your reverb and each snare hit will sound different (often more interesting)—and with tonal material, such as guitar, any slightly out-of-tune notes will be covered up in the reverb tail.

Finding and choosing the right reverb is the most basic thing you will want to do—and Altiverb's browser makes this very easy. You can just click on a photo of a room or device to select an impulse response, or use the keyword search field to help you find items like 'metallic resonances'—and these load instantly. If you find one that is close to what you want, just hit the Similar button to be presented with several alternatives. Or hit the News button to see recently released IRs (posted free every month) and download these directly into the plug-in.

Clicking any of the various 'tabs' at the top of the main window makes different features available in the display. The new interactive equalizer has four parametric bands, and you can drop almost any sound file onto the IR import tab to use as an impulse response. If you use a recording of a piano chord, for instance, then clap your hands into Altiverb, out comes the chord! You can use the controls in the gated reverb tab to adjust the gate, using note values locked to your song's tempo. Any reverb tail can be reversed and handled the same way, and Altiverb's pre-delay can also be set up in this manner—so it's easy to add a slapback delay timed to your track, for example.

A click on the 'positioner' tab at the top of the Altiverb window reveals the stage positioning display—see Figure 9.54. This stage positioner allows you to accurately place your sound anywhere onstage. With a mono impulse response, when you enable the On/Off button at the top left, a speaker icon appears that you can drag around with your mouse to place your audio on the 'stage' of the sampled room. With stereo impulse responses, you get two speaker icons and you can mirror the placement of these by enabling the Stereo Link Edit button.

Altiverb can be thought of as an echo chamber into which you feed your audio via speakers placed inside the chamber. You retrieve the source audio together with the sound of this room reverberation via microphones placed elsewhere in the chamber. So, audio fed into Altiverb goes to its virtual speakers then is collected through its virtual microphones to produce the effected output.

You can intuitively adjust the distance between the virtual microphones and the virtual speakers by dragging the speaker icons around the display. When you use the positioner, this controls the timing and loudness of the direct sound, so 'Direct' in the time tab is automatically switched and on controlled by the positioner.

Figure 9.54
Altiverb positioner

Overall, Altiverb offers lots of great features for creative music making. The brightness algorithm, modulation, stage positioning features, tempo-locked gated, and reverse reverbs are all real winners—and you can create fantastic aural textures by just using any sound as an impulse response.

Avid Space

Space is a convolution reverb (see Figure 9.55) that comes with a set of presets covering all the popular reverb categories.

Four tabs at the top of the window let you choose what to display in the window below: the impulse response waveform; a picture of the location in which the IR was created or of the device used; the Snapshot page that lets you store up to 10 snapshots of all the settings; or a fourth page that lets you set the preferences.

Below the picture display are five more tabs that switch between different sets of controls for levels, delays, early reflections, reverb tail, and decay settings. Unusually, Space provides separate control over the delay time of the late reflections; it also offers plenty of controls for the reverb tail, and the Decay page has controls for the low-, mid-, and high-frequency elements of the IR, using a three-band EQ section with adjustable crossover frequencies. At the far left is a Reset button that resets all the parameters to their default values with controls for wet and dry levels and decay below this.

To the right of these controls and displays is a browser section with folders for the different categories such as Rooms, Chambers, Plates, and Digital Reverbs. The impulse response library has a good mix of real acoustic spaces, springs and digital reverbs, postproduction ambiences, and effects—although this is not quite in the same league as Altiverb's wonderful libraries!

Figure 9.55
Avid Space
convolution reverb

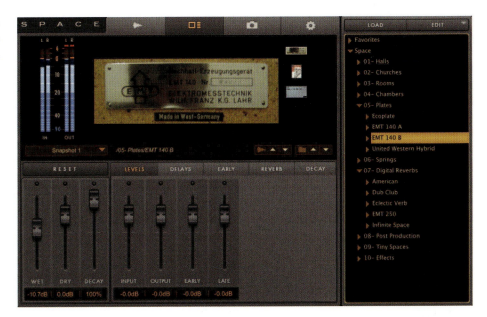

McDSP Revolver

McDSP's Revolver (see Figure 9.56) is a convolution reverb with dedicated and routable EQ, two stereo delay lines, a reverb decay crossover network, and specialized stereo imaging control. It comes with more than 300 impulse responses including typical acoustic spaces, rooms and halls, reverb plates, spring reverbs, and so forth. Even more are available to download from the McDSP Web site, and if you really want to 'roll your own', a set of tools is provided to help you to create impulse responses.

The user interface is a pleasure to use, with logical groupings of faders, easy-to-read legends, a large rotary control for the reverb duration, and a bank of eight buttons arranged vertically. The bank of buttons lets you select different pages of controls. The selected controls are then displayed in the lower half of the user interface, with the wet and dry mix controls always displayed at the left of each page.

A relatively large display is positioned in the upper right quadrant of the user interface with tabs across the top that let you display one of the following: the current impulse response; an image of where the impulse response was taken from; a graph that shows the responses of the various EQ sections; a flow diagram that not only lets you choose how the delay lines are routed in relation to the reverb, but also lets you choose where you want the EQ to be applied; or a system page that lets you control the tail cut (i.e., the level at which the reverb trails become truncated) and also has a latency mode control and a stereo mode control.

Figure 9.56
McDSP Revolver

The selection of presets has enough variety to suit most applications, and it makes a lot of sense to include Revolver if you decide to buy a bundle of McDSP plug-ins.

Pitch and Time Manipulation

It is often useful to be able to select an audio file or clip and change its pitch or its length/tempo. Pro Tools provides an AudioSuite plug-in, Time Shift, for exactly these purposes.

So why would you want to pay out extra for an AudioSuite time compression/expansion and pitch-shifting plug-in when Pro Tools already includes those tools? The reason is that the standard Pro Tools plug-ins have restricted capabilities compared with the 'premium' plug-ins that you pay extra for. These will allow you to make greater changes of duration and tuning (and modify longer audio files) with less distortion and fewer digital artifacts.

There are lots of alternatives to choose from. For example, SoundToys offers Speed, which lets you change the pitch or the length of your audio files by simply grabbing a knob and rotating it. Serato offers Pitch 'n Time, which lets you alter pitch and tempo graphically or numerically and has a Varispeed mode, where altering the tempo also alters the pitch. These have both been around for a long time and offer significant improvements compared with the basic Pro Tools plug-ins.

Avid X-Form

Avid's X-Form (see Figure 9.57) is one of the highest-quality duration-and-tuning tools available for the Pro Tools platform.

The user interface is simple enough to operate with everything laid out logically and labeled clearly. You can either use the rotary controls to set the time or pitch shift amounts or simply type in the amounts using your choice of Bar|Beats, minute:seconds, feet:frames, time code, or samples. The gain control lets you lower the input level by up to 6 dB to avoid clipping in the processed signal.

X-Form lets you alter audio files by even larger percentages with even fewer undesirable artifacts than most of its rivals: it can process files within a range of 12.5 to 800 percent of the original length or one-eighth to eight times the original duration. It can pitch shift audio up or down by three octaves, even on polyphonic material—and with proper formant correction so that the audio sounds completely natural.

X-Form provides a choice of two time-stretching/pitch-shifting algorithms: the Monophonic algorithm for working with solo instruments or vocals, and the Polyphonic algorithm for use on audio that contains chords—or a stereo mix. Also, when used with the transient sensitivity control, X-Form allows you to compress or stretch drum loops and other percussive material without losing crucial attack transients. I have compared X-Form with most of its competitors and it definitely sounds better.

NOTE

Part of X-Form's processing relies on separating 'transient' parts of the sample from 'nontransient' parts. Transient material tends to change its content quickly in time, as opposed to parts of the sound that are more sustained. Sensitivity is only available when polyphonic is selected as the audio type. For highly percussive material, lower the sensitivity for better transient detection, especially with the rhythmic audio setting. For less percussive material, a higher setting can yield better results. Experiment with this control, especially when shifting drums and percussive tracks, to achieve the best results.

Figure 9.57
Avid X-Form

AVID X-FORM VERSUS TIME SHIFT

I asked Avid to explain the differences between the standard Time Shift plug-in and the new X-Form plug-in. Here is what I was told: 'While both Time Shift and X-Form offer time compression/expansion and pitch shifting with high-quality results, there are a few key reasons why X-Form outshines Time Shift. The main thing that separates Time Shift from X-Form is the quality of the processed file. The iZotope Radius TCE engine used in X-Form is one of the best sounding TCE algorithms on the market, offering transparent time stretching and pitch shifting results. Also, X-Form is simpler to use and more effective than Time Shift when selected as the default TCE Trim tool because it has just one main Shift control. Another place that X-Form shines is when processing files using extreme settings for compressing or expanding the original file's length. The nature of X-Form's iZotope algorithm with its FFT-based processing enables the algorithm to achieve better results than the traditional "frame overlap" algorithms used in other time stretching products.

'Both Time Shift and X-Form offer phase coherent processing for processing multi-channel tracks. This is especially useful for stretching a multi-channel track while preserving the original audio file's surround "image". While both algorithms do phase coherence processing well, X-Form provides results that collapse the image less—resulting in a processed audio file that more closely resembles the original audio image. The Polyphonic algorithm in X-Form is able to preserve formants on polyphonic material, which is a feature that only Celemony Melodyne could do in the past. X-Form can preserve formants in Monophonic or Polyphonic modes while Time Shift can only preserve formants in Monophonic mode.'

Recording Equipment Emulation

We have now reached a point where just about every piece of equipment that you would find in a well-equipped recording studio has been modeled in software and made available as a plug-in for Pro Tools.

There are even virtual alternatives to the musical instruments—and even to the musicians and singers, but let's not go there just now. . . . Instead, let's take a look at the kinds of equipment that would be used to record, say, a guitar player. First of all, if it's an electric guitar, you need an amplifier and speaker.

Using software emulations of these, all you need is a suitable interface for Pro Tools with an instrument level jack for the guitar and once the guitar has been recorded into Pro Tools you can process this at your leisure using the plug-ins.

ENGL Guitar Amplifiers and Speaker Cabinets

ENGL guitar amplifiers and cabinets were first developed in Germany in the mid-90s to provide high-performance rock and metal guitar tones. Brainworx has developed software versions of the E646 VS Limited Edition (see Figure 9.58) and E765 Retro Tube (see Figure 9.59) models and has licensed these to Universal Audio for its Powered Plug-ins platform.

Each of these plug-ins is a faithful digital recreation of the original tube amp together with an FX rack that has a noise gate and a vintage delay, and a power amp simulator with an onboard power soak. The plug-ins both have 64 recording chain presets that provide enormous flexibility. Captured through Brainworx's ultra-rare Neve VXS72 console, these presets let you audition your sounds through a variety of vintage mics, ENGL cabinets, and outboard gear, including hardware emulations from Millennia, SPL, and elysia.

Figure 9.58
Universal Audio
ENGL E646 VS
Limited Edition

Figure 9.59
Universal Audio
ENGL E765 Retro
Tube

All of these features can be used or bypassed individually according to your preference. So, for example, you could bypass the plug-in's preamp section and record from your guitarist's own preamp if you prefer, but still use the plug-in's power amp simulation, FX, and recording chains.

The main attraction of these plug-ins is, of course, the sounds that a guitarist can get from them—and they provide a welcome alternative to the many available Fender, Marshall, and Vox simulations. Don't forget to try other instruments through these processors, though. The effect on a TR808 bass drum or a Wurlitzer piano can totally transform these sometimes bland-sounding instruments into exciting sounds that 'sit' much better in your mixes!

Millennia NSEQ-2

Millennia Music and Media Systems makes some of the finest microphone preamplifiers and equalizers that money can buy. One of its most popular models is the NSEQ-2. This is a twin-channel parametric EQ unit that features both vacuum tube and discrete J-FET solid-state signal paths that can be

Figure 9.60

Universal Audio
Millennia NSEQ-2

selected using a front-panel switch. Brainworx
has reproduced this in software and licensed it to
Universal Audio for its Powered Plug-ins platform—
see Figure 9.60.

The best way to appreciate the potential of this
plug-in is to download and install the demo
software from Universal Audio's Web site and try out
the presets—see Figure 9.61. These demonstrate
just how creative you can get using a high-quality
device with these features—going far beyond what
you can achieve with the basic EQ plug-ins supplied
with Pro Tools.

API Vision Channel Strip

The API Vision Console Channel Strip plug-in for
the UAD platform is based on API's flagship console
found in studios and sound stages across the
globe—see Figure 9.62.

The plug-in includes five of the API modules from
the Vision console: the 212L Preamp, the 215L
Sweep Filters, the 550L EQ, the 225L Compressor/
Limiter, and the 235L Gate/Expander.

01 Female Vocals Tube
02 Female Vocals J–FET
03 Male Vocals Tube
04 Male Vocals J–FET
05 Bright Acoustic Guitar Tube
06 Bright Acoustic Guitar J–FET
07 DI Bass Tube
08 DI Bass J–FET
09 Mix Start LR Tube
10 Mix Start LR J–FET
11 Mix Start MS TUBE
12 Mix Start MS J–FET
13 Room Drums TUBE
14 Room Drums J–FET
15 Snare TUBE
16 Snare J–FET
17 MetalliKick TUBE
18 MetalliKick J–FET
19 HiHat Sweetener TUBE
20 HiHat Sweetener J–FET
21 Synth Bass Booster TUBE
22 Synth Bass Booster J–FET
23 Solo Guitar De–Harsher TUBE
24 Solo Guitar De–Harsher J–FET
25 Vocal Group Widener TUBE
26 Vocal Group Widener J–FET
27 FM Synth Widener TUBE
28 FM Synth Widener J–FET
29 909 Beat Booster TUBE
30 Wide Piano MS TUBE
31 Piano LR TUBE

Figure 9.61

Universal Audio
Millennia NSEQ-2
presets

There are plenty of useful presets provided by five leading practitioners—see
the accompanying screenshot. My personal favorites are those developed by
Nathaniel Kunkel—especially Upright Bass and Male Vocal. All in all, you get a
lot of bang for your buck with this plug-in—highly recommended!

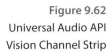

Figure 9.62
Universal Audio API
Vision Channel Strip

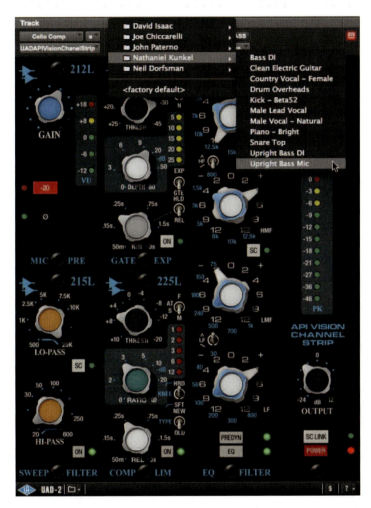

FROM THE MANUAL

'The tone of the 212L mic preamp has its roots in the classic API 2488 series all-discrete recording consoles, best known for the famed "LA" sound. The API 212L incorporates the API 2520 op-amp and the same circuit as the legendary API console input modules dating from the 1970s. This mic preamp articulates high frequencies with great detail, while delivering the big sounding, warm bottom end that API is famous for.

(Continued)

'With identical features as the modern API 550B EQ, the API 550L (L for "long" frame) is a continuation of the 550A EQs that have played a major role in the history of record making, but with an additional filter band and several new frequencies. The 550L artfully blends the past with the present, and is only available in modern API consoles. Making use of API's "Proportional Q" innovation, the 550L intuitively widens the filter bandwidth at minimal settings and narrows it at higher settings without the need for additional bandwidth controls.

'Ideal for almost any application, the widely versatile API 225L Compressor's auto-output level remains at unity regardless of the threshold or ratio settings. This feature allows for real-time adjustments without the need for changing the output level. Both New ("feed-forward") and Old ("feed-back") methods are selectable via the front panel, providing two choices of gain reduction. Soft provides a more subtle compression resulting in a natural sound, while Hard results in a sharp knee type with a severe limiting effect.

'The 235L Noise Gate/Expander is one of the fastest noise gates available. The API 235L can reduce noise in any type of program without losing any part of the source. Its extreme flexibility and superb sound make it ideal for all recording or mixing studio applications. The Expander function uses a 1:2 ratio, allowing the signal to "sneak up" to the full signal level without any loss of "under threshold" vocal or percussion nuances. Setting the threshold in the Gate function to the desired level, then switching to the Expander mode is the perfect workflow.

'The API 215L is a unique passive, sweepable cut filter, designed specifically to contour the sound in a way that preserves the natural tone of the signal. The 215L is a low pass filter with a slope of 6 dB per octave, and a high pass filter with a slope of 12 dB per octave. The filters are isolated from each other with the same discrete transistor buffer used in the famous 550 series equalizers.'

Studer A800

The Studer A800 multichannel tape recorder, first introduced in 1978, was regarded as the finest machine of its type ever built. Universal Audio has meticulously modeled the characteristics of the entire audio signal path, and of four different tape formulations (3M 250, Ampex 456, BASF 900, and Quantegy GP9). All three tape speeds—7.5, 15, and 30 inches per second—are provided and the head-bump effects and high-frequency shifts are modeled for each of these.

The plug-in (see Figure 9.63) can be inserted on mono tracks, on group busses, or even on whole mixes as a stereo insert.

Figure 9.63
Universal Audio
Studer A800

If you have the inclination, you can even tweak the lineup and other parameters that are revealed when you click 'Open' just above the Studer legend. Of course, there are plenty of great presets provided to get you started with most styles of popular music—from jazz to reggae, from rock to hip-hop—and the venerable Studer A800 has been used to record more classic pop records than most!

Avid Reel Tape Saturation

Reel Tape Saturation (see Figure 9.64) lets you make recordings with Pro Tools that sound as though they were recorded onto analog tape by emulating the tape saturation and compression that happens when you record with Ampex 456 or Quantegy GP9 tape using a 3M M79 or Studer A800 multitrack tape machine.

A bias control simulates the effect of under- or over-biasing the tape machine and a cal adjust control simulates the effect of three common calibration levels. The plug-in also emulates the artifacts that occur at the different tape speeds (7.5, 15, or 30 IPS) and even lets you add tape hiss.

For example, if you have recorded a drum kit and then decide that you would like to get the 'fatter' sound of drums recorded to an analog multitrack tape recorder, you have two choices. You can either insert a Reel Tape Saturation plug-in onto each drum track to allow individual adjustment or bus all the drum track outputs to an auxiliary track and just use one Reel Tape Saturation plug-in to affect the sound of the whole kit.

Figure 9.64
Avid Reel Tape
Saturation

Avid Reel Tape Delay

If you are really serious about simulating the sound of a tape machine, you will want to recreate all the kinds of delay and distortion effects created by the tape machine itself. Reel Tape Delay (see Figure 9.65) allows you to do this.

As Avid's Web site explains: 'Psychedelic, Dub, and Electro music styles are hotter than ever, and the warm, ear-pleasing sound of tape-based delay has never been more relevant. Reel Tape Delay emulates analog tape echo and delay effects with expertly modeled frequency response, noise, wow and flutter, and distortion characteristics.'

Reel Tape Delay can reproduce wow and flutter fluctuations in tape speed—whether small or large—and can even simulate the effect of adjusting tape speed while recording or playing back. Delays are fully adjustable over a wide range and the bass and treble controls built into the echo/delay feedback loop let you create a variety of effects, from subtle to extreme.

Figure 9.65
Avid Reel Tape
Delay

Avid Reel Tape Flanger

The Reel Tape Flanger (see Figure 9.66) produces the popular flanging effect that many producers and remixers like to use to spice up their mixes. This effect was first used on a popular hit record, 'Itchykoo Park' by the Small Faces back in the late 1960s, by a friend of mine called George Chkiantz who was the house engineer at Olympic Studios at that time. This effect and the so-called automatic or artificial double-tracking, abbreviated as ADT, was also used on various recordings by the Beatles and other popular groups around the same time. It has always been difficult to simulate this effect electronically, as the original effect was created manually using analog tape recorders with all their quirky characteristics.

According to Avid: 'The reason why most flanger effects sound unconvincing is because the studio technique that they imitate is dependent on so many real-world factors. Reel Tape Flanger thoroughly models the tape flanging process and hardware characteristics, and provides an arsenal of controls for shaping the output in real time. The results are as close as you can get to the real thing—and a lot easier to use! The sonic characteristics of frequency response, noise, wow and flutter, and distortion are captured in detail and applied to a variety of tape speed and delay control options. Infamously difficult manual flange and artificial double-tracking (ADT) techniques are simple to re-create. Simply adjust the variable delay manually (or use the LFO and Depth controls) and listen to the captivating frequency sweep and "crossover" cancellation effects heard on so many classic recordings. You also get easy access to tape speed fluctuation and feedback controls.'

The Reel Tape Flanger sounds very convincing and is extremely easy to use, with all the controls clearly and informatively labeled, and a selection of presets is provided to get you started.

Figure 9.66
Avid Reel Tape
Flanger

Ampex ATR-102

The Ampex ATR100 series of tape recorders was introduced at the spring 1976 AES Conference in Los Angeles. This was produced in two-track and four-track versions with interchangeable head blocks to cater for both 1/4" and 1/2" tapes. The next-generation ATR-102 was released in 1978, followed by the 16-track ATR-116 and the 24-track ATR-124. Despite general agreement that the ATR-124 was the best-sounding 24-track ever made, this did not achieve commercial success. However, the ATR-102 went on to become the one of the most popular choices to record final mixes onto.

Many engineers consider the ATR-102 the best-sounding tape machine for final mixdown, and Universal Audio has modeled this tape machine meticulously in its UAD Ampex ATR-102 plug-in—see Figure 9.67. Back in the '70s and '80s, it was also quite common for tape machines to be used to create slapback echo effects, and the UAD Ampex ATR-102 is all set up to do this using a simple set of controls, while extending the delay times beyond what would have been possible with the original machine.

A popular custom head block modification for the ATR-102 allowed the use of 1" tape to provide even higher fidelity than with 1/2" or 1/4", so all three tape head widths are accurately modeled and selectable in the plug-in. The UAD Ampex ATR-102 also models seven popular magnetic tape formulas—each with its own subtle sonic variation, distortion onset, and tape compression characteristics. All four tape speeds in the original hardware are modeled, including 3.75, 7.5, 15, and 30 inches per second, together with each speed's distinct frequency shift, head bump, and distortion characteristics. As you might expect, higher speeds have higher fidelity while 3.75 IPS has a distinctively 'lo-fi' character.

Four calibration levels are available, each with different tape response characteristics for a given level into the recorder, and the plug-in simulates the behavior of the transformers in the original hardware circuit that 'color' the sound. These transformers can be optionally disabled in the plug-in for a 'cleaner' sound. The plug-in also models the hum, hiss, wow, flutter, and crosstalk characteristics of the original hardware and these noise components can also be individually disabled, adjusted, or exaggerated for creative purposes.

UAD Ampex ATR-102 includes lots of great presets from prominent ATR-102 users including Chuck Ainlay, Richard Dodd, Buddy Miller, Mike Poole, and others. Nevertheless, if you are really determined to set things up just the way you want them, the UAD Ampex ATR-102 includes the full suite of tools required to manually calibrate the recorder. These tools include a tone generator (with multiple test tones and levels), a distortion meter with digital readouts, and a full suite of magnetic reference laboratory (MRL) alignment tapes, which are used to calibrate the playback electronics—paradise for tweakheads!

You can always use the plug-in as an individual mono or stereo insert effect or as an auxiliary group effect applied to a group of drums, guitars, or whatever. But, more typically, you will use the UAD Ampex ATR-102 as the last stereo insert on your master fader (or possibly the second-to-last insert before a brick-wall processor) so that you can add the signature sound of this classic analog device to your mixes.

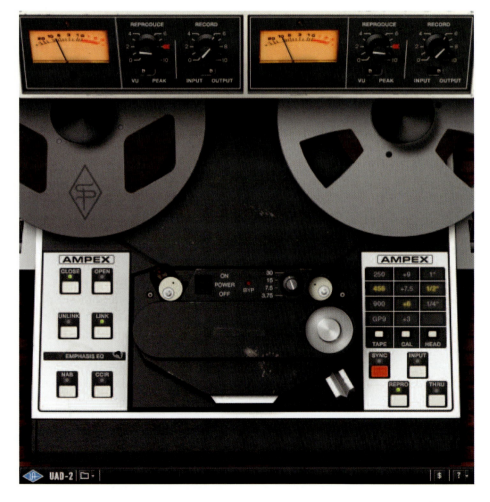

Figure 9.67
Universal Audio
Ampex ATR 102

Utilities

Just one final section in this chapter, for utilities—those useful little devices like guitar tuners that do just one simple job but do it well.

Avid Utilities

Avid's InTune (see Figure 9.68) has preset tunings for mandolin; ukelele; four- and five-string banjo tunings in various popular variations such as Irish tenor banjo; six-, seven-, and eight-string guitars; and four-, five-, and six-string bass guitars. Very useful for players of fretted instruments—I play banjo, mandolin, ukulele, six-string guitars, and four-string bass guitar myself, for example.

Figure 9.68
Avid InTune Tuner

Avid also offers a much better metronome plug-in than the standard Pro Tools Click plug-in: the Metro plug-in (see Figure 9.69) includes all the popular metronome sounds that people ask for time and time again, such as the Urei click, the MPC click, a rim shot to use as a click, a shaker to use as a click, a woodblock to use as a click, and so forth. Even better, it lets you add extra sounds on subdivisions of the bar so that you can have more interesting and rhythmically supportive click patterns to play to.

A third utility, MasterMeter (see Figure 9.70), is useful when you are making 'hot' mixes that sound loud. 'Hot' mixes can often cause distortion due to signal peaks above 0 dB that occur between samples. These may play back

Figure 9.69
Avid Metro

without distortion in the studio environment, but when the same mixes are played through a consumer CD player, the digital to analog conversion and oversampling processes can result in distortions becoming audible. By running your mix through the MasterMeter, you can check the time code positions within your mix at which these troublesome peaks occur so that you can take appropriate action to avoid these.

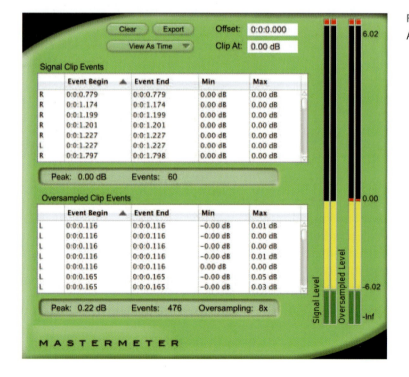

Figure 9.70
Avid MasterMeter

Summary

These days it seems that we are spoilt for choice when it comes to choosing among so many suites of signal processing plug-ins available from so many manufacturers. And, truth be known, you can get great results, both technically and creatively, with most of these. After all, this is mostly down to the workman, not the tools, assuming that the tools are all capable of processing audio at high quality (which most are).

Nevertheless, several do stand out from the crowd:

- FabFilter is not only pushing the edge technically with its designs, but has managed to come up with some of the best user interface features—and without copying older devices.

- When it comes to delay and modulation effects, SoundToys is head and shoulders above most of the competition.

- Avid's own Reel Tape Suite and its Pro series of dynamics processors feature excellent designs with pristine audio quality.

- Both Sonnox and McDSP have winners among their offerings, such as the Oxford Supresser and the ML4000 Multiband Dynamics.

- Flux Verb V3 is probably the best of the algorithmic reverb types, while Audio Ease Altiverb sets a standard for convolution reverb that will be hard for any competitor to equal or surpass.

- Universal Audio's recreations of analog devices are stunningly good, and most engineers 'of a certain age' are going to feel very comfortable using these—especially the Studer A800 and the venerable Ampex ATR-102 emulations.

in this chapter

CHAPTER 10

Mastering and Delivery

What Is Mastering?

When a final mix is recorded in a suitable stereo (or multichannel) format, this is often referred to as the *master mix*. Typically, this master mix goes through a further mastering stage—often in what is commonly referred to as a *mastering studio* under the supervision of a mastering engineer.

This stage, en route to the pressing plant where CD discs will be replicated for commercial distribution, is more correctly referred to as *premastering*—as it involves preparing the mixes so they are ready for the mastering stage at the pressing plant at which the premaster tapes or discs are used to create the *glass masters* used in the replication process.

But almost everyone refers to the premastering stage as the *mastering stage* and to the premastering engineer as the *mastering engineer*—as I mostly will in this book.

What Does the Mastering Engineer Do?

Basically, the mastering engineer takes the final mixes that will be distributed for sale to the public and adjusts these to suit the distribution medium.

Some producers also ask the mastering engineer to make creative changes to the mixes during the mastering process to enhance the music in various ways.

The final mixes may arrive at the mastering studio in a variety of formats. The preferred format for the digital age would be 32-bit or 24-bit WAV files created at the sample rate used for the project—with a maximum peak level of −3 dB so that the mastering engineer can adjust the final levels secure in the knowledge that these source files do not have clipped waveforms.

In practice, many mastering studios are also equipped to work with older formats, including DAT, Sony 1630/1610 with U-Matic VCR, 1/4" tape, and 1/2" tape.

For distribution as WAV files via the Internet, mastering may simply involve a final quality control check. But just about any other form of distribution, such as MP3 files for the Internet, vinyl discs, CDs, or the various surround formats on DVD, will require special attention to compensate for the peculiarities of the medium.

> **NOTE**
> There is a trend for some mix engineers to provide several two-track 'stems' containing mixes of, say, the rhythm section, the vocals, and perhaps brass, strings, or other instrumental components. The mastering engineer can then choose the final balance between these elements to accommodate the preferences of the client. When mixing for surround, six or more tracks may be created at the mix session, and if three stems are created this could result in 18 or more tracks for the mastering engineer to deal with.

Each mix should start and end cleanly, either abruptly or with appropriate fade-ins and fade-outs. Unless you are sufficiently experienced at 'topping and tailing' audio recordings and fading audio in and out, these tasks are best left to the mastering engineer. And it is always a good idea to include a little extra audio before and after the music starts and finishes just in case the mastering engineer needs to analyze this to obtain a 'noise profile' that can be used to remove similar noise from the wanted material, or to allow for last-minute adjustments to the fade times, or whatever.

The mastering engineer can often remove pops, clicks, and other noises, including broadband noise, during the mastering session using specialized hardware and software 'tools' available for these purposes. It is also possible to make 'repairs' to damaged tracks during the mastering session by copying material from an undamaged section and pasting this to replace the damaged section. However, if 'major surgery' of this type is required, you may need to take your recordings to an engineer who specializes in rescuing damaged recordings or cleaning up old material before you arrange the mastering session.

During the premastering process, the sequence of tracks for an album is finally decided on and decisions are made about how long the gaps between tracks will be. The average gap would be around two seconds. But if a track follows a long fade-out, the gap could be reduced considerably so that this following track comes in almost immediately. Some albums even have no gaps between tracks, or have crossover fades so that you hear the next track fading in as the previous one fades out.

If the mixes have been created at around the same time by the same mix engineer in the same studio, they will probably sound similar tonally. If not, the mastering engineer can apply EQ or multiband processing to compensate for mixes that sound too dull or too bright.

One of the most important processes in premastering is to make sure that the subjective loudness levels of all the tracks are about the same—apart from any tracks that are supposed to sound quieter or louder than the rest—and that the loudness levels of all the tracks are comparable with those of commercial releases within the particular genre.

Having adjusted the levels of the individual tracks, the mastering engineer then needs to set the overall level. This will be dictated by the level of the highest peak, even if this level is only reached during one of the tracks. When this peak 'hits the end stops' at 0 dBFS, you cannot increase the overall level without clipping the waveform—unless you use compression or limiting to reduce the level of the highest peaks so that the average level can be increased without clipping any waveforms.

With careful use of compression and limiting, peaks can be reduced without too much loss of quality, or sometimes even producing substantial subjective improvements in the listening quality. Once these peaks are 'tamed', the average level of all the tracks can be raised so that normal playback levels produce loudness levels comparable with other recordings of the same genre.

Unfortunately, many of today's mastering studio clients are demanding that mastering engineers apply unreasonable amounts of compression and limiting, using increasingly powerful multiband dynamics processors, in foolish endeavors to make their CDs sound the loudest when played on radio or on TV. This results in recordings that have very little dynamic range left and that often contain clipped waveforms. The high playback levels that are achieved can cause serious problems with consumer equipment or with broadcast equipment that is not designed with these levels in mind—resulting in even more distorted sound.

Classical, jazz, and the various acoustic music genres generally have a much wider dynamic range than pop or rock records. But when it comes to much of today's hip-hop, rap, and other mostly synthesized forms of popular music, there is very little dynamic range left—they stay loud from beginning to end without any letup! As a listener, I find many of these recordings to be virtually unlistenable. My ear becomes fatigued after hearing this kind of stuff for as little as 30 seconds—let alone a whole album's worth. The problem is not the music—it is the poor sound quality that is the result of too much compression and limiting. These processes take away the natural dynamics of musical sounds, replacing these with something that literally has 'squashed' waveforms—with the transients removed or reduced to a minimum. So the music just does not sound as good or as pleasing to listen to.

As Bob Katz says: 'The loudness war may have begun with analog records, but the current problem is many decibels worse than it was in analog. LPs were mixed largely with VU meters, which created a degree of monitoring consistency, but today's peak level meters give entirely too much more room for mischief, and today's digital limiters provide the tools to do the mischief. The net result: great consistency problems in CD level. The peak meter is currently being seriously misused. Remember that *the upper ranges of the peak meter were designed for headroom, not for level.*'

Are You Experienced?

Before deciding to master any recordings yourself, you should consider whether you have sufficient experience to successfully make the judgments that a mastering engineer, with fresh ears and many years of focused experience in this area, can make with confidence.

For example, although I have mastered a few albums for commercial CD release in the past, I recommend that most of my clients take the mixes that I have produced to a proper mastering studio with an experienced mastering engineer—unless it is a very low-budget project.

On the other hand, when it comes to preparing WAV, QuickTime, or MP3 files for Internet delivery I do feel more confident to handle the mastering myself. The WAV files may only need trimming or fading, for example, and it is easy enough to choose sensible settings when preparing audio for the various Internet file formats then audition these online to make sure they have encoded successfully.

Also, the tools and techniques mastering engineers employ can be used in a variety of other postproduction situations so it is worth learning about these techniques in anticipation that they will prove useful sooner or later—as you become more experienced.

And the more you know about these processes and the more experience you have using these yourself, the better you will be able to communicate with mastering engineers whenever the situation arises.

Tools of the Trade

First of all, you need to make sure that you can hear everything 'properly' before you start your mastering session. For this you need a suitable room with a high-quality monitoring system that is capable of reproducing the entire audio spectrum without compromise and is correctly set up and interfaced with your audio equipment.

The Room

The room has to be suitable for mastering. It should be designed such that any standing waves, particularly at low frequencies, are minimized—or you will end up boosting or cutting at these lower frequencies for the wrong reasons. The room should be, say, 20–30 feet long to provide sufficient bass response and to allow sufficient room for the speakers to be placed correctly and far enough from the walls to avoid proximity effects—unless they are mounted on the walls using soffits. The room should also be very quiet—with no noisy equipment fans, whirring hard drives, and the like. There should be nothing between the monitors and the mastering engineer to disturb the sound field created by the loudspeakers, and reflections from the room boundaries should be minimized. So there are lots of important requirements here that are unlikely to be satisfied in anything other than a dedicated, acoustically designed, mastering studio room.

The Monitoring System

Near-field monitors should not be used for mastering: you need full-range monitors that have something like a 12" or larger woofer—6", 8", or even 10" speakers are too small to reproduce the longest wavelengths at the lowest frequencies. It's a matter of the physics of sound reproduction.

If you try to master on near-field monitors, especially in a room without proper acoustic treatment, there is a good chance that you will overcompensate for the lack of bass and end up with bass-heavy mixes. Or you will overlook problems like rumble and other low-frequency noises.

It is possible to use sub-woofers with near-fields to extend the low frequency performance, but most experts will tell you that this is not as accurate for mixing music in stereo as using properly designed full-range monitors. If you must use a system with sub-woofers, then use a pair of these—one for each side of the main stereo speakers—rather than the single LFE sub-woofer used for most 5.1 surround setups.

Most near-field monitors will also compress high-level transients when monitoring at the louder playback levels, which makes these speakers unsuitable for judging the correct amounts of compression to apply during the mastering process. Near-fields can also be deceptive when you are judging reverberation levels and stereo separation.

> ### TIP
> The monitors that I see in the best mastering studios are usually ATC self-powered models such as the 100s or 200s, PMC passive monitors with Bryston power amplifiers, or B and W 602 passives with chord amplification. And the larger mastering studios usually have 5.1 surround versions of these monitoring systems.

Monitoring Levels

Ideally you will monitor at similar levels to those of your listeners to make sure they hear the balance that you want them to. An awareness of the Fletcher-Munson loudness curves will help you to understand why this is. The main thing to realize is that the ear responds differently to lower and higher frequencies as you raise the overall monitoring level—so you are likely to add too much bass if you monitor too low and add too little bass if you monitor too high.

A good monitoring level to aim for would be between 80 and 90 dB SPL—85 dB is the standard used for film dubbing, for example. But if you know that typical playback levels will be much lower than this, if the music will primarily be heard via radio or TV broadcast in the home, for example, then it would be wise to monitor at this 'target' level instead.

I find it very useful to check mixes from time to time at quite low levels, between 55 and 65 dB SPL, for example. At such low levels you will hear the most prominent elements within the mix and if anything that should be audible is totally lost or if anything sticks out too much, these things will be much more noticeable at these low levels.

Stereo Balancing

Don't make the mistake of thinking that you can judge the correct stereo balance by simply watching the left and right channel meters! You can only make this judgment using your own ears. So you must make sure that you are sitting in the monitoring 'sweet spot' and that the speakers are set up and adjusted correctly and wired to work in phase. Any level imbalances between the left and right channels will upset these judgments, as will any wrong positioning of the speakers or the listener.

Ideally, the speakers should be arranged to form an equilateral triangle with the listener positioned in the 'sweet spot' at the apex of the triangle with equal distances between the speakers and from each speaker to the listener. This arrangement causes the adjacent sides of the triangle to be angled at 60 degrees to each other. It is very important to make sure that these distances are the same, or the position of the center image will be shifted to one side or the other.

Operating Levels and Metering

CD mastering back in the 1980s involved analog processing of analog tape mixes that were then converted to digital audio. Consecutive sample count methods of overload detection used by the Sony PCM 1610/1630, DMU-30 meters, and DTA-2000 analyzers worked well with this type of audio, but these designs are not adequate for the digitally processed audio that mastering studios typically have to deal with today.

The standard meters in Pro Tools and other DAWs and digital mixers simply detect and display the levels of each sample—a process that tells you nothing about numbers of consecutive 0 dBFS samples. It is better to use meters that count the 'overs' that occur when the signal tries to exceed 0 dBFS resulting in clipping of the waveform. If these 'overs' last long enough to become audible or there are too many of them, the engineer can then take appropriate action to reduce levels.

Engineers also need to be aware that the peak value of a digitized sine wave, depending on its frequency and phase, may be up to 3 dB lower than the actual peak value of the analog signal that is reconstructed from this digital signal when it is played back on a CD player, for example. This is simply due to the nature of the A/D conversion process and is not revealed by most meters currently in use. The temptation for the unwary mastering engineer is to use this 3 dB as additional headroom into which the digital gain can be increased. The problem here is that the peak value of the reconstructed analog signal coming out of the D/A converters will then be 3 dB above 0 dBFS.

To make matters even worse, the expensive D/A converters used in professional mastering studios, unlike those in more typical replay equipment, may actually play back these higher levels without fully revealing the distortions associated with these '0dBFS+' levels. So the mastering engineer may remain unaware of these hidden distortions that inevitably lead to increased listener fatigue.

The reasons for this are explained in some detail in a paper written by Søren Nielsen and Thomas Lund and presented at the 107th AES Convention in 1999 titled 'Level Control in Digital Mastering', and also in a follow-up paper in 2000 titled 'OdBFS+ Levels in Digital Mastering'. These, and other similar technical papers, can be downloaded from the TC Electronic Tech Library at www.tcelectronic.com/TechLibrary.

According to Thomas Lund: 'In our tests we haven't found a single professional or consumer CD player that doesn't significantly distort when subjected to 0 dBFS+ signals. When peaks reach +3 dBFS, most players distort more than 10 percent. Many also latch up and take a while to get out of distortion mode again—so a period of time after a peak has occurred will also be distorted. Distortion, which cannot be removed, is the price paid for trying to adjust the end listener's level control by making a "hot" CD. However, not all the distortion is recognized in the studio because it does not show on the meters and is not heard before the signal goes through a reproduction chain. Therefore, an additional penalty is added: unpredictable reproduction quality due to exhausted headroom in D/A converters, in sample rate converters, and in data reduction systems such as MP3 codecs.'

So the safest plan is to stay well away from 0 dBFS when working with standard peak meters to avoid inadvertently clipping waveform peaks—because the way that these meters are designed makes it more difficult to see when clipping occurs. 'Sample based peak meters require that a headroom of at least 3 dB is maintained in order to prevent distortion and listener fatigue,' says Lund, adding that 'Hyper-optimized audio creates distortion and listener fatigue on CD, Film and Broadcast Commercials.'

Various third-party meters are available for Pro Tools systems that can check for inter-sample peaks, and Pro Tools itself now includes the MasterMeter plug-in that was developed by Trillium Lane Labs.

For more information, read what Bob Katz has to say on his Web site at www. digido.com/bob-katz/level-practices-part-1.html or in his book *Mastering Audio* (Focal Press).

Always Use Your Ears!

All this reminds me of one of the most important lessons I ever learned. I had recently graduated from Salford University with a BSc degree in electro-acoustics and I was working as a trainee audio engineer for Standard Telephones and Cables (STC), learning how to design microphones and other acoustic transducers under the supervision of a very experienced senior engineer called Dennis (who, naturally, we all nicknamed 'the menace'). We worked in a small laboratory with an anechoic room for acoustic tests and several workbenches packed with unbelievably expensive Brüel and Kjær test equipment including real-time analyzers that cost almost as much as a house! One day Dennis came over to me as I worked at the bench making some sophisticated measurements and said that he was about to introduce me to the best set of audio test equipment in the world—and he grabbed me by the ears!

Dennis insisted that I take a set of headphones and use my ears to listen to the changes in sound quality as he increased the signal level being fed into the transducer under test, pointing out how the sound got 'fatter' with extra harmonics appearing as the device started to overload. Then he carefully explained that despite however much the test equipment in this state-of-the-art laboratory cost, it could not even get close to matching the analytical power of the human ear/brain combination. Until that message really sank in, I had been feeling in awe of the expensive gear that I was using to test the microphones and had been assuming that this equipment would tell me everything I needed to know. Thanks to Dennis, the point was made, and not lost on me, that audio engineers have the best audio test equipment in the world attached to the sides of their heads—and this is what they should rely on first!

Although I moved into music production shortly afterward, I never forgot this lesson, for which I truly thank this early mentor—thanks Dennis! Now every time I see a VU meter or a PPM, or a real-time frequency analyzer, or whatever, I think of Dennis and start listening even more—and that is what you should do as well. Always remember that meters and analyzers can 'lie'—or not tell you 'the whole truth and nothing but the truth'. If they are not correctly calibrated or they are not designed appropriately or used intelligently, they may fool you into thinking that everything is okay when really there are problems lurking, either immediately or further down the line. So never forget to use your ears to check audio quality, despite whatever your meters or analyzers may (apparently) be telling you!

Cables and Connectors

It is definitely worth using high-quality cables and connectors to hook all your audio equipment up. I use solid silver, oxygen-free cables between my Bryston 4B power amplifier and my ATC SCM20 near-fields and can confirm that this makes a big difference to the clarity of the sound—even your grandmother could easily tell the difference! I use similar high-quality cables for both the analog and digital connections between my Pro Tools interface, a Yamaha DM1000 mixer, and the Bryston 4B.

Using higher-quality cables for the analog connections makes the biggest difference, but replacing the AES, SPDIF, and even the word clock cables with oxygen-free versions with gold connectors makes additional, if smaller, improvements to the stereo imaging and overall sound quality.

Signal Processing

Most professional mastering studios have a high-quality reverberation and effects unit, such as the Lexicon 960L or the TC Electronic System 6000 or one of the many fine reverberation plug-ins that are now available. And if analog equipment is used, a selection of high-end equalizers, compressors, and limiters from manufacturers such as Universal Audio, Prism Sound, Neve, Manley, and others can be found in most mastering studios.

The Digital Audio Workstation

You also need to decide on which DAW to use for mastering. Since the mid-80s, Sonic Solutions has led the field for CD mastering and currently offers its SonicStudio soundBlade system—www.sonicstudio.com/sonic/products/sonic_productoverview together with its suite of NoNoise audio restoration software tools—www.sonicstudio.com/sonic/products/sonic_nnproductoverview.

Another leading system is the SADiE Mastering Suite, which includes a bundle of iZotope signal processing. These systems offer integrated PQ encoding, extremely flexible crossfade editing, and other features especially useful for CD mastering.

Magix Sequoia (only available for Windows) has been highly rated by a number of mastering engineers. Steinberg's Wavelab software (which can also handle multiple audio tracks), available for both Mac and PC, can also be used for mastering audio. Other systems available include Pyramix, Audiocube, and DSP Quattro.

Recognizing the increasing need for specialized mastering tools to be made available for Pro Tools HD systems, third parties such as Universal Audio and McDSP offer plenty of classic compressors and multiband dynamics tools; iZotope offers suites of audio restoration and mastering tools; Sonnox offers audio restoration tools, limiters, and maximizers; Blue Cat offers advanced metering; and CEDAR Audio offers advanced restoration tools.

Can Pro Tools Be Used for Mastering?

Assuming that you have a suitable room with suitable monitoring, can Pro Tools be used for mastering? After all, it can easily be used to order tracks, balance levels, EQ, compress and limit, do crossfades and edits, de-noise, and make other repairs to the audio. But Pro Tools cannot do PQ encoding or produce DDP-formatted output. And vinyl mastering is obviously completely out of the question without a lathe . . .

So how important are these things? If you are running a professional mastering studio that services commercial record labels, these considerations will, of course, be very important. But if you are only mastering for Internet delivery, for example, you needn't worry about vinyl and CD.

You may also wish to use mastering techniques when finishing audio for a variety of purposes during the postproduction of sound for radio or TV broadcast, video or film soundtracks, DVD-Audio, or other formats—and Pro Tools is ideal for all of these applications.

Mastering

The Workflow

I recommend that you sort out the track ordering at the outset of your mastering session because this will affect the spacing that you choose between tracks, which in turn depends to some extent on the way the tracks start and finish. If the tracks need to be topped and tailed or faded in or out, you will often have to make judgments about how best to do these things based on what is happening with the tracks that precede and follow the one that you are dealing with. I usually do the level balancing next, followed by any equalization that may be necessary, although the order in which you do these things is not so critical.

Track Ordering and Spacing

When preparing tracks for an album, getting the running order to sound good can be very important to the success of the album. This is similar to the way a bandleader chooses to order the music for a live performance. The idea is to make the listening session as inviting and as interesting as possible and to show off all the music selections in the best possible way.

One approach is to start with the best, most up-tempo, or most catchy song to try to hook the listener from the outset. This might help when someone is auditioning lots of CDs, just playing the first track to get an idea of what the album might be like.

On the other hand, you could take the opposite approach, more as in a live performance, where you build up to a climax and either finish off with your best, brightest-sounding piece so that the audience goes away feeling very satisfied, or add one more piece after this, by way of an encore, which could be a beautiful slow piece that relaxes the listener at the end of the session.

It is often a good idea to place the tracks in order of increasing tempo to help achieve a build-up effect, but this very much depends on what you have to work with. If you have an equal mix of fast and slow tracks, for example, you could alternate these, or you could keep all the fast ones together and all the slow ones together.

Another possibility is to order the tracks according to how loud or how densely textured they sound, the main consideration being to make the album flow properly from one selection to the next.

Another thing to consider is the spacing between the tracks and whether to crossfade between selections. If one track has a long fade-out, you might want to start the next track almost immediately after the end of the fade. After a loud, abrupt ending you might want to wait two or three seconds before starting the next track.

Some musical styles sound good with crossfades between the tracks so the listener gets a continuous listening experience from beginning to end of the album.

Topping and Tailing

Here we are talking about removing audio material from the top, or beginning, and from the tail, or end, of each track. Many recordings will have extra sounds just before the mix is supposed to start and as it is finishing or has just finished that need to be cleaned up.

For example, if you have recorded 'live' overdubs to a click, the click may have leaked through the musician's headphones so that it is audible just before the music starts and just after it ends—or on quiet beginnings or endings. This should ideally be sorted out during the editing sessions before the original mix sessions, but if these things have been overlooked, it falls to the mastering engineer to resolve them.

You should also strive to make things sound 'natural' and 'believable' wherever possible. For example, if you remove the intake of breath immediately before a singer sings the first note, or the string or fret noise before the guitarist plays his first chord, this often won't sound 'right'. Similarly, if you cut everything 'dead' at the end, this will often sound unnaturally abrupt.

You should make sure that you don't cut off any decaying sounds from cymbals, pianos, guitars, or any sustaining instruments. And it can be good to let quiet sounds remain audible, such as the pianist releasing the sustain pedal or the percussionist putting her instruments down: this all makes the recording sound more natural and 'believable'.

> **TIP**
> Like all 'rules', there will be times when the artistic effect that the artist and/or producer wants or needs will require that these extra 'noises' or sounds are removed.

Fade-ins and Fade-outs

Even if the music actually starts and finishes very abruptly, you should consider putting very short fade-ins and fade-outs at the beginning and end of the audio—this can make the transitions from the silence between the selections into the music sound much smoother, even though these fades can be very short.

Often, you will hear hiss and noise become more audible as the music gets quieter and stops at the end—or even at the beginning of the music in some cases.

This can be tape hiss from tape recordings or it can come from guitar or other instrument amplifiers. For example, my Wurlitzer and Fender Rhodes Suitcase electric pianos are 'prime offenders' that often have hums, buzzes and other 'grungy' noises audible even when no notes are being played.

All of these things have to be taken into account when you decide on how best to fade—what shape to make the fade and how long the fade should be.

Of course, there are times when you will want to make a short fade-in or fade-out because this produces the best effect.

It's all a matter of judgment—and this is what the mastering engineer is good at. He or she has the quiet acoustically designed environment with the high-quality full-range monitors and no other distractions—which is not always the case for the mix engineer.

Equalization

The overall tonal balance of each recording should normally showcase the low, mid, and high frequencies equally. Of course there will be exceptions to this rule, but it is generally true. To get a good idea of what a good tonal balance sounds like, you should listen to a selection of successful recordings within the genre you are working with to see how the tracks you are mastering compare. However, there are going to be times when a producer or client has deliberately aimed for something that sounds very different, in which case you must discuss the situation with your client before making drastic changes.

You need to be very careful when applying EQ changes during mastering—it is all too easy to end up inadvertently altering some of the important internal balances within the mix, such as the vocal, bass, or drum kit levels—so you should usually be thinking in terms of small adjustments rather than extreme changes.

It is very tempting to simply boost or cut the bass frequencies if the level of the bass guitar is too low or too high, for example. However, this is likely to affect the sound of the bass drum and even of the guitars, vocals, and other instruments. Here's where you need to try using narrow-bandwidth EQ filters to be more precise, or to use multiband processing techniques that can bring up the levels of quieter sounds or reduce the levels of louder sounds.

If you need to remove low-frequency rumble, noises at particular frequencies, high-frequency noises, or hiss, you may get good results by using high-pass, low-pass, or notch filters, or by cutting at the ear's most sensitive frequencies between 3 and 4 kHz. However, it is all too easy to end up adversely affecting the parts of the mix that you want to preserve. It is often better to use the various specialized noise reduction and removal tools instead.

If you know that the audio will only ever be played on small computer speakers and the like, you might want to roll off the lower frequencies that will never be heard anyway, to avoid problems. But if you are delivering full-bandwidth WAV files to a customer who would then replay these on a high-quality system, you should not do this.

> **TIP**
> In general, while EQ-ing during mastering, you need to be fairly cautious—and don't forget to constantly make comparisons with and without your changes.

Brainworx Mastering EQ

The Brainworx bx_digital V2 plug-in (see Figure 10.1) is a state-of-the-art digital mastering equalizer and processor that provides precise 11-band equalization and M/S (mid-side) processing of single tracks, subgroups, or summed signals—letting you add presence and transparency to mixes, create impressive lows, correct problems, and emphasize the character of individual instruments and entire mixes.

Brainworx developed this plug-in for the UAD platform to emulate its bx1 and bx2 hardware mastering units. Like the hardware, the bx_digital V2 plug-in works in any of three modes: as a stereo/dual-mono EQ; as an M/S matrix equalizer for applying level and EQ changes separately to mid and side signals;

and as an M/S recording processor for converting mid and side mic signals into conventional stereo. The bx_digital V2 plug-in package also includes a bx_digital V2 mono version—perfect for mix equalization of sources like vocals, kick drum, and other typically mono signals.

The bx_digital V2 processor uses a built-in M/S matrix that separates stereo signals into mono sum and stereo difference signals that provide separate control of the mono and stereo elements. So users can simultaneously cut high frequencies of a mix to reduce the 'essing' of the lead vocals, while boosting high frequencies of harmony instruments, like guitars, keyboards, and pianos. Or effortlessly separate low-level signals to enhance the clarity and punch of a mix.

The bx_digital V2 is one of the most versatile mastering plug-ins available, with its 11-band equalizer, M/S de-esser, mono maker, and intelligent bass and presence shifters, plus its extra M/S features—including individual pan controls for M and S, and stereo width control.

The User Interface

The upper section of the user interface contains all the equalizer controls together with the M/S switching, stereo width, balance, gain, and pan controls.

The middle section, above the frequency response graphs and the various meters, contains the 'Shifting' EQ section with its four processors:

- Presence Shift boosts at 12 kHz and simultaneously cuts at 6 kHz. It has a range of ±12 dB and puts 'air' into the sound without exacerbating sibilance.

- Bass Shift boosts at 63 Hz and simultaneously cuts at 315 Hz. With a range of ±12 dB, Bass Shift can fatten the bottom end or push the low midrange without booming out the low bass.

- De-esser is a two-channel dynamic equalizer with an adjustable frequency range of 20 Hz to 22 kHz, and an adjustable threshold range from 0 to -60 dB.

- Mono-Maker forces all frequencies from stereo to mono, and is adjustable from 20 Hz to 22 kHz—perfect for tightening up the low end in any mix. (Mono-Maker is typically set between 60–100 Hz.)

The bottom section of the user interface has two frequency response graphs, vertical pre-EQ, post-EQ, and output meters, and horizontal meters for balance and correlation.

Figure 10.1
Universal Audio's
Brainworx bx_
digital V2 plug-in

Normalization versus Loudness Matching

I almost never use normalization to change the levels of my mixes. The reason for this is quite simple: normalization cannot do what I want it to do when mastering an album, which is to make sure that all the tracks play back at approximately the same loudness levels.

Normalization looks for the highest level peak in the audio file and moves the level of all the audio in the file upward until the highest peak just reaches 0 dBFS—or some other preset level such as −0.3 dBFS or even −3 dBFS.

> **NOTE**
> Normalization takes no account of the average level of the audio in the file—and the loudness on playback depends mostly on the average level of the audio, not the level of occasional peaks.

Compression and limiting do allow you to change the loudness level by reducing the peak to average ratio of the audio and then applying makeup gain to increase the average level of the audio—which makes the audio louder on playback.

However, ultimately the human ear is the only true judge of loudness, so you have to make careful manual adjustments to levels to match loudness—which is part of the art of the mastering engineer!

Level Balancing to Achieve Loudness Matching

Balancing the levels from track to track on an album requires careful listening to each track while making comparisons and judgments.

You can get lucky with a set of recordings that are all quite similar in terms of arrangement, instrumentation, mix style, and so forth. In this case you would choose the loudest track and adjust the levels of the rest to match this— auditioning short sections from each recording in turn and bringing its level up to match that of the loudest.

> ### NOTE
> This loudest track is the one track that I might be tempted to use normalization with, raising its gain so that its maximum just reaches the maximum level I have decided on for the album. I recommend that you set this maximum level to −3 dBFS to try to avoid any problems with clipped audio by keeping well away from 0 dBFS.

It can often be a lot more complicated than this, especially when all the tracks sound radically different to each other and when they have quiet introductions, middle sections, or endings. If this is the case, you are simply going to have to spend time carefully listening to the various different sections of each track.

It is also very important to get the transitions between tracks to sound 'right'— which also depends on the track ordering and spacing, topping and tailing, and fade-ins and fade-outs.

> ### TIP
> It should not really need saying (but I'll say it anyway) that you should keep your monitor gain at one constant level during your level balancing session—otherwise you will run the risk of becoming confused about the levels within the mix you are working on.

Compression

Compression reduces the dynamic range of music, changing the balance between the high-level, mid-level, and low-level components of the mix—which can make the mix sound 'punchier', especially when you apply makeup gain to raise the average level of the compressed mix.

> **TIP**
> If you want to raise the average level of the mix without otherwise affecting the mix, you should use a limiter, not a compressor.

However, before you start heavily compressing all your mixes, be aware that compression can adversely affect the stereo image and depth, the ambience, and the internal balance—bringing forward instruments that were previously intended to be heard in the background, for example.

> **TIP**
> The best time to change the mix is, of course, during the mixing session—not at the mastering stage.

Shadow Hills Mastering Compressor

Lots of compressors are available in AAX formats for Pro Tools, several of which are optimized for mastering, such as Universal Audio's emulation of the Shadow Hills Mastering Compressor—see Figure 10.2. This offers flexible stereo or dual mono operation and allows the user to control music dynamics in two stages—first with an optical section, then with a discrete (VCA) section. Both sections can be bypassed, effectively providing three compression types—optical, discrete, and combined—in a single unit.

The optical section is highly program dependent, with basic threshold and gain settings, and a two-stage release time akin to the classic LA-2A, so you can use the optical section to gently tame your dynamic range, and bring up the overall level. The discrete section takes over where the optical section leaves off, and provides a more aggressive sound with precise control over gain, threshold, ratio, attack, and release. You can also choose between three output transformers: nickel, iron, and steel. By switching the selected transformer type in and out of the signal path, you can hear the different frequency response, distortion characteristics, and transient limiting of each transformer—going from 'clean' (nickel), to 'colored' (iron), to 'dirty' (steel).

Mastering applications demand the most subtle uses of the compressor. So, for example, you might dial in 1 dB of optical gain reduction and 2 dB of discrete gain reduction; select a ratio of 1.2:1, an attack time of 30 milliseconds, and a recover time of 0.1 seconds; set the 90 Hz side chain filter in, and select the transformer matrix according to your preference. On the other hand, you might bypass the optical and discrete sections altogether and only pass signal through the transformer matrix, just for the colorations this imparts!

Figure 10.2
Universal Audio's
Shadow Hills
Mastering
Compressor

Multiband Compression

If you want to change the mix more radically you can use multiband compression to 'beef up' instruments that fall mostly into particular frequency ranges such as the bass drum, the snare drum in the midrange, or the percussion instruments that have lots of high frequencies.

Talking about high frequencies: it used to be the case when tape was used as the recording medium that the higher frequencies would be reduced in level on playback by the deficiencies of the medium such as tape saturation and compression. Digital recordings don't suffer from these deficiencies, so today's recordings are often too bright, and can be more fatiguing to listen to as a result.

Multiband dynamics processors can be used in mastering to counteract this situation by allowing the mastering engineer to boost low and high frequencies when these are at low levels while keeping these frequencies flat, or even applying a cut to the high frequencies when these are at high levels in the mix.

Adjusting the spectral balance using multiband processors also helps to make the mix sound louder without increasing the peak levels.

> ## TIP
> Don't forget to make A/B comparisons of the processed and unprocessed audio at equal loudness levels, or the nonlinear response of your ears at different loudness levels is likely to trick you into preferring the processed audio—remember Fletcher and Munson!

Limiting

Using a limiter or maximizer, it is common practice to set a ceiling of −0.3 dBFS. This helps to avoid inter-sample peaks clipping consumer D/A converters.

Although there are various designs used in competing products, typically a maximizer is a type of limiter that uses multiband processing. Examples include iZotope's Ozone maximizer and Avid's Maxim—see Figure 10.3.

Figure 10.3
Avid Maxim Peak
Limiter

Alternatives include Universal Audio's precision maximizer, which increases apparent loudness without reducing dynamic range, and Sonnox Inflator, which also does not use compression—instead changing the relative probability of samples in the audio such that there is a greater probability of larger values than in the original audio. The Sonnox Oxford Transient Modulator can also be used to add 'life' to recordings and to reduce peaks so that you can increase overall level.

Sonnox Oxford Inflator

There is a strong trend in today's popular music toward producing mixes that are as loud as they possibly can be without overloading the playback equipment. Compressors and limiters are the most commonly used tools that can do this trick, and specialized versions of these have been developed in both hardware and software. Nevertheless, if the material has previously been compressed, there may be nowhere to go if you still want to increase the loudness. And there can be various undesirable effects such as 'pumping' or 'breathing', or a dulling of the sound, or a loss of percussive quality due to the flattening of transients.

Sonnox Oxford Inflator (see Figure 10.4), originally developed as a signal processor for Sony's highly regarded Oxford mixing console, can increase the loudness of almost any audio material, regardless of the level of prior compression or the amount of dynamic range still available. It does this by changing the relative probability of the audio samples so that there's a greater likelihood of larger sample values than in the original audio. So this method does not suffer from the previously mentioned drawbacks. The Inflator not only enhances loudness, but also increases the dynamic impact and imparts a 'warmth' to the processed sounds—similar to that of valve-based audio equipment.

Figure 10.4
Sonnox Oxford Inflator

The user interface is very straightforward, with faders for input and output level, effect amount, and curve—all with associated meters. Increasing the effect parameter 'fattens up' the sound while the curve parameter can add 'life'— making hi-hats 'sizzle' more, for example. The Band Split function processes low-, mid-, and high-frequency bands separately to prevent prominent frequencies in any band interacting with other bands to produce distortion.

Sonnox Transient Modulator

The Sonnox Transient Modulator (see Figure 10.5) lets you put life back into dull or flat-sounding recordings and mixes, without the unwanted changes in overall timbre associated with multiband compression techniques. You can use the transient modulator to sharpen up the sounds of percussive instruments or to soften any overly percussive-sounding acoustic musical instruments. It allows you to radically change the dynamics of the instruments, accentuating or flattening attacks and transients to bring sounds forward or push them back in level. It can also be used to increase or reduce the effects of ambience or to produce rounded and dynamic percussive effects. And because it can be used to reduce the levels of very short transient peaks, you can also increase the overall level—as with a limiter.

Figure 10.5
Sonnox Transient
Modulator

The threshold control causes the process to operate only on audio above its set level, ignoring all signals below that level.

With the ratio control set to zero, no processing takes place. Positive values for ratio increase the gain during transients while negative values do the opposite—reducing the gain. So, for example, if a transient during the attack portion of a snare sound is 10 dB above the average level and the ratio is set to + 1, the gain would be increased by the same amount, that is, 10 dB, while the transient lasts—making the attack of the drum much sharper and louder.

569

On the other hand, with a negative ratio of -1 the drum attack transient would be reduced by 10 dB, bringing it back to the average level of the signal and, consequently, completely removing the transient—making the attack of the drum sound much duller and softer.

By carefully reducing the overshoot parameter the effect can be 'focused' onto shorter sounds such as those made by small bells or similar small percussion instruments. Low values for the recovery parameter allow the effect to be applied to most, or all, the transients. Longer values reduce the amount of effect that is applied, depending on the rate at which transients occur in the audio. So, for example, if there were some spill from busy hi-hats in between snare hits, setting a long recovery value would prevent the extremely short transients from the hi-hats from triggering the effect.

The rise time parameter alters the response to short transients. With the control set at minimum all transients, however short, will be processed. As you increase the rise time, the shorter transients will be progressively ignored. So you can use this parameter to prevent the transient modulator from acting in response to very short, inaudible transients, for example.

The dead band control can also be used to prevent the transient modulator from acting in response to small level changes or insignificant transients. Increase this from zero only if unwanted processing becomes noticeable, particularly during quiet sustained passages. The dead band control may also be used to produce dramatic effects by applying the transient modulator solely to the loudest transients in the audio material. In this case it's best to set low thresholds and high ratios to get the maximum effect, before progressively increasing the dead band to exclude smaller events from the effect.

When used with negative ratios, the transient modulator can increase the loudness of audio by reducing very short transients that would otherwise cause overloads. In this case, very small overshoot and recovery values should be used with a minimum rise time setting, in order to catch the fastest transients only. A negative ratio coupled with a suitable dead band setting should be set to reduce these transients by the required amount, allowing the overall level of the audio material to be increased before limiting occurs.

Metering

Metering is very important for mastering engineers—providing essential visual feedback about levels and other attributes of the audio. Pro Tools 11 has a lot more metering options than previous versions, but dedicated metering plug-ins offer more advanced features that can be particularly helpful for mastering engineers.

A spectrogram that graphs energy levels across the frequency spectrum provides a visual check on what you are hearing, so it is useful to keep an eye on one of these while working on EQ settings, or you might insert one at the beginning and one at the end of your mastering chain so that you can see the changes you have made. A vectorscope and a correlation meter can also provide additional useful feedback during mastering.

Two sets of third-party meters that I have had the opportunity to work with are Blue Cat's DP Meter Pro and iZotope's Insight.

Blue Cat's DP Meter Pro

Blue Cat's DP Meter Pro (see Figure 10.6) is a unique audio analysis tool that combines flexible and customizable audio meters with advanced side

Figure 10.6
Blue Cat's DP
Meter Pro

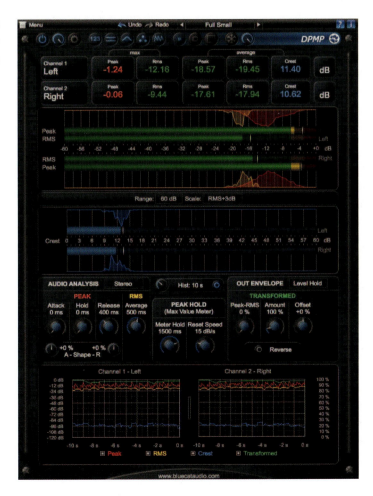

chain control capabilities that work together with its MIDI and automation outputs. Almost every aspect of its peak, RMS, and crest factor meters can be customized.

> **NOTE**
>
> The crest factor, also known as the *peak over average ratio* (PAR), is the difference in decibels between the peak and the RMS levels. A higher crest factor means more dynamics, while lower values mean more compression, and sometimes distortion: 0 dB will be reached with a pure sine wave, while a usual value for crest factor on mastered music is in the 6 to 15 dB range (from pretty compressed to pretty uncompressed).

Large colored statistics displays provide an overview of a track, while histograms give you a precise measurement of the dynamic range of the track and the graphs show the evolution of the audio envelopes over time, displaying peak levels in red, RMS levels in yellow, and the crest factor in blue. A mid-side switch lets you verify mono compatibility problems and check the 'stereo-ness' of your audio sources very easily.

> **NOTE**
>
> Blue Cat's DPMP is not a loudness meter: it is an audio level meter that measures physical values as opposed to perception-related values.

At the top of the interface, below the main toolbar, you will find various buttons that can be used to activate or deactivate each view—statistics, levels and crest factor meters, histograms, analysis controls, and the history view. Three buttons let you choose from small, medium, or large user interface sizes, a button that freezes all meters and graphs, and a brightness knob to control the brightness of all the meters and graphs.

Below these, the statistics view shows, for each available channel, the maximum and average values for levels and the average value of the crest factor. Values are displayed with colors that correspond to the scale chosen in the level meters view. See Figure 10.7.

Figure 10.7
Blue Cat's DP Meter
Pro statistics view

The level meters view displays peak and RMS levels with their histograms—peak histogram in red, RMS histogram in yellow. See Figure 10.8.

Figure 10.8
Blue Cat's DP Meter
Pro level meters

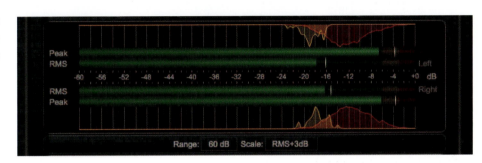

Below the level meters are crest factor meters for each channel with associated histograms—see Figure 10.9.

Figure 10.9
Blue Cat's DP Meter
Pro Crest factor
meters

The controls view lets you adjust the audio analysis, metering, and envelope generation parameters—see Figure 10.10. The main parameters you will use to control your meters are the peak release time (to control how fast the peak meter decreases); the RMS average (to control how fast the RMS and crest meters react to signal changes); and the history length (Hist.) to define the histograms' memory. Other parameters provide even more advanced control: attack controls how fast the peak meter reacts to signal increases; hold controls how much time the peak meter will hold its current position before falling down, when the level is decreasing.

A drop-down menu at the top right of the audio analysis section lets you switch between Stereo (left/right) and Mid/Side mode. In Mid/Side mode, the mid channel represents the mono part of the signal, and side the stereo part. This is useful to monitor how wide your mix is and to check that there is no mono compatibility (drops in the mid level shows that left and right channels may be out of phase).

Figure 10.10
Blue Cat's DP Meter
Pro controls view

The history view displays the envelopes for peak, RMS, and crest factors as well as the transformed envelope—see Figure 10.10.

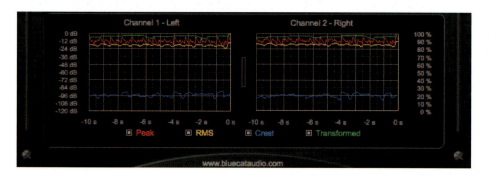

Figure 10.11
Blue Cat's DP Meter
Pro history view

iZotope Insight Meters

iZotope's Insight meters (see Figure 10.12) can display up to five modules within its user interface window: the spectrogram, spectrum analyzer, loudness history, sound field, and levels.

At the bottom of the Insight plug-in's window are two buttons to let you open the Preset Manager and Options menu as well as select which meters you wish to view at any time.

Figure 10.12
iZotope Insight
meters

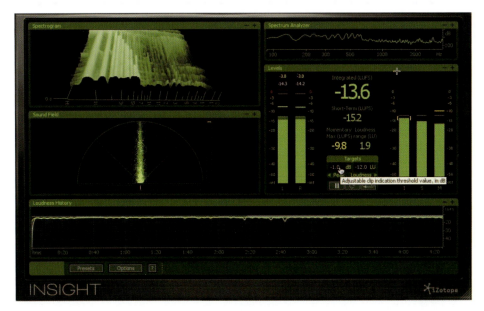

Level Meters

Insight has various level meters to monitor the incoming levels of your audio channels—see Figure 10.13. At the left of the display, the main left and right level meters display combined instantaneous (true peak) and averaged (RMS) levels. The lower, bright bar represents the average level (RMS) and a higher, dimmer bar represents the peak level. There is also a moving line above the bar representing the most recent peak level or peak hold.

> **NOTE**
>
> The peak meters are fast meters that measure true peak analog waveform values. If you are tracking your audio for possible clipping the true peak meters are appropriate. RMS (Root Mean Square) is a software-based implementation of an analog-style level meter. The RMS meters display the average level calculated over a short window of time. K-System K-12, K-14, and K-20 meters are also provided as options.

In the central area, the integrated loudness is shown as an LUFS value at the top of the display with the short-term loudness LUFS value below this. The maximum value of the momentary loudness in a given period of calculation is shown as an LUFS value in the momentary max readout in the central area of the display, with the loudness range LU value to the right of this. At the bottom you can enter targets for the maximum peak level in dB and for the maximum loudness level in LU.

The three meters at the right of the display show instantaneous, short-term, and momentary loudness levels. The meters labeled 'M' show momentary loudness, measured over the course of 400 ms. Short-term loudness, calculated over the course of three seconds, is displayed in the loudness meter labeled 'S', as well as in the short-term loudness readout. Integrated loudness is displayed in the loudness meter labeled 'I' as well as in the larger integrated loudness readout. This is an infinite average and generates a single loudness calculation for the total calculated period or program. This value is most commonly enforced by loudness standards.

Figure 10.13
iZotope Insight
level meters

Sound Field

Insight features a vectorscope that provides a view of the stereo image as well as meters that illustrate the channel correlation and balance—see Figure 10.14. The vectorscope and the surround scope are both available from the sound field partition.

A vectorscope juxtaposes the two channels of a stereo signal on an x-y axis in order to display the similarity or difference between the two channels. A mono signal will produce a straight vertical line while signals with a wider stereo image will produce more horizontal shapes. A vectorscope can be used while mixing or mastering to monitor the overall stereo width of audio. Additionally a vectorscope can immediately alert you to potential issues of phase cancellation. This is useful when placing microphones in a stereo pair, inspecting how various stereo signals combine, and ensuring your audio will translate to mono playout when necessary.

Figure 10.14
iZotope Insight
vectorscope

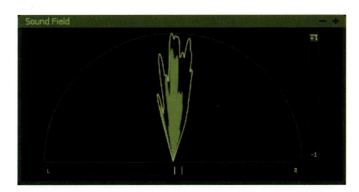

Below the vectorscope is a balance meter that shows the overall balance between the left and right channels of your mix. A lighter bar illustrates the balance in real time while a slower dimmer bar follows behind the real-time calculation to be more readable. A correlation meter is also provided immediately to the right of the vectorscope display. This allows you to check that the left and right signals will sum to mono without any cancellation of frequencies.

The surround scope allows you to easily visualize how the audience will perceive the collective surround channel's levels during playback. The surround scope offers a stylized display of the amplitude of your surround channels that stresses the spatial relationship of the tracks while illustrating each track's level—displaying each individual surround channel's presence relative to the others. While the amplitude envelopes display the levels of individual channels, the balance indicator illustrates the average surround location of your audio. The surround scope also monitors the phase relationship between neighboring audio channels and displays an alert when a negative correlation or phase cancellation takes place.

Loudness History Graph

Insight features a loudness history graph that allows you to monitor loudness trends over time and to retrospectively inspect any loudness issues in your mix. The timescale used by the graph resizes in real time so the graph automatically zooms to the current running time utilizing all available space.

The loudness history graph can plot and display short-term loudness (drawn in white), momentary maximum loudness (drawn in grey), and integrated loudness values (drawn in green). See Figure 10.15.

> **NOTE**
> The grey indicator line superimposed on the loudness history graph represents the loudness target value set in the levels partition. When the integrated loudness value exceeds the loudness target value, the integrated loudness is drawn in red to bring your attention to a potential loudness violation.

Figure 10.15
iZotope Insight
loudness history

The loudness history graph is useful for not only retrospectively diagnosing any issues with loudness in your mix, but also for immediate feedback on the loudness trends of your audio for monitoring while mixing. For example, if your integrated loudness is slowly increasing, this trajectory will be immediately apparent in the graph.

Spectrum Analyzer

A spectrum analyzer measures amplitude across the frequencies that encompass the spectrum of human hearing. The vertical axis represents amplitude while the horizontal axis represents frequency.

A spectrum analyzer can be used when mixing or mastering to help identify frequencies that you hear within your audio. This can be a useful alternative to using your ear while sweeping through frequencies using a parametric equalizer to identify frequencies of interest.

Insight's spectrum analyzer (see Figure 10.16) can be set to display both peak hold (shown as a bold curve) and average spectra (shown as a darker, thinner curve) simultaneously.

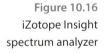

Figure 10.16
iZotope Insight
spectrum analyzer

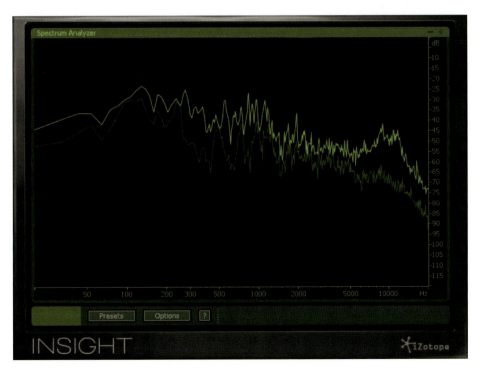

Spectrogram

A spectrogram allows you to visualize frequencies as they occur over time using a graphic representation of your audio. The vertical axis represents frequency while the horizontal axis represents time. Amplitude is displayed as color intensity in a 2-D spectrogram and as height in a 3-D spectrogram. Insight allows you to display either 2-D (see Figure 10.17) or 3-D spectrograms.

> **NOTE**
> As opposed to a spectrum analyzer that is often monitored in real time, a spectrogram is typically used to analyze the frequencies within your audio retrospectively.

Figure 10.17
iZotope Insight 2-D spectrogram

The spectrogram can also display various audio streams simultaneously for analysis in both 2-D and 3-D modes using iZotope's Meter Tap plug-ins.

Meter Taps

You can route audio from any track or bus in your mix to Insight's spectrogram using the Meter Tap plug-in installed with Insight. Meter Tap plug-ins can be inserted anywhere in your session and can send streams of audio to Insight's spectrogram. In the example shown in Figure 10.18, I have inserted a Meter Tap plug-in on a bass guitar track and have labeled this as 'Bass'.

Figure 10.18
iZotope Insight
Meter Tap plug-in

When Meter Tap plug-ins have been individually labeled, the corresponding names will appear for selection in the spectrogram's Meter Tap Selection window—see Figure 10.19. In this example, I have routed a bass and two lead guitar tracks to the spectrogram, which is displayed in 3-D here. The frequencies from these three tracks are shown in different colors, so they can be identified. Also, note that amplitude is displayed as height in the 3-D spectrogram.

Figure 10.19
iZotope Insight 3-D
spectrogram

NOTE

Left-clicking and dragging the 3-D spectrogram allows you to move it within its bounding partition in Insight. Right-clicking and dragging allows you to freely rotate the 3-D spectrogram. Additionally, your mouse wheel allows you to easily zoom in and out of the 3-D spectrogram. Together these functions allow you to easily adjust the orientation of the 3-D spectrogram to best suit your needs.

TIP

Meter taps allow you to analyze various audio streams and present these together in useful ways. For example, you might evaluate a signal before and after processing by placing meter taps before and after a plug-in. Or balance a drum mix by placing a meter tap on each track that makes up your composite drum stem, then compare levels at given frequencies.

iZotope Ozone 5 Advanced Mastering Suite

Ozone 5 Advanced is a suite of essential mastering processors available from within a single plug-in and also provided as separate, individual plug-ins—providing a complete mastering system to extend the capabilities of DAWs such as Pro Tools.

In the lower section of the 'full' plug-in (see Figure 10.20), buttons and controls are provided to enable/disable and modify the dynamics, equalizer, harmonic exciter, stereo imager, reverb, and maximizer modules.

NOTE

Ozone uses analog modeling to give each of the mastering modules a smooth natural sound. For example, the equalizer recreates the soft limiting exhibited by a vintage valve equalizer, while the harmonic exciter mimics the musically pleasing harmonic saturation of a vacuum tube component.

Ozone 5 comes with an extremely comprehensive set of presets to cover all sorts of situations, subdivided into folders for general purpose mastering, genre-specific mastering, instruments and busses, post and broadcast, special effects, and utility categories—see Figure 10.21.

Figure 10.20
iZotope Ozone 5
Advanced complete
mastering system

Figure 10.21
iZotope Ozone 5
Advanced general
purpose mastering
presets

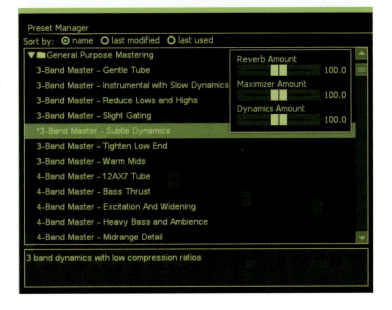

Ozone 5 Advanced also has a Meter Bridge window that can be opened using
the large button positioned just underneath the standard ozone meters. This
meter bridge provides a spectrogram, a spectrum analyzer, a vectorscope, and
level meters—all the tools you need for visualizing changes made during the

mastering process, troubleshooting problematic mixes, and comparing your mixes to reference tracks (see 10.22).

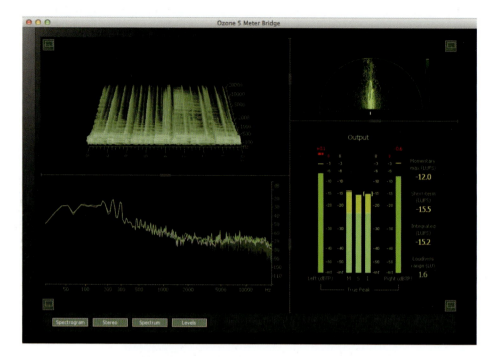

Figure 10.22
iZotope Ozone 5
Advanced meter
bridge

Ozone actually includes two 'paragraphic' equalizers (parametric equalizers presented in a graphical way), each of which has eight adjustable filter bands that can be used to boost or cut frequencies. Ozone has a multiband, multifunction dynamics processor module and a harmonic exciter module that has a selection of exciter presets modeled on tubes, triodes, and tape saturation. The stereo imaging module can be used to adjust the perceived width and image of the sound field.

Ozone also has a reverb module. Although reverb is normally added during the mixing stage, there are plenty of situations in which reverb needs to be added during mastering—for example, to cater for a last-minute change of mind by the producer, or when working with stems containing separate vocals and music, or to smooth out variations between tracks. Mastering reverb can be used as a way to unify the collection of recordings.

If the mix engineer has done a good job with imaging and reverb on the individual tracks and as a result you have a cohesive sense of space, you probably won't need to add any additional reverb to the final mix. In some cases, however, a little mastering reverb can add an overall depth to the sound.

Sometimes the mixes have been cut too short, and the mastering engineer doesn't have a long enough audio tail to do the required fade-out. A slight application of reverb at this point can help extend the natural decay of the last chord of the music, for example, allowing for a more comfortable fade-out.

Ozone provides both acoustic space and plate 'studio-style' reverbs that you can apply to your mixes. The reverbs use hybrid processing, utilizing both convolution and advanced algorithmic technology.

The final module in the signal chain is the maximizer—see Figure 10.23. Ozone's maximizer has five unique limiting modes: IRC I, II, and III use transparent, psycho-acoustically advanced algorithms; 'Hard' uses a brick wall-style algorithm; 'Soft' is more gentle and 'smeary'.

> **NOTE**
>
> When working with IRC algorithms, selecting the 'Intersample Detection' option allows Ozone to intelligently predict the behavior of the analog signal reproduced for the listener and prevents any inter-sample clipping from occurring in the analog domain.

Figure 10.23
iZotope Ozone
5 Advanced
maximizer module

The window at the top of the maximizer module can display the signal spectrum, the gain reduction trace meter, or the dither shape display. The gain reduction trace is a scrolling meter that displays the incoming signal's waveform with a superimposed tracing that illustrates the amount of gain reduction taking place in real time. The gain reduction trace can help you to set attack and release controls appropriately and monitor the envelope of gain reduction. Below this window are various sliders and other controls:

The threshold control determines the point at which the maximizer will begin limiting. Turning down the threshold limits more of the signal, which in turn will create an overall louder mix. In other words, by turning down the threshold you limit the dynamic range of the mix, and the maximizer automatically adds gain proportionally to maximize the output level.

Margin determines how much to boost the output signal after limiting. If the margin is set to 0 dB, the signal will be boosted all the way up to 0 dB. If the margin is set to −0.3 dB, makeup gain will be applied until the output signal is at −0.3 dB. Note that in Soft mode, the level may be allowed to cross the margin setting, while in Intelligent or Hard modes the margin serves as an absolute stop point.

The next slider controls either the release or the character according to which mode is active. When the maximizer is set to Soft or Hard mode, the release slider lets you set a release time for the limiting. In general, more extreme limiting will benefit from longer release times. The IRC modes provide intelligent release control (the release time is automatically varied depending on the audio material)—so the release slider is replaced by a character slider. This control allows you to modify the behavior or 'character' of the limiter by controlling how fast and aggressive it is.

The transient recovery feature allows you to fine-tune the shaping of transients before limiting takes place helping to preserve sharper sounds like drums while still optimizing loudness. The higher the amount of transient recovery set the more pronounced the transients will be after the limiting process.

The ozone limiter defaults to 100 percent stereo linking, which imposes one limiter across the stereo image. Enabling the stereo link feature allows the left and right channels to be limited independently.

When the dither type is set to 'None', no dither or word length reduction is applied. When the type is set to MBIT+, Type 1, or Type 2, dithering is applied and your audio will be reduced in bit depth to the number of bits selected in the bit depth control.

> **NOTE**
>
> Dither is applied independently of the maximizer and is not bypassed when the maximizer module is bypassed.

All of Ozone 5 Advanced's modules are capable of delivering the very highest-quality results. The way they can all be accessed from within one plug-in that also has the benefit of the meter bridge adds a further level of convenience.

Audio Restoration Tools

Mastering engineers often use audio restoration tools, an obvious application being to restore older recordings that have been copied from vinyl discs, tape or cassette recordings, or whatever.

Software is available that can reduce or remove the sound of tape hiss from tape and cassette recordings and reduce or remove the snaps (clicks), crackles, and pops that you typically hear when you play back vinyl discs. Noise removers can take away broadband noise while notch filters can remove specific frequencies. Declickers can get rid of clicks, pops, thumps, and any other small clicking sounds while decracklers can get rid of scratches and groups of clicks. Hum removers can get rid of ground loop hum, rumble, and other low-frequency noises.

Noise suppression techniques are also needed to clean up noisy dialog for film production, suppress ambient noise for live TV and radio broadcasting, revitalize sound effects libraries, and enhance speech for forensic audio investigations.

One of the most exciting recent developments for audio restoration is Celemony's Capstan software, which, amazingly, can remove wow and flutter from old recordings made using tape, compact cassette, wax, shellac, or vinyl! See www.celemony.com/en/capstan. Handsomely priced at just 3,790 euros, the software can also be rented for 199 euros per five-day period.

Three of the leading audio companies developing audio restoration tools are Sonnox in Oxford and CEDAR in Cambridge in the United Kingdom, and iZotope in Cambridge, Massachusetts in the United States.

Sonnox offers Restore—a suite of three plug-ins including the Oxford DeClicker, which incorporates DePop, DeClick, and DeCrackle; Oxford DeBuzzer hum and buzz removal; and Oxford DeNoiser, which includes a separate DeHisser.

CEDAR Audio offers its Studio Suite for Pro Tools in AAX format comprising four plug-ins: Adaptive Limiter, Debuzz, Declip, and DNS One noise reduction.

iZotope's RX Advanced software has the most comprehensive suite of audio restoration tools available.

CEDAR Studio Suite for Pro Tools

Available as a range of RTAS, AAX, and AudioSuite plug-ins for the Mac and PC, the CEDAR Studio Suite for Pro Tools allows you to eliminate a wide range of common problems and significantly improve the sound quality of your audio.

The Adaptive Limiter offers peak and oversampling modes. Employing a unique algorithm developed by CEDAR, it calculates a continuously varying EQ profile that constrains the peak level of the output while retaining the integrity of the input. The result is a perfectly controlled signal that remains much more natural than audio processed using conventional limiters.

Debuzz does exactly what it says it does, and does it better than just about anything else out there. It works with fundamentals up to 500 Hz, handles harmonics effectively, and leaves virtually no audible traces of its work!

Declip allows you to identify and remove most instances of clipping in a single operation, effortlessly removing clipped samples and reconstructing the original signal, improving clarity, reducing or even eliminating distortion, and restoring the original dynamic range of the damaged audio. Two clip removal algorithms are provided—one optimized for light clipping, and one optimized for heavy amounts of clipping. This ensures that you remove the problem without causing damage to the genuine signal.

DNS One is regarded by many professionals as the best tool for film and broadcast work when any type of noise suppression is required. Rumble, hiss, it takes these in its stride—and the sounds of air conditioning can at last be silenced!

> **NOTE**
> CEDAR Audio (www.cedaraudio.com) is the world's only company dedicated solely to audio restoration and speech enhancement for film, post, TV, and radio broadcast, CD and DVD mastering, libraries and archives, and audio forensic investigation.

Adaptive Limiter 2

The Adaptive Limiter is a loudness maximizer with advanced resampling and noise-shaping capabilities, peak and oversampling modes—ideal for mixing, final mastering, and other creative duties. The user interface offers a graphical display with large meters below this displaying peak input and output levels—see Figure 10.24. Rotary controls are provided for spectral characteristics (the degree to which the profile can change from one frequency to the next) and temporal characteristics (the rate at which the profile can change over time) at very low frequencies (LF) and very high frequencies (HF). If you set the spectral control to its zero position, the Adaptive Limiter acts as a single-band limiter. Increasing the control allows the Adaptive Limiter to respond accordingly across the spectrum of frequencies to a greater or lesser degree. In action, the Adaptive Limiter produces louder, cleaner mixes than conventional compressors—mixes that sound smoother and more three-dimensional.

Figure 10.24
CEDAR Adaptive
Limiter 2 AAX
Native plug-in

Debuzz

Buzzes and hums can creep into the highest-quality recordings and live audio, but can be eliminated quickly and easily using Debuzz—see Figure 10.25. This powerful but simple processor removes unwanted signal components across the entire audio spectrum, and is capable of removing all manner of buzzes and hums with fundamentals as high as 500Hz. Debuzz will also track wandering tones and, unlike traditional filters, can successfully restore your audio without unwanted side effects such as limited bandwidth or the hollow sound introduced by comb filters.

Figure 10.25
CEDAR Debuzz AAX
Native plug-in

Declip

Declip uses a de-clipping algorithm that identifies and removes most instances of clipping in a single pass. As with the other CEDARTools plug-ins, you first select the audio region that you wish to process in the Pro Tools Edit window, then select 'Declip' from the AudioSuite menu to open the plug-in. See Figure 10.26.

In the plug-in window you can view a visual indication of the density of the audio signal's sample values. If the Signal Analysis window displays 'hard' vertical edges (as shown in the extreme example depicted later in this chapter), this shows that the signal has been clipped at that sample value. You can then use Declip to remove the offending samples and reconstruct the signal.

The Declip screen comprises four areas: the upper toolbar, the signal analysis display, the de-clip detection and removal controls, and the I/O metering with its associated output gain control.

The signal analysis display represents the signal as the likelihood of a given sample value (the vertical or sample likelihood axis) plotted against sample value (the horizontal or sample value axis). The sample value axis shows the input sample values as a proportion of full-scale digital. Sample values that occur frequently will have greater values on the vertical axis than those that appear on few or no occasions. The sample likelihood axis is shown in a logarithmic form that allows you to see instances of slight clipping much more clearly and easily than would be possible on a linear axis. The sample density curve is drawn from this data and is shown in the signal display area.

When the audio is not clipped, the signal display shows samples smoothly distributed both vertically and horizontally. When the audio is clipped, the sample density curve has truncated edges. To fix this, you move the clip detection markers to points between the damaged data and the clean data to define the area containing the clipped samples that will be removed and replaced with repaired audio.

Clipping is displayed as peaks at one or both sides of the signal display. Looking at Figure 10.26, these peaks are at the left-hand side. The purple and green vertical lines that can be seen at the left and the right, respectively, are the clip markers. You can point, click and drag these lines, use the two rotary clip detection controls, or type numerical values to set the clip detection values.

Any samples lying outside these (i.e., further from the center line) are removed from the signal and replaced with 'good' signal generated by the algorithm. Two clip removal algorithms are provided: heavy is a general purpose algorithm suitable for most clipping problems, and will correct badly clipped signals; light is suitable for less damaged signals.

When you have adjusted the various parameters to your liking, the output gain control then allows you to fit the de-clipped audio into the available headroom before you press the Render button.

Figure 10.26
CEDAR Tools Declip
AudioSuite plug-in

DNS One

DNS One AAX is a fully automated implementation of CEDAR's DNS designed specifically for Pro Tools users. It's ideal for all studio environments, and especially for postproduction for film, TV, and video. It is particularly useful in postproduction to suppress background noises such as traffic sounds or aircraft fly-bys or to reduce excessive reverberation from audio recorded in a large studio or soundstage that should sound as though it was recorded in a smaller room or other acoustic space. Suppressing reverberation can also be useful to increase intelligibility of dialog tracks.

The original CEDAR DNS1000 became a studio standard for eliminating background noises such as traffic and air conditioning. DNS One AAX now offers the same performance as its hardware siblings, the DNS1500, DNS2000, and DNS3000, with a new user interface that allows you to control hundreds of channels of DNS (software and hardware, in any mix you choose) from a single Pro Tools host.

You control DNS One AAX using the dedicated DNS Control System software that runs on Windows 7 and on Mac OSX 10.6.8 onward. More than just a GUI, it supports both AAX Native and AudioSuite and allows you to control as many instances of DNS One AAX as your host system can support, plus up to 126 DNS2000s and/or DNS3000s. It is also fully integrated with Pro Tools automation and its hardware control surfaces such as the Icon, D-Control, and D-Command. The DNS Control System is completely cross-compatible between DNS One AAX, the DNS2000, and the DNS3000, so users can install DNS One AAX alongside existing hardware units, creating sessions on one and later, if desired, recalling them on either of the others.

TIP

The DNS One can also be used in music production to clean up noisy tracks or to remove tape hiss from older recordings. I used the DNS One very successfully to remove all the tape hiss from an early '60s album that I had digitized from vinyl to prepare for digital remastering. I had de-clicked and de-crackled these previously and I could hear, buried in the hiss at low level, but still audible in quiet passages, some annoying traces of noise modulation. Removing the hiss using DNS One removed all this noise modulation as well!

> **NOTE**
>
> Inevitably, some of the high frequencies are lost during this process, and, at times, I still found myself preferring versions that still had audible tape hiss in the quiet passages.

Operating the DNS One

A row of buttons near the top of the DNS One's user interface lets you choose a frequency range to work with—see Figure 10.27. Choices include low, 20 Hz–400 Hz; mid, 200 Hz–6 kHz; high, 4 kHz–18 kHz; low + mid, 20 Hz–6 kHz; mid + high, 200 Hz–18 kHz; and full range, 20 Hz–18 kHz.

Figure 10.27
CEDAR DNS One
AAX Native plug-in

For example, tape hiss is usually in the mid + high range, or sometimes just in the high range. Whichever frequency range you choose, this is split into six frequency bands, each of which has a gain fader.

Start by choosing the full range, then move all six faders to the bottom of the range. Raise the level control at the left of the window until the noise disappears. This gives you the approximate setting for the level.

Next, move the band gain faders back up to the 0 dB settings so that no processing is applied in any frequency band. At this point, you move each fader down in turn to discover by trial and error which have the greatest effect on removing the noise.

If you find that the leftmost faders reduce the noise while the rightmost faders do not, you should switch to a lower frequency range using the radio buttons, until you find a range in which lowering all six faders helps to reduce the noise: this procedure helps you to identify the frequency range across which most of the noise is distributed.

At this point, you can further adjust the level control so that you have reduced the noise without introducing audible artifacts.

NOTE

When you are happy with the results, it is usually best to record the processed audio to disk so that you can disable the plug-in. To do this, route the output of the auxiliary track that is monitoring the processed audio from the DNS One to the input of an available audio track.

iZotope

iZotope Inc., a leading audio technology company, offers its spectral audio editing and repair suite, RX—see www.izotope.com for more information on RX 3 and RX 3 Advanced.

The standard version of RX 3 features de-noise and de-clip modules, parametric EQ, drawing and selection tools, MBIT+ dither, audio recording and monitoring, SRC sample rate conversion, waveform statistics, and more. RX 3 Advanced features de-clip, de-click, time and pitch, and channel operations modules and lets you clean up dialog in real time with the dialog de-noiser, remove or reduce reverb from audio using the de-reverb module, and monitor your audio and loudness compliance using Insight, iZotope's comprehensive metering suite. RX 3 Advanced also includes iZotope's Radius—world-class time stretching and pitch shifting.

NOTE

Insight is iZotope's essential metering suite, bundled with RX 3 Advanced as an additional plug-in. With its extensive set of audio analysis and metering tools, Insight is perfect for visualizing changes made during mixing and mastering, troubleshooting problematic mixes, and ensuring compliance with international broadcast loudness standards like BS.1770–1/2/3 and EBU R128. Fully customizable and scalable, Insight allows you to visually monitor all relevant information from your mono, stereo, or surround mix in a convenient floating window.

Used everywhere from Hollywood postproduction studios to leading sound editing facilities, iZotope RX 3 transforms troubled recordings into professional-grade material with remedies for noise, clipping, hum, buzz, crackles, and more.

Designed to be the complete toolbox for cleaning and repairing audio, RX 3 is ideal for audio restoration and archiving, recording and mastering, broadcasting and podcasting, sample library creation, video production, forensics, and any application that demands spotless audio.

RX 3 Advanced works as a stand-alone application or as a suite of plug-ins for use in a host, including many popular DAWs and NLEs. Supported plug-in formats include AAX for Pro Tools 11, RTAS, AudioSuite, VST and VST 3, and AudioUnit (AU).

RX3 Advanced Standalone

RX3 Advanced standalone version offers a complete set of tools for repairing audio—including various processing modules together with a range of tools for selecting, isolating, and viewing audio—all accessible from one window. See Figure 10.28.

An overview of the audio file's entire waveform appears above the main spectrogram display. This provides a handy reference point when zooming and making audio selections in RX. To the right of the spectrogram/waveform display are the amplitude ruler for the waveform, a frequency ruler for the spectrogram, and a color map for the spectrogram.

The main spectrogram display, which occupies the largest part of the window, shows a range of frequencies (lowest at the bottom of the display,

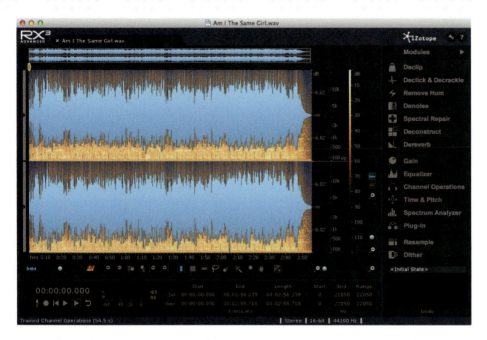

Figure 10.28
iZotope RX3
Advanced stand-
alone

highest at the top) and shows how loud events are at different frequencies. Loud events will appear bright and quiet events will appear dark. Using the spectrogram, you can see at a glance where there is any broadband, electrical, or intermittent noise, for example. The spectrogram display also features a transparency slider that lets you superimpose a waveform display over the spectrogram, allowing you to see frequency and overall amplitude at the same time. This can be invaluable for quickly identifying clipping, clicks and pops, and other events.

One of the key benefits of RX is that it has a set of tools for selecting audio not just by time, but also by frequency. This allows you to work with sounds that fall in only part of the frequency range at a given point in your project. This is very helpful for isolating and repairing a variety of audio problems including intermittent background noises.

In addition to the spectrogram/waveform display, RX offers a highly configurable spectrum analyzer (see Figure 10.29), which displays the FFT information around your playhead or selection. Use it to get a detailed side view of the spectrogram and accurate details about tonal peaks.

Figure 10.29
iZotope RX3
Advanced spectrum
analyzer

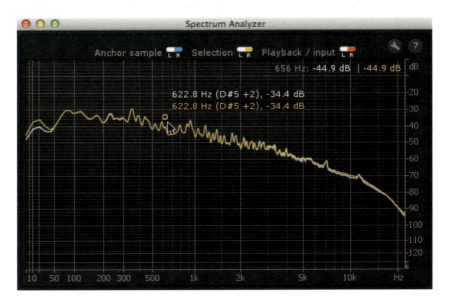

Press Alt/Opt-D to bring up the Waveform Statistics window—see Figure 10.30. Waveform statistics are important for managing levels, discovering amplitude irregularities, honing in on distortions, and ensuring compliance with modern loudness laws. You can get details about the peak and RMS levels of your selection, as well as quick measurements of BS-1770 loudness statistics from this useful utility window.

Figure 10.30
iZotope RX3
Advanced
Waveform Statistics
window

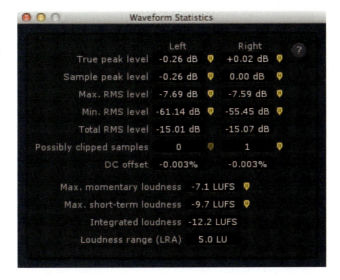

Modules are accessible from the list at the right-hand side of the user interface. Repair modules include de-clip, de-click, and de-crackle, remove hum, de-noise, spectral repair, deconstruct, and de-reverb. Utility modules include gain, equalizer, channel operations, time and pitch, spectrum analyzer, plug-in, resample, and dither.

After processing your audio selection with RX's modules, you can save and export your audio, or make use of RX's batch processor to organize and save your files. RX also features iZotope's 64-bit SRC and MBIT+ dithering modules to further prepare your files, enabling you to deliver the highest audio quality possible.

De-clip

De-clip repairs digital and analog clipping artifacts that result when A/D converters are pushed too hard or magnetic tape is oversaturated—see Figure 10.31.

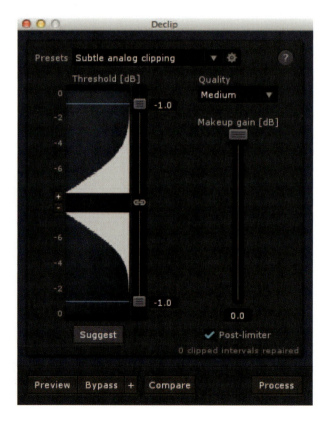

Figure 10.31
iZotope RX3
Advanced de-clip
module

De-clip can be extremely useful for rescuing recordings made in a single pass, such as live concerts or interviews, momentary overs in 'perfect takes', and any other audio that cannot be rerecorded.

De-clip will process any audio above a given threshold, interpolating the waveform to be more round. Generally, the process is as easy as finding the clipping you want to repair, then setting the threshold just under the level where the signal clips.

> **TIP**
> You can find clipping by either listening for the distortion that clipping causes or finding the high harmonic overtones of distortion in the spectrogram.

De-click and De-crackle

De-click and de-crackle is the main tool to use for restoring old vinyl and other recordings that have degraded because of clicks, pops, or crackles—see Figure 10.32.

This module has three tabs: De-click, which takes out the most prominent clicks, De-crackle, which will remove a stream of lesser clicks, and Interpolate, which can be used to fix a single click.

Figure 10.32
iZotope RX3
Advanced de-click
and de-crackle
module

Deconstruct

Deconstruct lets you adjust the levels of tones and noise independently in your audio—see Figure 10.33.

Figure 10.33
iZotope RX3
Advanced
deconstruct module

This module will analyze your audio selection and separate the signal into its tonal and noisy audio components. The individual gains of each component can then be cut or boosted individually.

This can be especially powerful with a range of audio files and applications, from increasing the 'breathiness' of a flute, to removing noisy distortion buried among tonal content. The deconstruct module may produce better results than the de-noise module when the noise is highly time variable, such as residual vinyl noise after de-clicking/de-crackling has been applied.

As opposed to de-noise, which separates signal from noise based purely on frequency magnitude, deconstruct analyzes the harmonic structure of a signal independently of level. It does not matter if a tonal signal like hum is quiet or prominent in a signal, deconstruct will always consider it tonal and adjust its gain accordingly.

> **TIP**
>
> Not all distortion is caused by clipping. If there is a distortion that can't be tackled with the de-clipper, try looking for a bright region of noise obscuring tones in the middle of the spectrogram. You can also hunt for it by making a frequency selection from 500 Hz to 5 kHz. After you make the selection, play back your frequency selection. Move it around or adjust the top and bottom of the selection until you hear exactly where the distortion is located. Once you have the distortion in your signal selected, use the deconstruct tool to lower the level of the noise components compared with the level of the tonal components.

Time and Pitch

The Time and Pitch module has two pages that you can access using the tabs at the top of the window: Stretch and Shift or Pitch Contour—see Figure 10.34.

Time and Pitch uses iZotope's sophisticated Radius algorithm to give you independent control over the length and pitch of your audio. It is useful for retuning audio to fit in a mix better, or adjusting the length of audio to deal with BPM or time code changes.

iZotope Radius is a world-class time-stretching and pitch-shifting algorithm. You can easily change the pitch of a single instrument, voice, or entire ensemble while preserving the timing and acoustic space of the original recording. iZotope Radius is designed to match the natural timbres even with extreme pitch shifts.

The most important choice to make when using Radius is between Mix and Solo modes. You should use Solo mode only when processing a single instrument with a clearly defined pitch. The human voice is a good candidate for Solo mode, as are most stringed instruments, brass instruments, and woodwinds. For all other types of source material, Mix mode will usually offer better results. Also note that when a single instrument plays more than one note at once, such as a chord played on piano or guitar, mix mode is a better choice.

Time and Pitch's Pitch Contour page can be used to change the pitch of a selection to quickly correct small pitch variations or gradual pitch drifts over time—see Figure 10.35.

Figure 10.34
iZotope RX3
Advanced Time
and Pitch Module
Stretch and Shift
page

The pitch contour changes pitch by continuously changing the playback speed of the audio. The effect is similar to speeding up or slowing down a record or tape deck while it is playing back. Because the pitch contour uses resampling to synchronously change time and pitch, it cannot be used to adjust pitch without also adjusting time.

The horizontal axis shows the length of your current selection. If you have no selection, the horizontal axis represents the entire length of your file. The vertical axis shows the amount of pitch shifting that will be applied. A curve through the top half of the display will create a higher shift in pitch and shorten the audio correspondingly. A curve through the lower half of the display will create a lower shift in pitch and lengthen the audio correspondingly.

You can correct a gradual pitch drift over time by adjusting the points at the far left or right of the display, drawing a straight sloping line from the beginning of your selection to the end. These points are locked to the vertical axis. Clicking on the contour display will create a new pitch node. You can create up to 20 pitch nodes to achieve very complicated pitch shifts.

Figure 10.35
iZotope RX3
Advanced Time and
Pitch Module Pitch
Contour page

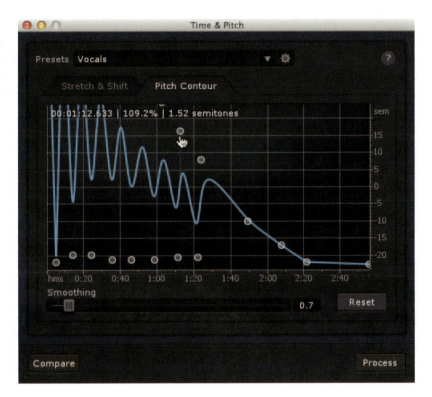

Equalizer

RX includes a six-band parametric equalizer module with four adjustable notch filters and two adjustable pass filters—see Figure 10.36.

The EQ is useful for manually shaping the overall sound of a file or selection. EQ can often be a simple first step to preparing a file for restoration, and can be used for cutting harsh high frequencies, removing rumble from dialog, steeply high passing out wind noise from a location recording, or cutting distortion overtones to increase the intelligibility of a voice.

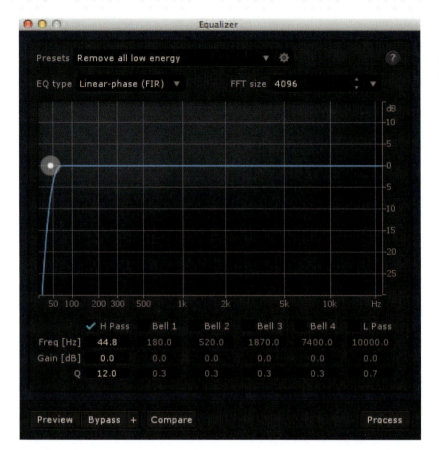

Figure 10.36
iZotope RX3
Advanced equalizer
module

Channel Operations

The channel operations module can be used to correct a number of issues with the stereo relationship of two channels, including variable level problems, phase imbalance, timing, and stereo noise in an otherwise mono signal. Four tabs along the top of this module let you choose which page to display in the window below—Mixing, Phase, Azimuth, or Extract Center.

Many restoration tasks require a simple rebalancing of levels. The Mixing page provides specific control over both left and right signal and balance levels. This simple operation can be used to downmix stereo material into mono, invert waveforms, transcode left/right stereo into mid/side, subtract a center channel, and much more.

The controls provided in the Phase page allow you to balance asymmetric waveforms by rotating signal phase, for example. Asymmetric waveforms such as dialog, voice, and brass instruments can occasionally occur in audio. Making the waveform more symmetrical gives the signal more headroom.

The Azimuth page provides control over left and right channel gain and delay—see Figure 10.37. Azimuth adjustment can help repair stereo imbalances and phase issues that can occur with improper tape head alignment or other speed-related issues.

Figure 10.37
iZotope RX3
Advanced Channel
Operations Module
Azimuth page

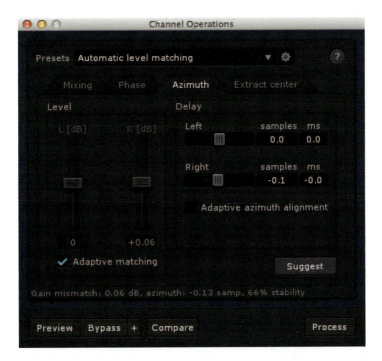

The Extract Center page lets you preserve or remove the center of a stereo file—see Figure 10.38. Extracting the center will retain the center of a stereo field and attenuate everything on the sides, such as signals panned to the left or right. Center channel extraction will preserve a stereo image if the side channels are retained. This can make it more desirable in some cases than mid-side encoding (which would sum left and right hard pans into one channel).

When the signal you want to preserve is even in both channels and the noise is uneven between channels, extracting the center can remove a lot of noise. For example, a mono record transferred to a stereo tape would have side channel noise that would be suppressed by extracting the center channel. In this case, make sure that the 'Keep Center' option is selected.

If you want to preserve the wide stereo information and remove the center information, you can keep the sides of the signal instead. In this case, make sure that the 'Keep Sides' option is selected. This is useful for karaoke-style removal of vocals from a song, especially because the process results in a coherent stereo image.

A pop-up selector lets you choose between two different algorithms: true phase, which cancels the center with phase information and retains the original panning of the sides, and pseudo pan, which extracts the side information and artificially stereoizes it into two channels. The reduction strength slider lets you control the level of the preserved signal. Lower settings will keep more information; higher settings will discard more information.

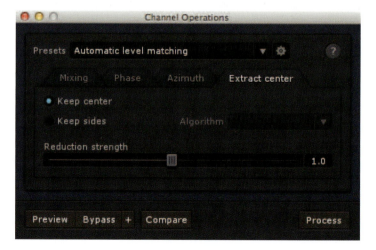

Figure 10.38
iZotope RX3
Advanced Channel
Operations Module
Extract Center page

Resample

The resample module allows you to convert an audio file from one sample rate to another—see Figure 10.39. Sample rate conversion (SRC) is a necessary process when converting material from one sampling rate (such as studio-quality 96 kHz or 192 kHz) to another rate (such as 44.1 kHz for CD or 48 kHz for video).

iZotope SRC's steep low-pass filter guards against aliasing artifacts and the post-limiter option can be used to limit the output level to below 0 dbFS to prevent any clipping from occurring.

Figure 10.39
iZotope RX3
Advanced resample
module

Plug-in Hosting Module

RX3 Advanced supports the use of AudioUnit (AU), VST, and DirectX plug-ins. These can be selected using the plug-in hosting module—see Figure 10.40.

NOTE

Only one plug-in may be loaded at a time. However, with the use of presets, multiple settings and presets may be recalled quickly in order to move between plug-in instances. When your plug-in is configured the way you

(Continued)

would like it, select 'Add Preset' from the small Preset drop-down arrow. Once your preset is named and saved, you can then assign that plug-in and preset to a keyboard shortcut with the Set Preset Shortcut feature. This keyboard shortcut will recall not only the plug-in settings, but the plug-in instance itself. As such, if a preset is saved with Plug-in 1, and Plug-in 2 is currently loaded into RX's plug-in window, pressing the preset keyboard shortcut will re-instantiate Plug-in 1 and recall the exact settings when the preset was made. This can allow for very quick editing, processing, and recall of plug-in instances and settings, providing a quicker workflow than traditional DAW track/mixer environments.

TIP

With a plug-in loaded into RX's plug-in module, you can make use of the same audio selection tools and preview and compare options that are available for other RX modules. This can allow for very detailed processing and greater accuracy when working with your existing plug-ins, giving you audio selection options unavailable in a traditional DAW setup.

Figure 10.40
iZotope RX3
Advanced plug-in
module hosting
Zynaptiq's Unveil

RX 3 Advanced Plug-ins

Seven RX 3 processors are available as AAX Native plug-ins: de-clipper, de clicker, de-crackler, de-reverb, de-noiser, dialog de-noiser, and hum removal. All of these are also available in AudioSuite format together with an additional plug-in—spectral repair.

> **NOTE**
>
> The deconstruct module from the standalone version is not available as a plug-in, neither are any of the utility modules.

RX3 Plug-in Presets

RX 3 Advanced plug-ins share several common features and controls. The plug-in preset system uses the same presets as in the RX 3 Advanced application, helping to provide a fluid workflow between environments.

Every control you move is kept in the plug-in history, and a flexible array of configuration options help you work the way you want to. From the preset manager, you can select from default presets and presets you have saved.

Clicking the Add button adds the current RX settings as a new preset. You can type a name and optionally add comments for the preset. Selecting the History button opens the History window. This view gives you a list of all of the actions you've performed inside the plug-in, allowing you to step back to previous settings and undo changes.

De-crackler

A continuous stream of quiet clicks is called *crackle*. Crackle ranges from the quiet but pervasive crackle of old degraded recordings to the smacks and clicks present in close-miked vocals. The de-crackle module can easily remove continuous clicks running through a recording—see Figure 10.41.

De-clicker

You can use the de-clicker module to remove a variety of short impulse noises, including clicks caused by digital errors, mouth noises, interference from cell phones, or any other audio problem caused by impulses and discontinuities in a waveform. See Figure 10.42.

Figure 10.41
iZotope RX3
Advanced
de-crackler

Figure 10.42
iZotope RX3
Advanced de-clicker

De-clipper

Clipping is a common problem that occurs when a loud signal distorts on input to a sound card/converter, mixing console, field recorder, or other piece of equipment. The result is overload distortion. This clipping can usually be seen most easily from RX's waveform display, where it appears as 'squared-off' sections of the waveform.

RX has a dedicated de-clipper tool that can in many cases rebuild the squared-off peaks caused by clipping and restore the recording to a natural-sounding state—see Figure 10.43.

Figure 10.43
iZotope RX3
Advanced
de-clipper

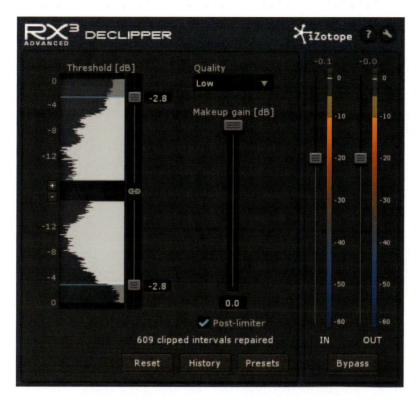

De-noiser

Unlike hum and buzz, broadband noise is spread throughout the frequency spectrum and not concentrated at specific frequencies. Tape hiss and noise from fans and air conditioners are good examples of broadband noise. In RX's spectrogram display, broadband noise usually appears as speckles that surround the program material. RX's de-noiser module is the go-to tool for removing noises of this type—see Figure 10.44.

Dialog De-noiser

The dialog de-noiser tool sets a new audio standard for real-time dialog treatment. When unwanted background noise and electrical interference threaten the quality of vocal recordings, the dialog de-noiser replaces the need for costly and time-consuming ADR. See Figure 10.45.

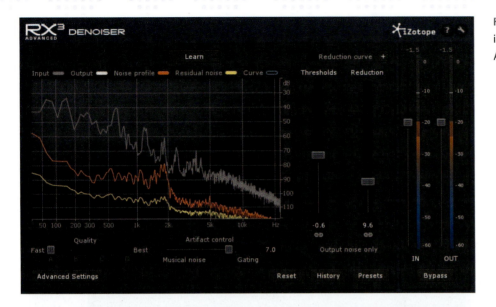

Figure 10.44
iZotope RX3
Advanced de-noiser

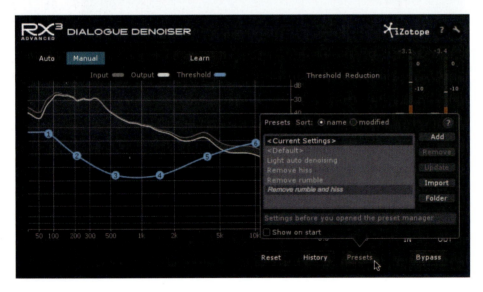

Figure 10.45
iZotope RX3
Advanced dialog
de-noiser

De-reverb

Reverb is the echoing decay of a sound in a space. The right amount of reverb in a recording can make a sound feel like it's sitting in a natural space. Too much reverb in a recording, however, can interfere with intelligibility. Reverb can interfere with other processing as well, including dynamics and mixing.

The de-reverb module—see Figure 10.46—gives you control over the amount of reverb in your audio. You can use it to subtly push reverb lower, remove it completely, or even enhance it.

Figure 10.46
iZotope RX3
Advanced de-reverb

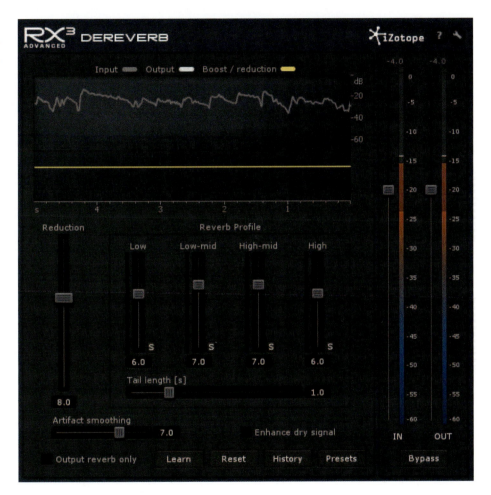

Hum and Buzz

Hum is usually the result of electrical noise somewhere in a recorded signal chain. It can usually be heard audibly as a low-frequency tone, usually based at either 50 Hz or 60 Hz depending on whether the recording was made in Europe or North America.

RX's hum removal module lets you set the primary frequency of the hum and lets you control suppression of up to seven harmonics above the low frequency—see Figure 10.47.

Spectral Repair

The spectral repair module intelligently removes unwanted sounds from a file (treating areas selected in the spectrogram/waveform display as corrupted audio that will be repaired using information from outside of the selection) with natural-sounding results—see Figure 10.48.

When used with a time selection, it can provide higher-quality processing than de-click for long corrupted segments of audio (above 10 ms).

When used with a time/frequency, lasso, brush, or wand selection, it can be used to remove (or attenuate) unwanted sounds from recordings, such as squeaky chairs, coughs, wheezes, burps, whistles, dropped objects, mic stand bumps, clattering dishes, mobile phones ringing, metronomes, click tracks, door slams, sniffles, laughter, background chitchat, digital artifacts from bad hardware, dropouts from broken audio cables, rustle sounds from microphone movements, fret and string noise from guitars, ringing tones from rooms or drum kits, squeaky wheels, dog barks, jingling change, or just about anything else you could imagine.

Sometimes a recording may include short sections of missing or corrupted audio. These are usually very obvious to both the eye and the ear! RX's spectral repair can be used to resynthesize gaps in audio, sometimes of up to a half second long or more, by using information around the gap to fill in the missing information using patterns and advanced resynthesis.

Intermittent noises are different than noises like hiss or hum; they may appear infrequently and may not be consistent in pitch or duration. Examples include coughs, sneezes, squeaks, footsteps, car horns, cell phone rings, and so forth. Because noises like this are often unpredictable, they usually need to be removed manually.

RX's spectral repair tool provides a number of different modes that allow you to select and remove intermittent noises and replace them with resynthesized content based on the surrounding audio:

- Attenuate mode removes sounds by comparing what's inside a selection to what's outside of it. Attenuate does not resynthesize any audio. It modifies dissimilar audio in your selection to be more similar to the surrounding audio.

- This mode is suitable for recordings with background noise or where noise is the essential part of music (drums, percussion) and should be accurately preserved. It's also good when unwanted events are not obscuring the desired signal completely. For example, this mode can be used to bring noises like door slams or chair squeaks down to a level where they are inaudible and blend into background noise.

- Replace mode can be used to replace badly damaged sections (such as gaps) in tonal audio. It completely replaces the selected content with audio interpolated from the surrounding data.

- Pattern mode finds the most similar portion of the surrounding audio and uses this to replace the corrupted audio. Pattern mode is suitable for badly damaged audio with background noise or for audio with repeating parts.

- Partials + Noise mode is the advanced version of Replace mode. It restores harmonics of the audio more accurately with control over the harmonic sensitivity parameter. This mode allows for higher-quality interpolation by explicit location of signal harmonics from the two sides of the corrupted interval and linking them together by synthesis. Partials + Noise can correctly interpolate cases of pitch modulation, including vibrato. The remaining nonharmonic material ('residual') is interpolated using the replace method.

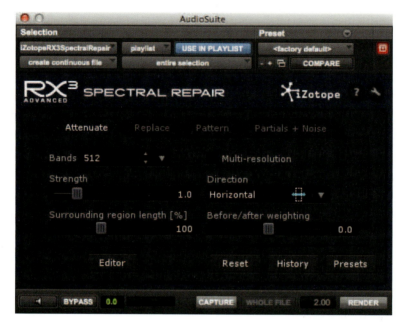

Figure 10.48
iZotope RX3
Advanced spectral
repair

TIP

To choose the different repair modes, click on the relevant tab near the top of the plug-in's window: Attenuate, Replace, Pattern, or Partials + Noise. The controls provided in the window will change according to the mode selected.

Zynaptiq Unfilter

Unfilter is a real-time plug-in that can be used to remove filtering effects, such as comb filtering, resonance, or excessive equalization—effectively linearizing the frequency response of a signal automatically. See Figure 10.49.

Unfilter automatically detects resonances, EQ, or filter effects—essentially any frequency response anomalies that aren't part of the original signal. You can then remove these by simply adjusting one control, leaving the rest of the signal virtually untouched—saving hours of work manually identifying and removing trouble peaks and notches.

Unfilter can function as an adaptive program equalizer that continuously adjusts itself while the audio is playing. This helps maintain a balanced spectrum when working with audio that changes over time, for example when streaming mixed music in broadcast applications. In addition, Unfilter incorporates a very useful free-form equalizer with various curve interpolation modes, from super-gentle slopes, through narrow notches, to brick-wall steps. It also includes a seven-band graphic equalizer, and has a super-steep 96 dB per octave shelving high-pass filter that (in the words of the Zynaptiq guru) 'removes low frequency rumble so efficiently you'll forget that there was ever anything down there'.

You can also apply the measured filter response from one recording to another—placing the two in the same acoustic 'world'. Or you can create room tone to fill editing gaps by applying a measured filter response to noise. You can also store filter responses to disk as linear- or minimum-phase impulse response files, building a library of 'colors' to have at your disposal. The fact that it can extract impulse responses in this way makes Unfilter a great tool for sound design, engineering and research.

Figure 10.49
Zynaptiq Unfilter

Zynaptiq Unveil

Unveil uses a model of the human auditory system to help work out which parts of the signal are reverb, effectively 'listening' to your signal much like a human would. These components are separated from your signal for processing using Zynaptiq's proprietary, artificial intelligence-based MAP (Mixed-Signal Audio Processing) technology. Unveil then makes it possible to either attenuate or boost reverb components within a mixed audio signal of any channel count, including mono sources, as well as to modify the characteristics of the reverb. See Figure 10.50.

So, Unveil can be used to alter the amount of reverb in a recording. You can turn a close-miked drum kit into a full-on room-heavy monster, move a sound back in the soundstage, or adapt the amount of reverb on different location recordings to match each other better. And Unveil lets you alter the reverb characteristics within a mixed signal, in real time even using mono source material.

Mix and mastering engineers can bring key mix elements into focus or move them to the background by attenuating or boosting perceptionally less important signal components—that is, unwanted audio components, commonly referred to as *mud*. Unveil can also be used to remove reverb and 'mud' from music recordings, so, for example, music producers can turn vintage drum loops rich in reverb into dry recordings, and vice versa.

Figure 10.50
Zynaptiq Unveil

Premastering for CD

What is commonly referred to as *mastering* actually involves two stages—the premastering phase in which aesthetic decisions about the sound of the mixes and consequent changes to these mixes are made, and the mastering phase proper in which the delivery media are created.

Sample Rates and Bit Depths

The final step in a CD mastering project is to create the 44.1 kHz/16-bit audio that will be used on the CD. Sample rate conversion should be the next-to-the-last process that you apply to your audio, while the last process should be bit depth conversion with the application of dither.

Recordings intended for CD release should ideally be made at 44.1 kHz, or a multiple of this such as 88.2 kHz, to avoid unnecessary rounding of calculations during the conversion process. There is no advantage to recording at 48 kHz instead of 44.1 kHz, and, in fact, there is a chance that some quality may be lost in the conversion process. If you want to use a higher sample rate, the best choice would be 88.2 kHz, rather than, say, 96 kHz, to minimize any quality losses due to rounding of mathematical calculations during the conversion process.

Most professional recordings these days are stored as 24-bit files (or, increasingly, as 32-bit files) on disk, so these have to be converted to 16-bit files before being sent to the pressing plant. This involves a significant loss of resolution, for example, going from the theoretically possible 144 dB of dynamic range provided by 24 bits down to the theoretically possible 96 dB of dynamic range provided by 16 bits.

Instead of simply throwing away (truncating) the lowest eight bits to reduce the bit depth, you should always apply a suitable dither algorithm—which will have the effect of extending the resolution of the 16-bit audio beyond the theoretical 96 dB range.

TIP

I have carried out extensive listening tests on 16-bit and 24-bit files. I do perceive the 24-bit files as being of higher quality, so I try to work with 24-bit (or 32-bit) files wherever possible. Having said this, the differences are not drastic in any way, and I would never hesitate to use a good-sounding 16-bit audio file, especially if a 24-bit version was not readily available.

NOTE

Bear in mind that increases in either the bit depth or the sampling rate result in corresponding increases in file sizes stored on disk. Double the bit depth, and the file is twice the size. Quadruple the sample rate and the file is four times the size. And don't forget that this is per audio channel, so multiply by two for stereo, by six for 5.1 surround format, and so forth. Perhaps even more important, higher bit depths and sample rates require more CPU processing, which is why the track counts in your digital audio software get subdivided as these are increased.

The 44,100 Hz sampling rate was originally chosen for the compact disc (CD) format by Sony because it could be recorded on modified video equipment running at either 25 frames per second (PAL) or 30 frame/s (using an NTSC *monochrome* video recorder).

The 48,000 Hz audio sampling rate was chosen as the standard to use with professional digital video equipment such as tape recorders, video servers, vision mixers, and consumer video formats like DV, digital TV, DVD, and films.

This rate was originally chosen because 48 kHz was readily derived by frequency division from the standard input frequencies that are used to derive television frequencies, and it readily synchronized with all video signals. Furthermore, it bore a simple relationship with the 32 kHz BBC/EBU sample rate—see more at: www.tvtechnology.com/opinions/0087/digital-audio-sample-rates-the—khz-question-/184354#sthash.O9LvnrXx.dpuf

> **NOTE**
>
> In practice, there are unlikely to be any audible differences between audio sampled at 44.1 kHz and audio sampled at 48 kHz. And any differences that you may hear are far more likely to occur as a result of differences between the analog electronics used in the converters, or because of other factors, rather than as a result of the different sample rates.

The 88,200 Hz sampling rate (double the 44.1 kHz standard) is used for convenience by some professional recording equipment when the destination is CD, because it is a simple multiple of 44,100 Hz that can easily be converted to 44,100 Hz.

The 96,000 Hz sample rate (double the 48 kHz standard) was developed specifically for use with DVD-Audio, HD DVD audio tracks, and BD-ROM (Blu-ray Disc) audio tracks.

The 176,400 Hz sampling rate (four times the 44.1 kHz standard) is used by HDCD recorders and other professional applications for CD production. This can also be easily converted to 44,100 Hz for CD.

The 192,000 Hz sampling rate (four times the 48 kHz standard) can also be used for DVD-Audio, HD DVD audio tracks, and BD-ROM (Blu-ray Disc) audio tracks.

> **TIP**
>
> I have carried out extensive listening tests with several high-quality converters and found that the Prism Sound Dream converters sounded better than everything else. I carefully compared the sound of these converters running at 44.1 kHz with several others running at 96 kHz and 192 kHz, and the Prism Sound converters sounded better at the

(Continued)

621

44.1 kHz/48kHz sample rates! There was no audible difference between the 44.1 kHz and 48 kHz sample rates. I also found it very difficult to perceive much, if any, difference between the Prism Sound converters running at 44.1 kHz and any of the converters running at any of the higher sample rates, although a couple of 'golden-eared' producers and engineers who joined me in these tests claimed that they could just hear the differences and preferred what they described as the 'airier' high frequencies at the higher sampling rates. The Apogee Rosetta, Lynx Aurora, and Crane Song HEDD 192 also offer extremely high-quality conversion and I could hear small differences in the sound quality of these, such as the Apogee converters which were less 'transparent' than the Prism Sound converters, adding a little subjectively pleasing 'coloration' to the sound. Nevertheless, they were all of a very high standard and I would be happy to use any of these.

NOTE

There are higher rates such as 352,800 Hz and 2,822,400 Hz used for super audio compact disc (SACD) and 5,644,800 Hz (eight times 44.1 kHz) used in some professional direct stream digital (DSD) recorders. You can buy music that has been recorded or remastered using high sample rates such as 352.8 kHz, and consumers can buy high-end audio converters such as the Eclipse 384 A/D and D/A converters (at around $7,000) or the Zodiac Gold D/A converter (at under $4,000) to listen at these high sample rates— but these will not concern most of us!

PQ Sub-codes

Each sector on an audio CD contains 98 control bytes identified by the letters P through W. These are often referred to as *sub-channels* or *sub-codes* and the most important of these are the first two, the P and the Q sub-codes. The P and Q sub-codes contain the track start and end times and the disc catalog number along with various other coded information.

For example, in the lead-in area the Q sub-code contains the table of contents while in the program area of the disc the P sub-code contains the start and end times of each track and the Q sub-code contains absolute and relative time code information. The Q sub-code also contains track and index marks, the emphasis, copy prohibit, and SCMS 'flags', the barcode info, and the ISRC code.

The track marks let your CD player cue to two seconds before the start of the audio material and the first index point within each track marks the beginning of the audio material. You can have up to 99 tracks on a CD, with numbers running from 1 to 99, and each track can have up to 100 index points within the track. Index 1 is where the audio starts in the track. If there is silence after the previous track ends and before the new track starts to play at Index 1, the start of this silence is labeled as Index 0. The other 98 index points are optional.

Older DAT and other digital recorders sometimes used a system of 'pre-emphasis' on recorded material with a corresponding 'de-emphasis' on playback. Pre-emphasis boosts the high frequencies prior to A/D conversion while de-emphasis removes the boost after D/A conversion. De-emphasis circuitry is built in to all CD players to provide compatibility with any material recorded using pre-emphasis. However, the emphasis bit must be set to 'on' in the track's Q sub-code so that the CD player will 'know' that it should use the de-emphasis circuitry while this track plays back.

Each track on a CD has a copy prohibit 'flag' bit setting in the track's Q sub-code to indicate whether copying the track is allowed. This copy prohibit bit was originally intended to prevent direct digital copying using DAT recorders—but virtually no recording equipment uses this today, so it has no effect in practice. Subsequently, the Serial Copy Management System (SCMS) was developed to prevent users from making a digital copy of a digital copy. The way this works is that the SCMS 'flag' bit gets incremented when the source material is digitally copied from one digital recording device to another—so you are allowed to make one digital copy of the source material using SCMS-equipped recorders, but you won't be able to make any further digital copies of that recording.

Checking this 'flag' on any CD tracks before you write the CD identifies these tracks as the first copies—so if you try to record these digitally from the CD onto an SCMS-equipped digital recorder it won't let you. If you use an SCMS-equipped recorder to make a copy of the CD, this will be the first digital copy and will have the SCMS flags set for each track by the system built into the recorder. Now, if you try to copy that recording digitally, SCMS will prevent you. Of course, if you want to allow digital copies to be made then don't set the copy prohibit or SCMS flags on the CD. Oh, and in case you were wondering, checking the SCMS bit will override the setting for copy prohibit.

The other bits, identified as R–W, are not normally used with audio CDs, but can be used to store additional information in these. An early extension of the basic CD format called 'CD + g' used these bits to add graphics and MIDI data for each track and some specialized CD + g players were manufactured that had video and MIDI outputs.

> **NOTE**
> Sub-codes can be inserted either at the mastering stage or at the disc pressing plant.

Media Catalog Numbers

If the disc is released commercially, you will almost certainly want to include the disc's Media Catalog Number, which is a unique identification number for the CD in the form of a UPC or EAN barcode. These code numbers are allocated by the EAN or UPC authorities and are normally supplied to the mastering engineer by the record label. Each consists of 13 consecutive digits that you can obtain from one or other of these organizations. When embedded into an audio CD, these are referred to as *Media Catalog Numbers* and are used to uniquely identify the product that is being sold—whether a single, an EP, or an album. Typically, the barcode will be printed on the back cover of CD, for example, where it can be scanned by a retailer at the point of sale. Independent artists can get these from digital distributors such as CD Baby, Tunecore, or Ditto Music.

International Standard Recording Code (ISRC)

The record label should also supply ISRC codes to identify each mix, although some labels are still not aware of the benefits of using ISRC codes to facilitate tracking usages of the sound recordings. An ISRC is an international standard recording code that can be assigned to each and any final mix to uniquely identify this mix. This code can then be used along with records kept by the record companies, royalty-collection organizations, and rights organizations to track the owners and rights holders who should be paid when the recording is broadcast publicly or exploited commercially. If you want to make sure that all the rights holders (the owners of the sound recording and the performers on that sound recording) get paid when the disc is broadcast, then you should definitely include an ISRC code. Independent artists can get these from digital distributors such as CD Baby, Tunecore, or Ditto Music.

NOTE

In 1989, the International Organization for Standardization (ISO) designated the International Federation of the Phonographic Industry (IFPI) as the registration authority for ISRCs. The IFPI is a not-for-profit members' organization registered in Switzerland that operates a secretariat based in London, with regional offices in Brussels, Hong Kong, and Miami. The IFPI, in turn, delegated part of the administration of ISRCs to several dozen national agencies, which allocate ISRCs to both record companies and individuals.

In the United Kingdom, anyone who owns the copyrights to sound recordings can join PPL, which collects payments for any public broadcasts or performances of commercially released sound recordings and shares this money between the rights holders and the performers on the recordings. There is no fee to join PPL. PPL issues rights holders a registrant code that will allow that label, company, or independent artist to assign ISRCs for its past, current, and upcoming recordings. You can apply here: www.ppluk.com/I-Make-Music/Why-Should-I-Become-A-Member/What-is-an-ISRC/.

In the United States, the appointed ISRC agency is the Recording Industry Association of America (RIAA). Recording rights owners based in the United States can apply for a registrant code through the U.S. ISRC Agency here: www.usisrc.org/applications/question/1. There is a one-time $80 application fee for the allocation of a registrant code. This registrant code is yours for life, and it will allow you to assign up to 100,000 ISRCs each year. The U.S. ISRC Agency has also allocated registrant codes for ISRC managers, such as ISRCmusicCodes.com, which they can use to assign ISRCs on behalf of clients or customers. ISRC managers can provide individual ISRCs to independent artists or to those who do not wish to manage their own ISRC assignment. ISRCs are issued as a part of the business arrangement between the company and an artist, and they are assigned to only those tracks that fall within this agreement. For more info, e-mail isrc@riaa.com.

ISRC Rules

The following information is taken from the *International Standard Recording Code (ISRC) Handbook*, third edition, 2009, originally published by the International ISRC Agency (IFPI Secretariat), London. This is used here under creative commons license. Minor modifications to and abbreviations of the original text have been made to aid clarity.

The original document is available at www.ifpi.org/content/library/isrc_handbook.pdf.

The ISRC Code

An ISRC is a unique identifier for sound and music video recordings where one, and only one, identifying code is allocated to each version of the recording.

The ISRC consists of 12 characters representing the country (2 characters), registrant (3 characters), year of reference (2 digits), and designation (5 digits).

The country code identifies the country of residence of the registrant and the designation code consists of five digits assigned by the registrant to each unique recording.

Registrant Code

The registrant code identifies the entity assigning the designation code in an ISRC. Because ISRCs are normally allocated at the point prior to the preparation of the final production premaster, the registrant code will normally reflect the original producer of the recording. However, if the producer of a recording sells the recording with all rights before its ISRC is assigned, the new owner should be considered the registrant when they elect to allocate an ISRC.

> **NOTE**
> The registrant code cannot be assumed to identify a current rights owner as the recording may have changed hands since code allocation. Additionally rights may vary territory by territory.

Year of Reference

The year of reference element identifies the year in which the ISRC is allocated to the recording—and this allocation will normally take place in the year in which the preparation of the final production premaster for the recording is finalized.

The year of reference element should always reflect the year of allocation of the ISRC. So, for example, a track that was originally released in 1996 but not assigned an ISRC until 2001 should have a 2001 year of reference ('01').

> **NOTE**
> The year of reference in an ISRC code does not have any copyright significance. The sound recording may be in the public domain, for instance, and the year in which the ISRC was allocated may not be the year in which the recording was made or in which it was first released (if it came out on vinyl back in the 1970s, for example).

When to Issue New ISRCs

The rules are that an ISRC code should be allocated to each separate, distinct recording and to any new or materially changed version of this. The basic principle is that a separate ISRC must be assigned to every different version of a recording where there has been new creative or artistic input—but you should not assign a new ISRC to an unchanged track if this is reused on a new album of recordings, for example.

A new ISRC must be issued when a sound recording is remixed or edited, if a new fade changes the length of a track by more than 10 seconds, or if a 'full restoration' of a historical recording is carried out. See later in this chapter for more detail.

Remixes, Edits, Session Takes

If multiple sound recordings are produced during the same recording session with or without any change in orchestration, arrangement, or artist and if they are preserved or turned into commercial products, each recording shall be encoded with a new ISRC.

A new ISRC shall be assigned to each remix, edit, or new version of a recording.

It is recommended that the registrant associate in its database the ISRC numbers of the original recordings used in the remixing.

Changes in the Playing Time

The playing time of a recording is an important characteristic as it is used for product design and also as a basic element for the calculation of fees by copyright authorities, broadcasting stations, and the owner of the rights.

If the playing time is changed a new ISRC shall be allocated.

The following rules should be applied in determining whether a new ISRC is to be allocated.

- A record begins with the first recorded modulation and ends with the last recorded modulation.

- Deviations in the playing time, resulting from different measuring methods or changes in fade and that have no influence on existing legal rights, should not result in the allocation of a new ISRC.

- When a change of duration is intended 'musically or artistically', a new ISRC should be allocated. The recommended threshold is 10 seconds.

Allocation of ISRC to Existing Recordings

Recordings that have not been assigned ISRCs should be provided with an ISRC by the present owner of the rights to such recordings prior to a rerelease. In these cases, the registrant code will be that of the present owner of the rights.

Restoration of Historical Recordings

Remastering and editing technologies (including re-pitching, re-equalizing, de-noising, de-clicking, etc.) offer many ways of processing historical recordings so that they meet contemporary quality standards.

When a full restoration of sound quality is carried out, the processed recording is to be considered a separate recording and thus obtain a new ISRC.

Remastering

When a track is remastered for the purpose of reproduction on a new carrier without restoration of sound quality, then no new ISRC is required.

NOTE
It is nevertheless the registrant's responsibility to decide where to draw the line between sound restoration (full remastering) and simple remastering.

Copyright Expired Recordings

An ISRC should be assigned to recordings even when copyright has expired.

A recording that is in the public domain in one country may be protected in another. Without the ISRC it would be impossible to track such a recording's exploitation. Even if rights in the recording have expired, obligations to authors and publishers may still exist.

And finally a situation could occur as it did in the United Kingdom in 1988, where expired rights could be resurrected by a term extension. Without the ISRC it would be impossible to track exploitation of recordings in these circumstances.

Associated Video Assets

As well as using the ISRC to identify sound recordings and music video recordings, the ISRC may be used to identify audio and audiovisual material:

- that is closely associated with a released sound recording or short form music video,

- that is released in association with it, and

- where a musical performance does not form a substantial part of the content (for example, interviews, documentary material, etc.).

Registrants taking advantage of this provision are required to ensure that such material can be clearly identified in metadata provided to repertoire databases.

Output Formats

The most common output format used to deliver CD production masters to pressing plants today is audio CD written to CD-R according to the Red Book standard using Disk-At-Once (DAO) mode. Typically, this will have both ISRC and UPC/EAN codes embedded in the PQ sub-channels. These discs are usually burned at low speeds, and with no other software drawing on CPU resources at the same time, in order to minimize the possibility of write errors. This type of disc may be referred to as a *premaster CD* or PMCD.

> **TIP**
>
> If you use a Plextor CD burner, most of these include Plextools CD error-checking software.

The disc description protocol (DDP) format is still sometime used for data supplied on Exabyte, or, increasingly often, on DVD, CD-ROM, or optical disc, or sent electronically via ftp directly to the plant. The DDP files are disc images used to create the glass masters used for CD replication (or duplication) at the pressing plants.

> **NOTE**
>
> Duplication is normally used for smaller quantities of CDs that are burned onto blank CD-R discs using multiple CD burners. Unit costs are higher than for replication, but turnaround times are shorter—typically two to three business days. Replication is used for larger quantities of CDs using a large-scale manufacturing process that involves making a 'glass master' from the supplied production master. The glass master is used to create a stamper that is loaded into an injection molding machine that replicates the CDs. Unit costs are lower, but turnaround times are longer—typically 7–10 business days.

The older Sony PCM 1630 format, time-coded DAT, or the short-lived Sony PCM-9000 optical disc formats are almost never used nowadays—although you may come across these formats when remastering older releases.

> **NOTE**
>
> Many mastering engineers believe that the Sony PCM 1630 and the DDP formats deliver higher-quality results than other formats—but Sony and Exabyte hardware is becoming much harder to find these days.

Premastering for Vinyl

Mastering professionally for vinyl usually involves making sure that there is not too much bass or too much high-frequency stuff in your mixes, that nothing is panned too widely, that the mixes are not over-compressed, and that the signal levels average at around 0 VU. You also need to bear in mind the running time per side: a seven-inch single always sounds best with a running time of about three and a half minutes, or maybe four minutes max, and a 12-inch LP running at 33 1/3 rpm lasts 20 minutes per side at best. This is why the 12" single format became popular during the disco era, as these could contain longer mixes and would play back at higher levels—the first major 'loudness wars' were fought back then! If you have recorded more than 40 minutes of music on your album, you are going to have to make some decisions if you want to release this on vinyl, either reducing the number of tracks or editing the existing tracks until they fit. Another option is to release a double album on vinyl. Also, whichever way you go with this, it is very likely that you will to have to rethink the running order of your tracks as well!

In Europe you can get low-budget vinyl packages, ordering, say, 250 seven-inch discs in printed jackets for a little more than £500, for example from Northern Record Pressing: www.nrpressing.com/?p=815. In the United States you can get 1,000 12" discs in color printed jackets for around $2,500, for example from Rainbo Records at www.rainborecords.com. For more options, go to Gotta Groove Records at http://gottagrooverecords.com/2009/07/every-vinyl-record-press-in-the-united-states/.

Quality Control

If thousands of discs will be pressed from your 'master' mixes, you should definitely consider carrying out serious quality control checks on your mixes before final approval for pressing is given.

This is one of the stages in your project that you ignore at your peril! There are plenty of stories circulating among mastering engineers of albums that were pressed with tracks missing, with the wrong mixes substituted, tracks in the wrong order, with the wrong labeling, with pops and clicks audible, with noticeable edits, with audible distortion, and so on and so forth.

> **NOTE**
> It is worth keeping in mind that to check one hour of audio will always take longer than one hour because sections will have to be re-auditioned if any problems are suspected or found, technical notes will have to be made, and discussions engaged in with the mastering engineer and/or the producer or mix engineer. This could even take as long as three or four hours, especially if it is a surround project.

Quality control checks are first the responsibility of the producer or client, then of the mastering engineer. When the project reaches the pressing plant, further, detailed quality control checks ought to be made. However, in practice, this is often no longer the case.

> **NOTE**
> At the larger mastering studios, there may be a separate QC engineer who checks everything very carefully after the producer and mastering engineer have carried out their own checks, giving the project an extra 'safety net' that helps to avoid expensive mistakes. This separate QC stage is even more important for surround sound projects because of the extra audio tracks, additional stereo mixes, stereo reductions, and so forth.

If problems are found, the mastering engineer needs to discuss these with the producer and/or the mix engineer to work out whether and how to fix these. Sometimes what sounds like an unwanted noise is deliberately placed as an element within a mix—but the mastering engineer cannot know this unless informed by those who do.

> **TIP**
> It can be useful to check for small ticks or noises using a pair of headphones, although the monitor speakers mastering engineers use should be sufficiently revealing to make these quiet sounds perfectly audible. Using headphones to check quality becomes even more important with surround sound mixes, where it is much easier to overlook noises on one of six or more speakers.

Finishing Up

After any mastering session you should make safety copies both for yourself and for your client.

It is wise to make versions of the finally edited, topped and tailed, faded, ordered, level balanced, and EQ-ed mixes without any mix bus dynamics processing and without any final normalization—saving the mixes as 32-bit files or as 24-bit non-dithered files.

NOTE

It is now possible to deliver 24-bit audio to end users—using SoundCloud, for example.

TIP

You could also provide a copy of these files to a mastering studio to make the final adjustments to the dynamics and the overall level for release on CD, vinyl, or other formats. If you have done this job well, you will save lots of time (and money) if the mastering engineer only has to check and approve your work before applying any final compression, limiting, and overall level setting.

The most common file formats are Waveform Audio File Format, commonly known as the WAVE file format or more often by its filename extension WAV, and the older Audio Interchange File Format (AIFF). Other formats you may encounter include Broadcast Wave (BWF), which is increasingly used by broadcast organizations, and Free Lossless Audio Codec (FLAC), which is sometimes preferred when supplying files via the Internet as the sizes of these data-compressed files can be as small as half those of a WAV file—and without any loss of quality (because they are, of course, 'lossless'!)

You can deliver your finished master files either to your client or directly to the pressing plant using conventional mail or other delivery services, or via the Internet using standard ftp or specialized services such as SoundCloud, Gobbler, or Drop Box.

Broadcast Wave and Cart Chunk Formats

The Broadcast Wave Format (BWF) is a file format for audio data that was developed by the European Broadcasting Union (EBU) as Version 0 in 1997 and revised as Version 1 in 2001 and as Version 2 in 2011—see https://tech. ebu.ch/docs/tech/tech3285.pdf from which the information here has been abstracted. It can be used for the seamless exchange of audio material between different broadcast environments and between equipment based on different computer platforms. As well as the audio data, a BWF file contains the minimum information—or metadata—considered necessary for all broadcast applications. The Broadcast Wave Format is based on the Microsoft WAVE audio file format, to which the EBU has added a 'Broadcast Audio Extension' chunk. BWF Version 2 is a substantial revision of Version 1, which incorporates loudness metadata (in accordance with EBU R 128). Version 2 is backward compatible with Versions 1 and 0. This means that software designed to read Version 1 and Version 0 files will interpret the files correctly except that Version 0 software will ignore the UMID and loudness information, which may be present, and Version 1 software will ignore the loudness information. The change is also forward compatible. This means that Version 2 software will be able to read Version 0 and Version 1 files correctly. The SMPTE Unique Material Identifier (UMID) is used by TV broadcast organizations.

> **TIP**
>
> Various free software applications are available that let you edit Broadcast Wave metadata, such as BWAV Writer, available at www.quesosoft. com, or Wave Agent, available from www.sounddevices.com/products/ waveagent/.

As use of the BWF has grown, so has the information that users wish to include with the payload—see https://tech.ebu.ch/docs/tech/tech3352.pdf from which the information here has been extracted. In 2003, the EBU created a chunk to carry Extensible Markup Language ('XML') data so as to prevent the need to define further chunks for each new requirement that arose and to encourage the integration of IT and broadcast applications. The EBU encourages the use of a uniform method of expression of identifiers carried as XML in WAV/BWF files. For example, unique identifiers such as the ISRC (International Standard Recording Code) provide the most appropriate method of ensuring that the file at hand is the expected resource (e.g., a clean version of the label copy supplied by the record company), which may share the same artist, title, and timing with other undesired versions (e.g., an instrumental version or a version with adult lyrics). With the right ISRC identifier embedded in a BWF file, a user can check that he has the correct file. The same XML technique could be used to carry any other custom identifier, such as a mix stem identifier.

The Cart Chunk WAV format was ratified as an AES Standard, AES46–2002, in 2002—see www.cartchunk.org from which the information here was abstracted. WAVE files have become a de facto standard for exchanging audio among diverse systems and applications in the radio business. The Cart Chunk standard allows users to embed the common labeling information that radio users need directly into a WAVE file. Based on the RIFF file format, the Cart Chunk format is a nonproprietary standard that allows additional metadata to be attached as an integral part of a WAVE file in the form of 'chunks', or integral units of data. WAVE files support nearly every imaginable audio data format. The WAVE file standard can support not only linear PCM, but MPEG layer I and II, MPEG-3, Dolby AC-3, and more. In fact, the MPEG data in an MPEG-3 WAVE file is bit for bit identical to MP3 files. WAVE files are simply convenient containers or *wrappers* for audio data. The .WAV extension does not determine what *kind* of audio data is in the file. The WAVE wrapper lets you also add labeling and *meta*data, such as the EBU Broadcast Extension Chunk (a.k.a. BWF)—or the Cart Chunk.

Mastering for iTunes, AAC, and MP3

The main points to observe when preparing MP3 or AAC files for distribution are to use the highest-quality source material (96 kHz sample rate, 32-bit or 24-bit files, ideally); to reduce the high-frequency content (to allow the codecs to produce higher-quality results throughout the rest of the spectrum); and to use a maximum value of −1 dBFS for the limiter.

TIP
It also makes sense to audition your audio masters on a range of likely playback devices, including iPods and their earbuds, speaker docks, home hi-fi speakers, and popular headphones.

Apple points out in its guidelines for mastering for iTunes that iTunes and all iOS devices have a playback volume controlling feature called Sound Check that lets listeners hear all their songs at approximately the same volume. According to Apple, 'It first determines the loudness of a track and then stores that information in the file's metadata. (Songs downloaded from the iTunes Store already contain this information.) The metadata is then used to raise or lower the volume of each track to prevent jarring volume changes while a device is shuffling songs.'

Radio stations use similar technology to control changes in volume between tracks played on air; MP3s have a technology called Replay Gain that can control volume changes; and the International Telecommunication Union (ITU) has specified similar technologies in its broadcasting standards (specifically BS. 1770).

As Apple further explains, 'The effect of Sound Check, as well as other volume-controlling technologies, is that songs that have been mastered to be too loud will be played back at a lower volume, letting listeners more easily notice any artifacts or unintentional distortion.'

The logical way to deal with this situation then becomes to 'always mix and master your tracks in a way that captures your intended sound regardless of playback volume'.

Apple provides a free set of mastering tools that can be downloaded at: www.apple.com/uk/itunes/mastered-for-itunes/. These include:

- Master for iTunes Droplet, which is a simple, stand-alone drag-and-drop tool that can be used to quickly and easily encode your masters in iTunes Plus format.

- afconvert, which is a command-line utility that can be used to encode your masters in iTunes Plus format.

- afclip, which is a command-line utility that can be used to check any audio file for clipping.

- Audio to WAVE Droplet, which automates the creation of audio files, in WAVE (Waveform Audio File) format, from any audio file (such as MPEG or CAF files) natively supported on Mac OS X.

- AURoundTripAAC, which is an AudioUnit (AU) plug-in that can be used to compare an iTunes Plus file to the original source audio file to check for clipping. (A free AudioUnit host application, AU Lab, is available as a download at apple.com/iTunes/Mastered for iTunes.)

> **NOTE**
> Apple's Round Trip plug-in has a useful Listening Test feature that lets you A/B test the source with the encoded audio, with a blind listening test to sort out the truly 'golden-eared' from the rest of us!

Useful as the Apple software is, it is a bit basic—relying on the command line interface for two of the items, small application 'droplets' for two others, and the AU format plug-in.

Sonnox offers two rather more sophisticated products—the Codec Toolbox and the Pro-Codec. These can be used to audition various codecs to help you choose the one that sounds best for the recording you are working with.

Sonnox Fraunhofer Codec Toolbox

The Sonnox Fraunhofer Codec Toolbox was designed by Sonnox in partnership with Fraunhofer, who originally created the MP3 and AAC formats that the majority of people listen to today online or on their iPod or MP3 player. It is very affordable, priced at just £35 in the United Kingdom, for example. The Codec Toolbox provides 16-bit resolution encoding of 16-, 24-, and 32-bit files at sampling rates of 48 kHz, 44.1 kHz, and 32 kHz and is available for Mac and Windows in AAX Native, RTAS, AudioUnits, and VST formats.

When you are preparing to create MP3 or AAC files, it is helpful to be able to hear exactly how your mix will sound when it is subsequently encoded for MP3 or AAC at different bit rates as you are choosing these settings (i.e., in real time, without having to encode the file then play this back). The toolbox lets you do this. You can also check that your mix isn't too loud and won't clip when played back. Other advantages are that it lets you audition codecs in real time (a huge time saver—no need to encode first then listen); you can batch and encode multiple files to save time; add metadata like track names and artwork; and use the codec toolbox to decode MP3 or AAC files to WAV or AIF.

The list of codecs available (see Figure 10.51) includes MP3 for general purpose usage; AAC-LC (AAC Low Complexity) for excellent audio quality at low bit rates such as ISDB television in Japan; HE-AAC (High Efficiency AAC) for good quality at bit rates of 32–48 kbps per channel (64 kbps stereo) as used for XM radio, mobile music downloads, and digital radio mondiale; HE-AAC v2 for good quality at bit rates of 16–24 kbps per channel (48 kbps stereo), as used for 3GPP music download, digital radio DAB+, and Internet radio streaming to mobile devices (e.g., iPhone); HD-AAC (High Definition AAC) lossless audio codec with an optional lossy core, used for music distribution and archival at roughly half the bit rate of a 16-bit uncompressed WAV file; Apple AAC for excellent audio quality at low bit rates, used for Apple iPod and iTunes with 256 kbps VBR in stereo.

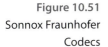

Figure 10.51
Sonnox Fraunhofer
Codecs

> **NOTE**
>
> The HD-AAC codec from Fraunhofer has a very clever feature; the single compressed lossless file includes a lossy core channel. It therefore acts as a lossless archival format, a lossless distribution format for the masters, and a final playback format for both lossless and lossy decoders—and all of this in a single file.

The Mac version of the Codec Toolbox also includes Apple AAC iTunes Plus (mastered for iTunes), which allows real-time audition of files produced under the Mastered for iTunes initiative that are destined for the iTunes store. It uses the same Apple codecs and resamplers that are used for the current iTunes catalog. It very specifically resamples to 44.1 kHz if necessary, and encodes as AAC-LC, 256 kbps, variable bit rate, and maximum quality, which are the settings used for the iTunes catalog. For online audition, the signal is then resampled back to the host DAW sample rate if necessary.

The Codec Toolbox consists of two applications—the Toolbox plug-in, for auditioning your mix in real time through various codecs, and the toolbox manager for encoding and adding metadata. A button at the top right of the Codec Toolbox's window launches the manager application from within the Toolbox plug-in.

Using the Toolbox plug-in in your DAW, you can easily audition your mix using a variety of different codecs and bitrates, in real time. The NMR indicator (noise-to-mask ratio) indicates possible audible differences between input and the codec output, so you can adjust the bit rate for your mix. The overs indicator shows if the process of encoding will introduce any clipping, giving you the chance to reduce the overall mix level slightly—see Figure 10.52.

When working on finalizing your mix or master processing, insert the Toolbox plug-in as the last insert on your master output. The Toolbox plug-in should be placed after your final limiter and dithering processor. Use the NMR meter to help assess the areas of your mix that may be affected by the effects of lossy compression. You may adjust areas of your mix or adjust your codec, mode, and bit rate until you find a suitable balance between audio quality and amount of data compression. (Remember: lower bit rates equate to greater data compression.)

Figure 10.52
Sonnox Fraunhofer
Codec Toolbox

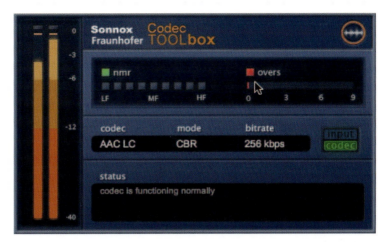

Figure 10.52
Sonnox Fraunhofer
Codec Toolbox

Play your mix or master from start to finish and pay attention to the bit stream overs meter. If this meter registers greater than 0 dBFS (red), it indicates that your mix or master has the potential to clip the DAC of the end user's playback device, leading to undesirable distortion.

> **NOTE**
> The sound of hard or soft clipping can be desirable for certain genres of music. However, in such cases, it is much wiser to impart this desired sound by using a suitable clipping processor prior to your final dither processing. This way, the clipped sound is faithfully reproduced, but, most important, no more is added by the different playback devices that may be used subsequently.

To accommodate for bit stream overs, simply reduce signal level prior to your final dither processing by enough to prevent the Toolbox plug-in overs meter from registering greater than 0 dBFS. When you have decided on suitable codec settings, and optimized your mix/master levels accordingly, bypass or disable the Codec Toolbox plug-in, then bounce to uncompressed WAV or AIFF and save the resulting file to your chosen folder.

The Toolbox Manager

The Codec Toolbox Manager (see Figure 10.53) is where the actual encoding or decoding is done. This has a 'Clip Safe' feature that will ensure clean encoded files, as the software auto-compensates for any 'overs' during the encode process.

You can open the Toolbox Manager either by clicking the manager icon at the top right of the plug-in or by opening the application directly from disk.

Figure 10.53
Codec Toolbox Manager window with folder browser at the left, file list in the upper central position, encode/decode section below this, the audio file playback audition section at the bottom, and the metadata editor at the right

The Folder browser at the left of the window is used to navigate folders on your file system containing supported media. Selecting a folder containing files supported by the manager will load them into the file list located to the right of the Folder browser. The file list provides a detailed view of files in the folder currently selected by the Folder browser.

Below the file list is the encode/decode section, which occupies the lower middle area of the Toolbox Manager window—see Figure 10.54. This provides information about the selected file, encoder settings, or decoder output format settings, output file name editing, and output file path selection.

If a WAV or AIFF file is selected, the codec settings selector will appear in the middle of the panel and the yellow Encode/Decode button at the bottom of the panel will be set to 'encode'. The output file name text field allows you to edit the name of the output destination file.

The manager can also decode compressed files for import into a host sequencer (or for general purpose decoding). To decode supported M4A or MP3 files, simply select them in the file list, choose WAV or AIFF, and press the DECODE button.

The audition section is situated at the bottom of the window. It contains controls for playing back audio files selected for audition in the file list.

Occupying the top right-hand area of the screen is the metadata editor. If a supported M4A or MP3 file is selected, any metadata it contains can be examined, edited, and written back to the file. If a WAV or AIFF file is selected, the section is used to edit the metadata that will be written when this file is encoded.

TIP
To save time, you can encode multiple files using the batch feature.

Metadata, such as artist, track name, and artwork, can easily be added to your files to be displayed on your MP3 player or iTunes.

NOTE
The Codec Toolbox Manager provides the ability to add and modify metadata in files with an 'mp3' extension, and some files with an 'm4a' extension. Adding and modifying metadata in m4a files compressed with

(Continued)

the Apple Lossless Audio Codec (ALAC) is not supported. The adding and editing of metadata in 'wav' and 'aiff' formats is not supported. The metadata contained within MP3 and m4a files is distinct from the audio data, and is stored in a separate format of its own. The manager supports the handling of two different metadata formats: ID3v2 format is typically for use in MP3 files. The ID3v2 standard is now in its fourth revision, but many players, including Windows 7 utilities, only support up to ID3v2.3. The manager can read and write metadata in all versions: 2.2, 2.3, and 2.4. iTunes Metadata is the standard developed and popularized by Apple, and found in m4a files. These two tag formats are widely supported, and can be parsed by most commercially available music player applications. Note that m4a files can contain metadata in the ID3 format, but MP3 files cannot contain metadata in the iTunes format.

Sonnox Fraunhofer Pro-Codec

The Sonnox Fraunhofer Pro-Codec plug-in is available for Mac and Windows in 32-bit RTAS format that works with Pro Tools 10. It is also available in 64-bit compliant AudioUnits and VST formats. The Pro-Codec allows you to audition, encode, and decode audio with codecs such as MP3 and AAC, making it possible to produce online-ready mixes in real time.

Additionally, the Pro-Codec Mac version enables engineers to use Apple's iTunes Plus codec for real-time auditioning, making it a powerful time-saving tool to efficiently mix directly for the Mastered for iTunes program. The exact same clipping behavior of the iTunes encoding chain can be monitored and levels corrected if necessary, saving time and ensuring high-quality output during the mixing process.

The Pro-Codec's extensive monitoring tools and built-in encoding and decoding features enable mixes to be optimized for specific target formats, with sample rates of up to 192 kHz encoded to 16-, 24-, or 32-bit files, ensuring maximum fidelity for final encoding and distribution.

By default the ONLINE mode is active, and the central panel of the GUI contains a large FFT display, showing the spectral content of the input signal in yellow, the DIFF signal in red (the difference between the input and output signals), and an NMR (noise-to-mask ratio) curve displayed in green.

Beneath that window five rows allow you to select up to five codecs for simultaneous export, in real time, to create encoded files. To the right of the codec list are three tabbed windows; shown here is the Bitrate Compression window that gives an indication of the data compression for the selected codecs—see Figure 10.55.

On the left-hand side of the window there is an input meter, a button to open the Export Settings window, and a total delay indicator that shows the total plug-in delay. On the right-hand side are controls for the monitoring and for the main display.

Figure 10.55
Sonnox Fraunhofer
Pro-Codec

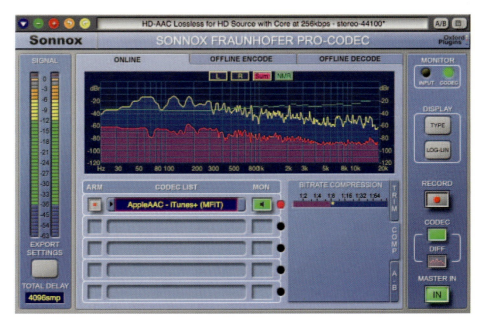

To encode a file in real time, you first have to record enable the codec. This is done by selecting the buttons to the left of the audition list. To record straight away, you can press the Record button when the host is playing back audio; this will write the output of your selected codecs to disk. You can also arm the codec when playback is stopped, so that codecs will start writing the instant you start host playback. Likewise, stopping playback will terminate the writing procedure.

| ✓ mp3 |
| mp3HD (Lossless to 16bits) |
| AAC–LC (Low Complexity) |
| HE–AAC (High Efficiency) |
| HE–AAC v2 (Parametric Stereo) |
| HD–AAC (Lossless to 24–bits) |
| **AppleAAC – iTunes+ (MFiT)** |

Figure 10.56
Sonnox Fraunhofer codecs

Also integrated into the plug-in are Fraunhofer's own AAC and MP3 codecs, including MPEG Surround and multichannel AAC formats for surround mixing. The full list of supported codecs includes MP3, AAC-LC, HE-AAC, and HE-AAC v2, MP3 Surround, MPEG Surround, AAC-LC multichannel, HE-AAC multichannel, and MP3-HD and HD-AAC lossless codecs. See Figure 10.56.

It is possible to encode files off-line using the Offline Encode tab—see Figure 10.57.

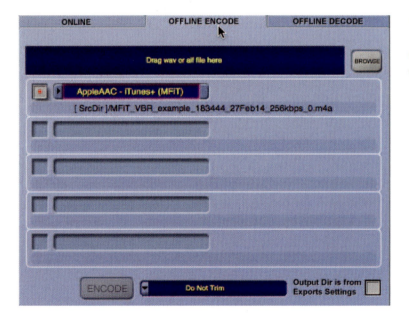

Figure 10.57
Sonnox Fraunhofer Pro-Codec Encode tab

The Sonnox Fraunhofer Pro-Codec can also decode compressed files for import into the host sequencer (or for general purpose decoding). To do this, use the Offline Decode tab—see Figure 10.58.

Figure 10.58
Sonnox Fraunhofer
Pro-Codec Decode
tab

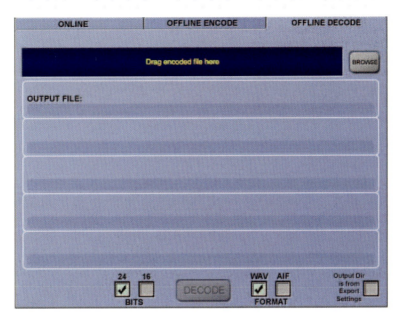

NOTE

The Pro-Codec has limited provision for ID3 metadata. The plug-in inserts the following metadata into an MP3 stream: date and time; codec type; bitrate and other settings; and a manufacturer message that reads 'Encoded using the Sonnox Fraunhofer Pro-Codec'. Currently, AAC metadata, for use with iTunes for example, is not supported.

Database Credits

It is increasingly important to get information about recordings into various databases that are now widely used for a variety of purposes. The major labels can be expected to be aware of all these, but if you are working with independent labels and artists, it would be wise to make your clients aware of these databases—and possibly to offer to submit the information on their behalf as one of your services.

iTunes, GooglePlay, HD radio, Amazon Cloud Player, and many other music services use metadata obtained from the Gracenote database about commercially released recordings: www.gracenote.com/company/faq/owner.

There are alternatives, including the All Media Guide/AMG Lasso database, which uses an 'acoustic fingerprint' system to recognize and identify media on CD, DVD, or in digital file formats such as MP and WMA. Once media is recognized, metadata from AMG's All Music and All Movie databases is delivered directly to the player: www.allmusic.com/product-submissions.

There are other similar services, including Muze, FreeDB, and MusicBrainz: http://search.muze.com/html/industry/labels.htm; www.freedb.org/en/faq.3.html; http://musicbrainz.org/doc/Beginners_Guide.

Another important database is used by the mobile phone/iPad app Shazam. Using Shazam, your mobile phone can recognize even a short section of any recording in its database when your phone can 'hear' this being played. Links are then provided to let you conveniently buy a copy from retailers such as iTunes or Amazon, or to post info about what you heard to Facebook or other social networking services: https://support.shazam.com/entries/23061432-How-to-submit-songs-or-albums-to-the-Shazam-database.

Like Shazam, another mobile phone/iPad app, SoundHound, can also ID a song title, lyrics, and artist information when it 'hears' a recording of this. And SoundHound goes even further: all you need to do is to sing, hum, or even whistle the tune! Singing or humming can be a bit hit or miss, and whistling can be a bit tricky, and Shazam seems to have the more accurate database at the moment, but SoundHound has lots of extra features including the ability to look up similar artists, look up other songs on the album, or the artist's entire discography, lyrics, videos on YouTube, tour dates, song album appearances, and much more: www.soundhound.com/index.php?action=s.faq.

Archiving

When your project is finally finished, you will probably want to wipe it from your hard disks to prepare these for your next project. You should already have at least two backups if you have been observing good practice throughout your project, but these may be on equally vulnerable hard drives. This is the point at which you should archive the project onto an archival medium that you can put into storage—at least until the equipment needed to restore the project from your archive needs to be replaced.

Archival media typically include tape-based technologies such as DLT and Exabyte, along with magneto-optical discs, DVD-R, and CD-R discs, that are guaranteed to last for many years—but will the machines to play these still be available in 10 or 20 years or more? Data storage technologies are changing so fast that you should definitely review the situation at least once a year, making new error-free copies to help prevent degradation or transferring your archived projects to newer formats as this becomes necessary.

More Info

Excellent sources of info about mastering include the various technical papers by Thomas Lund and others in the TC Electronic Tech Library at www.tcelectronic.com/TechLibrary.

Bob Katz's book *Mastering Audio* (Focal Press) is required reading for all mastering engineers and, indeed, for all audio recording engineers.

Bob Katz has also written the definitive book about mastering for iTunes. *iTunes Music: Mastering High Resolution Audio Delivery* (Focal Press) has some of the best explanations of the decision-making processes that take place while mastering audio that I have encountered. An essential handbook for anyone who wishes to learn more about mastering—and not only for iTunes!

From a practical perspective, Gebre Waddell's *Complete Audio Mastering* (McGraw-Hill Education) is the best all-around book on the subject that I have been able to find. Highly recommended!

Bobby Owsinski's book, *The Mastering Engineer's Handbook* (MixBooks), contains lots of tips and tricks along with interview material from leading mastering engineers that you may also find helpful.

The Bottom Line

The journey involves knowledge of and familiarity with a wide range of different processes; metering the audio to identify problems and confirm specifications; repairing and restoring damaged audio; premastering for CD and for vinyl, mastering for iTunes and MP3 delivery, and for delivery via broadcasts; getting the metadata sorted out, ISRCs attached, and release details into the databases; and finally archiving the project.

If you want to master an album for CD, vinyl, or other physical formats, and replicate thousands of copies, then you really should take your mixes to an experienced mastering engineer who can pretty much guarantee to get everything 'right'.

However, for lower-budget projects involving just a few hundred or maybe a thousand or so discs, or if you are releasing your product via the Internet or 'mastering' audio for other purposes, then you can achieve excellent results using Pro Tools and one or more of the software tools covered in this chapter.

The processes that may be applied during mastering include EQ, compression, multiband dynamics processing, making adjustments to the stereo image, adding harmonic 'excitement', maximizing, and limiting. Repairs and restoration may also be required. Many individual plug-ins are available to carry out these processes, notably from Universal Audio and Sonnox, along with suites of plug-ins that provide all these features, in particular from iZotope, which specializes in offering repair and restoration and general mastering tools.

Mastering for CD can be done 'in-the-box' using Pro Tools and the various plug-ins, but more control and features such as PQ encoding are available using Sadie or Magix Sequoia. If you are mastering for iTunes or for MP3, one or other of the Sonnox Fraunhofer Codecs is an essential tool to have at your disposal.

Index

Note: An italicized *f* following a page number indicates a figure on the corresponding page.

About the Author

Mike Collins is a studio musician, recording engineer, and producer who has worked with all the major audio and music software applications on professional music recording, TV, and film scoring sessions since 1988.

During that time, Mike has regularly reviewed music and audio software and hardware for magazines including *Future Music*, *Computer Music*, PRS for Music's *M* magazine, *Macworld*, *MacUser*, *Personal Computer World*, *Sound On Sound*, *AudioMedia*, *Studio Sound*, *Electronic Musician*, *EQ*, *MIX*, and others. Mike also writes industry news and technical reports for Pro Sound News Europe.

Mike has been writing for Focal Press since 2000. His first book, *Pro Tools 5.1 for Music Production*, was published in December of that year. A second book, *A Professional Guide to Audio Plug-ins and Virtual Instruments*, was published in May 2003. *Choosing & Using Audio & Music Software*, Mike's third title for Focal Press, was published in 2004 along with *Pro Tools for Music Production Second Edition*. Mike's fifth book, *Pro Tools LE & M-Powered*, was published in the summer of 2006. *Pro Tools 8: Music Production, Recording, Editing and Mixing* was published in the summer of 2009 and *Pro Tools 9: Music Production, Recording, Editing and Mixing* followed in the summer of 2011. *Pro Tools 11: Music Production, Recording, Editing and Mixing* was completed just in time for publication in December 2013.

In 2010, Mike was invited to join a team of audio transfer engineers at Iron Mountain's Xepa Digital Studios in Slough, near London, to help transfer a large part of Universal Music's back catalog of classic popular recordings from analog and digital tape copies of the archive stored in the United Kingdom to WAV files for archiving at Iron Mountain's secure underground facility in the United States. Using an Ampex ATR100 tape machine and Sonic Studio digital audio equipment, Mike personally transferred from 1/4" tape much of the Chess catalog along with many albums by Cat Stevens, Joe Jackson, Louis Armstrong, Quincy Jones, Barry White, Tricia Yearwood, and lots of other famous bands and artists from Universal's library of hit recordings from the past 60 years.

Throughout 2011, Mike regularly performed live as a duo with vocalist Aurora Colson at venues in and around London—see www.michaelandaurora.co.uk for more info. In the final quarter of 2011, Mike set up Rude Note Records—www.rudenoterecords.com—to release recordings he has produced, including Jim Mullen's solo jazz guitar album 'Thumbnail Sketches', together with three EPs and nine singles featuring Aurora Colson. Two collaborative albums featuring Mike with Jim Mullen, 'Blues, Jazz & Beyond' and 'Pop, Rock & Gospel', were released in December 2011 and February 2012, respectively. In April 2012, the label released Mike's remix of Vivienne McKone's song 'Everything Is Gonna Be Alright', along with a cover version of the Loose Ends hit 'Hangin' on a String', featuring vocalist Joanna Kay. These recordings are all available from the Rude Note Records Web site via Spotify, iTunes, Amazon, and various other online retailers.

Between May 2012 and May 2013, Mike worked on a new album with singer/songwriter David 'DaPaul' Philips, recording and mixing the album, playing guitars on various tracks, and coproducing most of the tracks with David. This gospel-influenced soul album, 'Soulful Spirit', was released in the summer of 2013 and is available at CD Baby—www.cdbaby.com/cd/dapaul2. This album reached #3 on the UK Soul Charts in December 2013.

From May 2013 until May 2014, all other projects were put on hold to allow Mike to complete two new books within this period—*Pro Tools 11: Music Production, Recording, Editing and Mixing* and this latest offering, *In the Box Music Production: Advanced Tools and Techniques for Pro Tools*.

Also active as a music technology consultant, Mike Collins often presents seminars and chairs discussion panels on Pro Tools, music production, music technology, music rights, and copyrights.

Contact Details

The author may be contacted via e-mail at mike@mikecollinsmusic.com. The author's Web site can be found at www.mikecollinsmusic.com and a professional profile is available at www.linkedin.com/in/mikecollinsmusic.